Enterprise Digital Reliability

Building Security, Usability, and Digital Trust

Manoj Kuppam

Apress®

Enterprise Digital Reliability: Building Security, Usability, and Digital Trust

Manoj Kuppam
Dallas, TX, USA

ISBN-13 (pbk): 979-8-8688-1031-2   ISBN-13 (electronic): 979-8-8688-1032-9
https://doi.org/10.1007/979-8-8688-1032-9

Managing Director, Apress Media LLC: Welmoed Spahr
Acquisitions Editor: Aditee Mirashi
Development Editor: James Markham
Coordinating Editor: Kripa Joseph

Cover designed by eStudioCalamar

Cover image by Freepik (www.freepik.com)

Distributed to the book trade worldwide by Apress Media, LLC, 1 New York Plaza, New York, NY 10004, U.S.A. Phone 1-800-SPRINGER, fax (201) 348-4505, e-mail orders-ny@springer-sbm.com, or visit www.springeronline.com. Apress Media, LLC is a California LLC and the sole member (owner) is Springer Science + Business Media Finance Inc (SSBM Finance Inc). SSBM Finance Inc is a **Delaware** corporation.

For information on translations, please e-mail booktranslations@springernature.com; for reprint, paperback, or audio rights, please e-mail bookpermissions@springernature.com.

Apress titles may be purchased in bulk for academic, corporate, or promotional use. eBook versions and licenses are also available for most titles. For more information, reference our Print and eBook Bulk Sales web page at http://www.apress.com/bulk-sales.

Any source code or other supplementary material referenced by the author in this book is available to readers on GitHub (https://github.com/Apress). For more detailed information, please visit https://www.apress.com/gp/services/source-code.

If disposing of this product, please recycle the paper

Table of Contents

About the Author

Manoj Kuppam, a seasoned reliability engineer with two decades of practical experience, has dedicated his career to cultivating software development skills globally. His research and academic contributions to reliability engineering have earned him widespread recognition. His innovative Site Reliability Engineering implementation framework has been particularly noteworthy, garnering numerous accolades.

Beyond his professional accomplishments, Manoj is committed to fostering intellectual growth and inspiring future generations. He actively mentors and promotes STEM initiatives in the North Texas region through Future City Organization, Tech Titans, and as a champion coach at Frisco ISD. His coaching and mentorship have been recognized by the Ministry of Science and Technology of the Government of Andhra Pradesh, India, with a special appreciation for his contributions.

Manoj relishes spending quality time with his family. He is a devoted father to his two sons and a loving husband to his wife, Anna, and their favorite time is to have brunches on the weekends.

Contributing Authors

Saurav Bhattacharya is a distinguished researcher and author with extensive expertise in account registration systems, digital identity, and cybersecurity. With a background in computer science from IIT Kharagpur and a career at Microsoft, Saurav has been instrumental in advancing technology solutions that address global challenges.

As the founder of an online security firm, SuperChargePlus, and president of the New World Foundation, he brings a wealth of knowledge and leadership to the peer-reviewed journal IJGIS.

Pradeep Chintale is a seasoned professional with over 18 years of experience in infrastructure automation and as a system analyst and cloud/DevOps engineer, specializing in the design, build, and operational support for application and infrastructure management. He holds a Bachelor of Science in Computer Science and a Master of Computer Applications from Mumbai University. Pradeep has been recognized for his significant contributions to the industry, receiving numerous accolades such as the Globee Award for Cyber Security 2024, the Industry Eagles Award for Innovation of the Year, and the International Achievers Award for Best Project of the Year. His expertise has also been acknowledged through his role as an industry judge for prestigious awards, including the Academy of Interactive & Visual Arts and the Globee Awards.

Pradeep is a prolific author and technical reviewer, having published the *DevOps Design Pattern* book globally and contributed to several academic journals and publications.

In addition to his professional achievements, Pradeep is an active member of the IEEE, where he serves on the Technical Program Committee and the Senior Member Panel. He is a board member at the New World Foundation and a member of the advisory boards for Harvard Business Review and Packt Publications. As a mentor, he actively participates in the European Startupbootcamp, guiding emerging startups toward success. Pradeep is also an accomplished inventor, holding several patents in the field of cybersecurity, focusing on AI- and ML-based solutions for audit, privacy, and risk assessment across various sectors. His work has been showcased at international conferences and the World Book Fair in Germany, London, and India, underscoring his influence in the global tech community.

 **Gaurav Deshmukh** is a highly skilled technology leader with over a decade of experience driving transformative software engineering initiatives. Throughout his career, he has held pivotal technical roles at prominent companies such as Guidewire, Cigna, Home Depot, American Agricultural Laboratory (AmAgLab), Tata Elxsi, and Amdocs. Gaurav's expertise encompasses a range of cutting-edge technologies, including cloud computing, cybersecurity, software automation, data engineering, and full-stack development with various programming languages and web technology frameworks. He employs his vast knowledge to create innovative solutions that optimize workflows and drive business growth. Gaurav holds both an MBA and a master's degree in Computer Science, with a focus on data warehousing and computer vision. He is dedicated to elevating the strategic role of software engineering

in delivering business value. As a distinguished leader, Gaurav can be reached at gauravkdeshmukh89@gmail.com to explore transformative technical initiatives.

Rajiv Avacharmal is a leading expert in the field of AI/ML risk management, with a particular focus on generative AI. With a distinguished career spanning over 13 years, Rajiv has held senior leadership roles at several multinational banks and currently serves as the Corporate Vice President of AI and Model Risk at a leading life insurance company. Rajiv's research interests lie at the intersection of AI/ML, risk management, and explainable AI.

Vishwanadham Mandala is a seasoned IT professional with over 20 years of experience in the industry, having worked with leading corporations such as Accenture, IBM, Oracle, Ciena, and Cummins. His extensive expertise spans big data, data engineering, cloud data engineering, AI and ML solutions, data streaming technologies, workflow orchestration technologies, data integrations, and technology leadership. Vishwanadham Mandala is committed to contributing to the growth of data science through impactful projects, book authorship, mentorship, and evangelization of cutting-edge technologies. His LinkedIn profile: https://www.linkedin.com/in/vishwanadh-mandala/.

Vishwanadham Mandala envisions leveraging his AI and ML expertise to drive impactful technological advancements.

Dr. Madan Mohan Tito Ayyalasomayajula is a dedicated researcher, author, and senior technology architect based in Texas. He is recognized as an industry expert with over two decades of experience in data architecture, distributed computing, cloud computing, big data, machine learning, and artificial intelligence. He holds a Doctorate in Computer Science from Aspen University and dual master's degrees from Osmania University. Dr. Ayyalasomayajula has played a pivotal role in architecting scalable data solutions and addressing the complex challenges of big data and AI. He is a senior member of IEEE and IET. He serves as a reviewer for prestigious journals and international conferences in AI and big data, contributing to the quality and relevance of cutting-edge research. His extensive academic and industry experience uniquely positions him to bridge theoretical knowledge with practical application, ensuring that innovative solutions effectively meet real-world needs. Dr. Ayyalasomayajula is also an active mentor to young IT professionals and shares his expertise through publications and industry events. As a thought leader in technology, he is committed to advancing the field of computing and fostering a culture of innovation and ethical AI practices, offering valuable insights into the future of technology and its impact on society.

Praveen Gujar is a distinguished product leader with expertise in enterprise data products in digital advertising and known for his transformative contributions to the tech industry, with a remarkable tenure at leading technology organizations such as LinkedIn, Twitter, and Amazon, where he has proven his ability to build large-scale enterprise products and drive significant business growth.

Fardin Quazi is a renowned expert in digital and business transformation within the healthcare domain, with 19+ years of extensive global experience in healthcare technology, management and admin solutions, robotics and intelligent process automation, AI/ML, and digital technology–based business transformation solutions. Fardin is working as Associate Director—Business Solutions, with Cognizant Technology Solutions, US Corp. He is a Certified Professional of the Academy for Healthcare Management, issued by American Health Insurance Plans. He holds an MBA in Information Systems and Operations and a bachelor's in Electrical Engineering. Fardin is volunteering as the Senior Vice President of Ethics Standards and Compliance at the New World Foundation and serves as an Editorial Board Member for the *International Journal of Global Innovations and Solutions*. He is currently living in Dallas, TX, with his family.

Harshavardhan Nerella is a distinguished cloud engineer with over seven years of experience, complemented by two master's degrees from prestigious universities in the United States. He has a robust background in cloud computing, cloud native solutions, and Kubernetes. He is deeply involved in research and technical community contributions. He has published research papers in esteemed journals, conferences, and authored articles featured in DZone's Spotlight section. His commitment to the field extends to his roles as a reviewer for various conferences and journals and as a judge for prestigious competitions such as Princeton Research Day and Technovation. Recognized as a Top Cloud Computing Voice on LinkedIn, he is also a highly sought-after mentor and interview preparation guide on ADPList, where he is ranked in the top 1% of mentors.

Anirudh Khanna is a distinguished technical and thought leader in backup and recovery, disaster recovery, and ransomware attack recovery. With over 15 years of experience, Anirudh has successfully led teams responsible for safeguarding data and ensuring business continuity for several Fortune 500 companies.

As a prolific author, Anirudh Khanna has published over 20 research papers in reputed journals and presented at more than seven international conferences. His deep expertise and commitment to advancing the field of data protection have been recognized with numerous prestigious awards, including Stevie, Globee, and Titan Awards for technological excellence.

For the past seven years, Anirudh Khanna has played a pivotal role in providing business continuity services for critical infrastructure at one of the largest utility companies in the United States. His dedication to maintaining essential services and operational integrity is reflected in his work. As a recognized leader, he is a Senior Member of IEEE and regularly reviews research papers for highly reputed journals and conferences.

Anirudh's extensive body of work, including articles and research papers, highlights his capabilities as a visionary in data protection and cyber recovery, cementing his status as a respected authority in the industry.

Sriram Panyam is a veteran software engineering leader with over two decades of extensive experience in developing technical platforms and organizations within major areas like large-scale distributed systems, cloud platforms, data analytics, SaaS products, and AI. Recognized for his innovative contributions and strategic leadership, Sriram has initiated programs globally, impacting billions of users. His tenure at top tech companies like Google, LinkedIn, and Amazon showcases his ability to handle complex, high-stakes scenarios, promoting a culture of innovation and growth. Notably, Sriram combines strategic vision with practical expertise, efficiently navigating teams through advanced technical initiatives. His notable strengths include mentoring and empowering engineering teams, fostering a spirit of innovation, particularly in large, slow-to-change environments. With a strong entrepreneurial spirit, Sriram has proven his capacity to motivate teams toward embracing and sustaining innovation. As a forward-thinking leader, he is dedicated to creating new technologies, empowering his teams, and developing the next generation of tech leaders.

With a master's degree in Computer Science from Indiana University Bloomington and multiple cloud certifications, **Ayisha Tabbassum** is an Onsite Lead for Cloud Operations and Multi-Cloud Architecture at Otis Elevator Co. She designs, automates, provisions, and secures Azure, AWS, and GCP infrastructure for various business domains and customer needs. She is also the founder and CEO of One Stop for Cloud, an Edtech company with the motto of providing simplified learning solutions for five major cloud platforms such as AWS, Azure, GCP, OCI, and IBM. She is a conference speaker on AI and cloud technologies. She has extensive work experience in using most sophisticated cloud platforms such as AWS, Azure, GCP, and IBM to create scalable, reliable, and cost-effective solutions. She is also responsible for reporting and addressing the security vulnerabilities in Azure Security Center, Wiz, and AWS Security Hub and designing and implementing policy add-ons to enhance security. In addition to her cloud engineering and architecture skills, she has a strong background in infrastructure automation and CI/CD application deployments, using technologies such as Git, GitLab, Jenkins, Ansible, Docker, Kubernetes, OpenShift, Dynatrace, Splunk, Prometheus, Grafana, SiteScope, Nagios, ELK, and Azure Monitor. She has applied these skills in diverse domains, such as ecommerce, retail, big data, and security, delivering high-quality solutions that meet business requirements and customer expectations. She is passionate about learning new technologies and staying updated with the latest trends and best practices in cloud computing and DevOps. She is also motivated by collaborating with cross-functional teams and stakeholders and contributing to the organization's goals and vision.

Parthiban Venkat is a lead data engineer with over a decade of IT experience, specializing in software development, data analysis, ETL processes, and cloud analytics. Having a strong focus on data warehousing and cloud migration, Parthi has successfully led key data engineering and migration projects across industries such as banking, healthcare, retail, hospitality, and gaming.

With a postgraduate degree in data science and machine learning, Parthi leverages advanced computation programming to design and implement scalable, data-driven solutions, delivering innovative strategies that enhance business performance, build reliable solutions, and drive digital transformation across diverse platforms. Parthi is passionate about applying cutting-edge technologies to solve complex data challenges.

About the Technical Reviewer

Sanyam Jain is a distinguished cloud security engineer with a deep-seated expertise in the cybersecurity domain. His unwavering commitment to safeguarding digital ecosystems is evident in both his professional achievements and contributions to the broader security community. Throughout his career, Sanyam has excelled in key roles within cloud security, security operations, application security, compliance, and security automation. He brings a comprehensive understanding of these areas, consistently developing and implementing robust strategies to protect critical infrastructure. His work ensures that enterprises not only meet but exceed their security objectives. Sanyam's technical proficiency is broad and deep, covering essential security disciplines such as network security, threat detection, data encryption, and access control. He is well-versed in leading cloud platforms, including AWS, Azure, and Google Cloud, enabling him to deliver security solutions that are both innovative and effective. His contributions extend beyond practical implementation. Sanyam's discovery of security vulnerabilities has been widely recognized and featured in esteemed publications such as Forbes, TechCrunch, ZDNet, Bleeping Computer, and over 40 other platforms. This recognition underscores his thought leadership and authority in the cybersecurity field. Academically, Sanyam holds a master's degree in Technology from BITS Pilani, where he graduated with distinction. His career is characterized by leading major projects that have significantly advanced enterprise security within the organizations he has served.

Acknowledgments

I would like to express my deepest gratitude to my wife, Anna, and my children, Bhavin and Vivin, for their unwavering support and understanding throughout this project. Their patience and encouragement allowed me to dedicate the time necessary to complete this work.

I am also immensely grateful to my talented co-authors for their invaluable contributions and dedication. Their expertise and hard work were instrumental in bringing this book to fruition. Additionally, I would like to thank all of our internal peer reviewers for their insightful feedback and suggestions, which helped to improve the quality of our work.

Introduction

In today's interconnected and data-driven world, ensuring the reliability of enterprise systems has become paramount. The hidden costs of unreliability, such as financial losses, reputational damage, and operational disruptions, have spurred organizations to prioritize reliability as a core business objective.

This book delves into the multifaceted landscape of enterprise reliability, exploring key concepts, metrics, design principles, governance models, testing strategies, and emerging trends. We will examine the distinction between DevOps and SRE and how they contribute to building reliable systems. It also explains the key parameters of performance, availability, resiliency, and scalability and their interlink with Site Reliability Engineering practice with examples from the real world.

PART I

Introduction

CHAPTER 1

Introduction

Authors:

Manoj Kuppam

Harshavardhan Nerella

Fardin Quazi

Reviewer:

Sriram Panyam

Reliability Engineering

Reliability engineering in modern computing relates to the practice of ensuring the software provides a dependable experience to the customer using the product. And often, this practice in the software world can be referred to as Site Reliability Engineering (SRE) as both focus on overall system reliability and share common goals. As large-scale systems stabilize their functionality and the focus changes to operational readiness and reliability, it makes sense to focus on improving the code and automation to free up time for the enterprises to explore new features. So, it is essential for organizations to care about reliability as a feature and the engineers embrace digital resiliency as a goal. This can include techniques, methods, and principles that drive the four major technical aspects, namely, performance, availability, resiliency, and scalability (*PARS* principles) of software reliability:

© Saurav Bhattacharya 2024
M. Kuppam, *Enterprise Digital Reliability*, https://doi.org/10.1007/979-8-8688-1032-9_1

Performance

This relates to the quickness of the system and its response time including any latency across the networks. A good performance leaves the customers and users with a high satisfaction score and avoids frustration.

Availability

It describes the uptime of a system and its availability for the customers to access and use it without failures as expected.

Resiliency

Resiliency is the system's ability to recover from failures or disruptions with least possible impact to the customers. This is typically handled by building highly available systems, embracing redundancy, predicting the point of failures, and proactively addressing the issues in the code.

Scalability

Scalability relates to the ability to handle workloads dynamically by provisioning the appropriate infrastructure capacity. Scalability ensures the systems are adequately designed to handle more users, data, or requests without compromising on the performance.

We will delve into each of these parameters in this book's chapters, and all the reliability engineering actions would be to achieve one of the PARS goals, hence resulting in better software reliability. Apart from the technical parameters, Site Reliability Engineering practice embraces an underlying operational characteristic that focuses on automation, documentation of runbooks, blameless postmortems, and continuous improvement:

Automation

Automating repetitive tasks and workflow-based action items not only saves time but also reduces human error, speeds up recovery time, and helps consistency. The DevOps practice has provided a platform for scripting deployments and reducing release timelines and provided an inspiration to automate configuration management, monitoring, event, and alert responses.

Runbooks

Standard operational procedures (SOPs) and runbooks provide detailed instructions that guide operations teams to react quickly and take actions to resolve the ongoing issues, perform maintenance tasks, and handle situations promoting knowledge transfer within the team.

Blameless Postmortems

Blameless culture focuses on finding the root cause of the problem and provides opportunity to prevent future occurrences of the issue without blaming the individuals.

Continuous Improvement

Reliability engineering is an iterative process. This ensures systems adapt to changing demands, user and systems behavior, efficient usage of infrastructure, and other parameters to constantly evolve and improve with time.

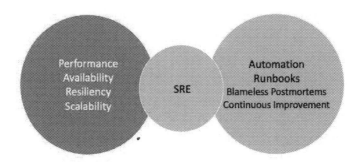

Figure 1-1. *The key technical and operational aspects of a reliability engineering culture*

Defining Reliability

Reliability engineering has become increasingly critical in modern software systems due to the distributed nature, cloud adaption, and higher failure points in the microservice architecture. To overcome the failures, system thinking and software engineering procedures combined with Google's SRE guiding principles have evolved with time to support different systems and applications with the best practices for modern reliability engineering. It is also important to note that reliability is a constant pursuit and goes beyond simply keeping the systems up and running.

With this context, reliability can be defined as

the ability of a system to consistently deliver its intended func-tionality with acceptable performance under varying conditions.

Imagine an objective of reliable transportation and a car that takes you from city A to city B and the customer expects this to be within 4 hours at an average speed of 60 miles per hour (mph). A reliable car would expect this goal to be accomplished consistently every time for a foreseeable future without fail and within the expected time of 4 hours and an average speed of 60 mph. If there is a flat tire situation, a spare tire would improve the resilience factor; if there is an engine failure, a second car or an alternative transportation would make the logistics highly available; if there is a known problem of closing the trunk, write down the instructions to safely close the trunk providing the runbook to handle the situation; and if there is a route that has higher traffic, a paid tollway may provide a longer but faster route to meet the timeframe, proving and providing the reliability in the goal of transportation.

The reliability engineering in software addresses the below questions:

- Does the system stand up to user satisfaction?

- What is the business impact of the system failure?

- Is our code written with compliance to the well-architected framework?

- Did the code handle applicable resiliency patterns to minimize the impact of failure?

- Did we document all the standard operating procedures in case of a failure?

- Is observability in place to get insights from monitoring tools?

- Are alerts set up to notify the support teams upon violations in the performance and failure events?

- Did we identify any opportunities to automate manual tasks?

- Can the systems scale to support the unexpected peak loads?

Each of these questions leads to better system thinking and operational guidelines, improving the overall stability of the software and infrastructure in place. These questions drive the SRE teams to build and maintain reliable systems that enterprises and users can trust and depend on.

In this section, we have understood the definition of reliability and touched up on the core characteristics of it from the technical and operational standpoint. We have also learnt that customer satisfaction is driven by consistent and dependable software, and it is beyond just being an up and available system. In this book, we will further explore each of the core topics and understand different metrics that must be measured and tracked, the cost of not following the reliability practices, impact of Site Reliability Engineering runbooks, application of the best practices to the futuristic technologies like artificial intelligence (AI), machine learning (ML) and blockchain, and observability principles.

Hidden Costs of Unreliability

In today's digital age, IT systems serve as the backbone of businesses across various sectors, driving operations, facilitating communication, and enabling data-driven decision-making. The reliability of these systems is not just a technical requirement but a critical business imperative. However, the concept of reliability extends beyond mere uptime; it encompasses the system's ability to perform consistently and predictably over time, ensuring business continuity and operational efficiency.

Understanding and addressing the hidden costs of IT unreliability requires a comprehensive approach that considers both the technical and business dimensions. It involves not only fortifying the IT infrastructure but also cultivating a culture of resilience and proactive management. By shedding light on these often-overlooked aspects, businesses can develop more robust strategies to mitigate risks and enhance their overall performance.

This chapter delves into the multifaceted nature of IT system unreliability, exploring its potential impacts on different aspects of a business. Through real-time examples and strategic insights, we aim to equip readers with the knowledge to identify, quantify, and address the hidden costs associated with unreliable IT systems, thereby fostering a more resilient and successful business environment.

Understanding IT System Unreliability

IT system unreliability is a multifaceted issue that goes beyond the occasional downtime or system crash. It encompasses any scenario where IT infrastructure fails to meet the set performance standards or expectations, affecting the smooth operation of business processes. Unreliability can manifest in various ways, such as frequent downtimes, slow system responses, inaccurate data processing, unable to scale, or inadequate security measures, each carrying its unique set of challenges and implications for the business.

A recent example of this is the **July 2024 incident involving Microsoft and CrowdStrike**, where a significant global IT outage linked to a CrowdStrike-related issue affected Microsoft's windows computers and servers, impacting critical sectors like airlines and banking. This event highlighted the various organizations leveraging one company's services and demonstrated the cascading effects of such outages across industries heavily reliant on these services.

Downtime is perhaps the most visible aspect of system unreliability. It directly halts business operations, leading to immediate revenue loss and customer dissatisfaction. However, other forms of unreliability, like slow performance and data inaccuracies, can be subtle yet equally detrimental over time. For instance, the July 2024 outage impacted about 8.5 million Microsoft devices globally, causing interruptions in various industries such as airlines and airports, public transit, healthcare, financial services, etc. This led to widespread productivity losses in workplaces worldwide, underlining how IT infrastructure failures can broadly impact day-to-day business operations.

Figure 1-2. *An analogy of impact of an unreliable code*

Inaccurate or missing data processing poses another critical challenge, as decisions made based on faulty data can lead to strategic missteps, financial losses, and erosion of customer trust. A notable example is the **2017 Uber driver payment miscalculation**. Uber had been incorrectly calculating its commission on New York drivers' earnings for over two years, leading to underpayment by an estimated $45 million. The issue arose not from inaccurate data but from the use of incorrect calculation methods. This incident underscores the importance of transparency and accuracy in data processing to maintain fair business practices and avoid costly errors.

Understanding these various dimensions of IT system unreliability is essential for developing effective strategies to mitigate their impact. It requires a comprehensive approach that includes regular system evaluations, auditing the access model, investment in robust technology solutions such as upgrading to newer hardware or adopting cloud computing, and fostering a culture of continuous improvement. By acknowledging and addressing the different facets of unreliability, businesses can enhance their resilience, maintain competitive advantage, and build stronger relationships with their customers and stakeholders.

Direct and Indirect Costs of Unreliability

The ramifications of IT system unreliability stretch far beyond the surface-level inconveniences, embedding themselves deeply within the operational and financial strata of an organization. These ramifications manifest as direct and indirect costs, each insidiously eroding the foundation of business efficiency, profitability, and reputation.

Direct Costs: The Immediate Financial Toll

Direct costs are the straightforward, calculable expenses that businesses incur when their IT systems falter. The most palpable of these is the loss of revenue. For instance, when an online retailer's website crashes, even briefly, the immediate loss in sales can be staggering. Amazon's 2018 Prime Day glitch [4], which lasted just an hour, is estimated to have cost the company $90 million in lost sales. This example underscores the tangible financial peril tied to system downtime.

Moreover, the costs associated with rectifying the issues—emergency technical support, overtime wages, and expedited parts or software procurement—can swiftly accumulate. But the financial bleed doesn't stop at repair bills; operational inefficiencies also take a toll. When the system glitches sidetrack employees, their diverted efforts translate to lost productivity, which, in essence, is money slipping through the organization's fingers.

Indirect Costs: The Stealthy Business Underminers

While direct costs punch holes in the budget, indirect costs subtly undermine the business's long-term health and market position. One such insidious cost is the erosion of customer trust. In an era where alternatives are just a click away, customers disillusioned by recurrent service interruptions are quick to jump ship, taking their loyalty and wallets to competitors.

The blow to a company's reputation from system unreliability can resonate far and wide, especially in today's digitally interconnected world. A tarnished reputation not only deters potential customers but can also devalue the company in the eyes of investors and partners.

Furthermore, the ripple effects of unreliable systems on employee morale and retention can be profound. A work environment marred by frequent IT disruptions can foster frustration and disengagement among staff, potentially leading to higher turnover rates. The hidden costs of recruiting and training replacements add another layer to the financial strain.

Lastly, the strategic blunders stemming from unreliable data or systems can lead to missed opportunities and misguided decisions, the costs of which may be incalculable but are undoubtedly significant.

In essence, the direct and indirect costs of IT unreliability weave a complex web of financial and operational challenges. Recognizing and addressing these costs is not just about fixing what's broken; it's about strategically investing in reliability to safeguard and propel the business forward in an increasingly digital world.

Understanding IT System Unreliability in Healthcare

IT unreliability in healthcare extends beyond downtime or system failures. For instance, downtime in healthcare IT systems can lead to delayed patient care, impacting patient outcomes. Slow systems can cause inefficiencies in patient data processing, leading to longer wait times and reduced patient satisfaction. Inaccurate data processing poses risks of misdiagnoses or incorrect medication, while inadequate security measures expose sensitive patient data to breaches, undermining patient trust and legal compliance.

Three Major Hidden Costs in Healthcare

The hidden costs of unreliability, though not directly affecting the bottom line, may have deeper impacts on a healthcare organization. This can be broadly categorized into three major areas: operational costs, reputational costs, and legal costs.

Operational Costs

Operational costs involve all expenses incurred by an organization to run its day-to-day activities. In the healthcare ecosystem, unreliability in a digital setup can increase operational costs in more than one way. Downtimes in the system, a common form of unreliability, can lead to service interruptions, such as delays in patient care. This not only affects the quality of services rendered to the patients but also overburdens the health practitioners, thus increasing the operational costs. Furthermore, system downtime translates into increased resource requirements for troubleshooting and rectification, which always include additional manpower and technological resources. This further adds to the upward surge of operational costs. In critical instances, this may also involve the loss of sensitive patient data and PHI, due to a dysfunctional system that may necessitate expensive data recovery efforts.

During the COVID-19 pandemic, hospitals faced a sharp decline in revenue due to the cancellation or postponement of elective procedures [1]. This led to increased operational costs as hospitals had to initiate layoffs, furloughs, and salary cuts to contain costs and maintain financial viability [1].

Reputational Costs

The reputation of a healthcare organization continues to be paramount in maintaining the confidence and trust that patients place in it. However, they can go a long way toward crippling that reputation in the form of sensitive data breaches. In the current digitally interlinked world, the news of such incidents can travel at lightning speed, thereby resulting in a substantial loss of trust among already existing and potentially existing patients. This can then be translated into less patient intake, directly hampering the revenue of the organization. Repairing a damaged reputation takes time and money through public relations campaigns and other damage control measures.

Healthcare organizations faced a reputational crisis during the COVID-19 pandemic. Despite their heroic performances, they were slammed by crises such as employee morale issues, conspiracy theories undermining community health, and exorbitant jury verdicts in medical malpractice cases [2]. These crises led to a significant loss of trust among stakeholders, affecting patient intake and overall reputation [2].

Legal Costs

Perhaps the gravest among them are the legal implications in case the system of digital healthcare is not reliable. Inaccurate processing of data would result in wrong diagnosis, inappropriate planning of treatment, and several other medical errors. This may result in litigation against the organization, with huge fines and penalties. In the event of a data breach resulting from a failure in the protection of patient data, healthcare organizations could be held accountable and face ensuing lawsuits. The resultant litigation, legal fees, and other penalties can add up to a significant cost to be borne by the healthcare organization.

Medical malpractice cases are a common source of legal costs in healthcare. For instance, a case involving a failure to diagnose led to a $950,000 recovery [3]. Such cases not only result in hefty fines and penalties but also necessitate additional resources for legal proceedings [3].

Conclusion

Throughout this chapter, we have explored the multifaceted nature of IT system unreliability and its pervasive impact on businesses. As illustrated through the notable incident of Amazon's 2018 Prime Day outage, the consequences of system failures extend far beyond immediate financial losses, permeating various aspects of business operations, customer relationships, and long-term strategic planning.

The direct costs, such as lost revenue and increased operational expenses, provide a tangible measure of the risks associated with system unreliability. However, the indirect costs—ranging from diminished brand reputation to eroded customer trust—often have a more insidious and lasting impact on a company's market position and growth prospects.

As businesses continue to rely heavily on IT systems for their core operations, the imperative to invest in robust, reliable infrastructure becomes increasingly evident. This investment is not solely about purchasing high-quality hardware or software; it encompasses a holistic approach that includes proactive monitoring, regular maintenance, disaster recovery planning, and fostering a culture that prioritizes IT resilience.

In conclusion, the hidden costs of IT unreliability underscore the need for a strategic, comprehensive approach to IT management—one that anticipates potential failures and mitigates their impact. By recognizing the broad spectrum of risks associated with IT system unreliability and adopting a proactive stance, businesses can safeguard their operations, protect their brand, and secure their competitive edge in an increasingly digital landscape.

The Intersection of DevOps and SRE

Site Reliability Engineering (SRE) and DevOps have emerged as two complementary disciplines in the ever-evolving landscape of software engineering, yet with distinct approaches to addressing the challenges of modern software delivery and operations. While they share some common goals, such as improving system reliability and efficiency with automation, they differ in their specific focus areas, metrics, and methodologies. Understanding these nuances is crucial for enterprises seeking to adopt the most appropriate practices to meet their unique requirements and drive digital transformation. SRE focuses equally on all aspects of software engineering from design to day-to-day operations, while DevOps focuses on getting the best software principles to implement the finished software product from source to its destination into a live environment.

SRE organization is a very complex structure, and driving it is less complex if you understand the road is not a straight path but has responsibilities across various components of the software system like

1. Ensure we have good knowledge of the road (historical performance data and knowledge of the systems).

2. We have an automobile that can navigate all terrains—team that understands and comprehends different layers of network, infrastructure, cloud, application code, DevOps pipelines, APIs, middleware, etc., and yes, I forgot some critical things like database and cache, and maybe more, but that's the point as SRE is an overall engineering duty and not just production support or application development specific role.

3. Instrument the monitoring agents/systems have good sensors to detect and collect right telemetry, so the car health is continuously monitored.

4. Implement good observability (dashboards and analytics) into the systems.

5. Ensure critical alerts are in place and autocorrection techniques are implemented with automation as an accelerator.

6. Perform blameless postmortems to reduce and avoid future repeats of the same issue.

To make this happen, we should understand that some of the duties of SRE overlap with DevOps and one can easily transition between jobs though SRE is often a lot more than just automation and needs a more comprehensive mindset than DevOps.

Site Reliability Engineering (SRE)

SRE is a discipline that originated at Google, combining software engineering principles with operations practices to ensure the reliability, scalability, and efficiency of large-scale distributed systems. SRE teams are responsible for designing, implementing, and maintaining systems that meet stringent service-level objectives (SLOs) and facilitate rapid innovation while minimizing operational overhead.

SRE Metrics and Focus Areas

SRE emphasizes the measurement and monitoring of key performance indicators (KPIs) and service-level indicators (SLIs) that directly impact system reliability and user experience. Some of the critical metrics measured in SRE include

- **Availability:** Measures the percentage of time a system or service is operational and accessible to users. For example, an ecommerce platform may have an SLO to maintain 99.99% availability during peak shopping seasons.

17

- **Latency:** Tracks the time taken for a system to respond to user requests or complete transactions. In a financial services application, low latency is critical for ensuring real-time trade execution and data processing. It is important to differentiate this with average response time (ART) as ART is response time taken in the context of a specific transaction in a span of time from sending the request to receiving the response, while latency is the delay or waiting time for the action to occur.

Latency is primarily influenced by factors like physical distance between sender and receiver, network congestion, and processing power of the systems involved. Average response time is affected by the factors like latency, server workload and processing speed, and the software application complexity.

- **Error Rates:** Monitors the rate of errors or failures occurring within the system or service. A content delivery network (CDN) may track error rates to identify potential issues with content caching or distribution.

- **Throughput:** Measures the number of successful transactions or operations processed per unit of time. For a high-traffic video streaming platform, throughput metrics are essential for capacity planning and ensuring a seamless viewing experience.

- **Durability:** Evaluates the system's ability to retain and retrieve data over time without loss or corruption. In healthcare applications, data durability is paramount for maintaining accurate and complete patient records.

SRE teams focus on optimizing these metrics by implementing robust monitoring and observability practices, leveraging automation, and adopting error budgets and risk management strategies. The primary goal of SRE is to strike a balance between innovation and operational stability, ensuring that systems remain highly reliable while enabling continuous delivery of new features and capabilities.

SRE Goals

The key goals of Site Reliability Engineering include

- Meeting and exceeding service-level objectives (SLOs) for system reliability, performance, and availability. For example, a cloud service provider may have an SLO of 99.99% uptime for their infrastructure-as-a-service (IaaS) offering.

- Minimizing toil (manual, repetitive tasks) through automation and scalable processes. SRE teams at a large ecommerce company may automate infrastructure provisioning and deployment processes to reduce operational overhead.

- Enabling rapid innovation and feature delivery by reducing operational overhead. A software-as-a-service (SaaS) provider may leverage SRE practices to accelerate the release of new features and capabilities to their customers.

- Fostering a culture of collaboration between software engineers and operations teams. In a large financial institution, SRE teams may work closely with developers to ensure that reliability considerations are integrated into the software development life cycle.

- Implementing robust monitoring, observability, and incident response practices. A global logistics company may adopt advanced monitoring and observability tools to gain visibility into their supply chain management systems and quickly diagnose and resolve issues.

DevOps

DevOps is a cultural and operational movement that emphasizes collaboration and communication between development and operations teams throughout the software delivery life cycle. It aims to break down traditional silos, streamline processes, and promote a shared responsibility for delivering high-quality software efficiently and reliably.

DevOps Metrics and Focus Areas

While DevOps encompasses a broad range of practices and principles, its metrics often focus on measuring the efficiency and velocity of the software delivery pipeline. Some of the key metrics measured in DevOps include

- **Lead Time:** Measures the time taken from code commit to deployment in production. For a software development team practicing agile methodologies, minimizing lead time is crucial for delivering value to customers quickly.

- **Deployment Frequency:** Tracks the number of successful deployments or releases to production within a given timeframe. A mobile app development team may aim for frequent deployments to rapidly iterate and incorporate user feedback.

- **Mean Time to Recovery (MTTR):** Measures the average time taken to resolve incidents or restore service after a failure. In a high-availability system, such as a telecommunications network, minimizing MTTR is critical to maintain uninterrupted service.

- **Change Failure Rate:** Monitors the percentage of deployments or changes that result in failures or incidents. A large enterprise software company may track change failure rates to identify and address bottlenecks in their release processes.

- **Defect Escape Rate:** Tracks the number of defects or issues that make it into production environments. A healthcare software provider may monitor defect escape rates to ensure patient safety and regulatory compliance.

DevOps teams focus on optimizing these metrics by implementing continuous integration and continuous delivery (CI/CD) pipelines, automated testing practices, and collaborative workflows between development and operations teams.

DevOps Goals

The key goals of DevOps include

- Accelerating the software delivery life cycle through automation and streamlined processes. A financial technology (FinTech) startup may adopt DevOps practices to rapidly iterate and deliver new features to their customers, gaining a competitive advantage in a fast-paced market.

- Improving collaboration and communication between development and operations teams. In a large manufacturing company, DevOps principles can help bridge the gap between software developers and factory automation teams, ensuring seamless integration of software systems with industrial processes.

- Fostering a culture of shared responsibility and accountability for software quality and reliability. A government agency may adopt DevOps to promote cross-functional collaboration and shared ownership of mission-critical applications.

- Enabling rapid feedback loops and continuous improvement through monitoring and metrics. A media streaming company may leverage DevOps practices to gather real-time feedback from users and quickly address performance issues or feature requests.

- Reducing lead times and increasing deployment frequency while maintaining high-quality standards. A gaming company may use DevOps to rapidly release new game updates and features, staying ahead of the competition and meeting the demands of their user base.

Table 1-1. *Table explaining how SRE and DevOps complement each other*

Aspect	SRE	DevOps
Focus	Stability of production environment	End-end software application life cycle
Team structure	Hybrid with SRE leads driving SRE objectives with all teams	Multidisciplinary with central DevOps team
Principles	High availability, scalability, performance, automation, and operation resilience and efficiency	Integration and automation with collaboration
Goal	Overall system reliability	Continuous integration and delivery

The Intersection and Differences Between SRE and DevOps

While SRE and DevOps share some common goals, such as improving system reliability and efficiency, they differ in their specific focus areas and the metrics they prioritize.

Areas of Intersection

Both SRE and DevOps share the following common goals and practices:

Promoting collaboration and breaking down silos between development and operations teams. In a large telecommunications company, SRE and DevOps practices may be combined to foster cross-functional collaboration between network engineers, software developers, and operations teams.

Emphasizing the importance of automation and continuous delivery practices. A cloud computing provider may leverage automation and continuous delivery practices to rapidly provision and deploy infrastructure resources and application updates.

Leveraging monitoring and observability tools to gain insights into system performance and reliability. A large retail chain may use monitoring and observability tools to track the performance of their ecommerce platform and inventory management systems.

Fostering a culture of shared responsibility and accountability for software quality and reliability. A financial services firm may adopt SRE and DevOps principles to promote a culture of shared ownership and accountability across teams responsible for critical trading and risk management applications.

Key Differences

Despite their commonalities, SRE and DevOps differ in the following ways:

Focus: SRE primarily focuses on ensuring system reliability, scalability, and efficiency, while DevOps emphasizes accelerating the software delivery life cycle and improving collaboration between teams. For example, in a large media company, SRE teams may focus on optimizing the reliability and performance of video streaming infrastructure, while DevOps practices are adopted to streamline the delivery of new features and updates to the company's streaming applications.

Metrics: SRE metrics tend to prioritize availability, latency, error rates, and service-level objectives, while DevOps metrics often focus on lead time, deployment frequency, and change failure rates. In a large ecommerce company, SRE teams may track availability and latency metrics for the company's online shopping platform, while DevOps teams monitor deployment frequency and lead times for new feature releases.

Scope: SRE typically operates at a system or service level, addressing reliability and scalability challenges for large-scale distributed systems. DevOps, on the other hand, encompasses the entire software delivery life cycle, from code development to deployment and operations. In a financial institution, SRE teams may focus on ensuring the reliability of mission-critical trading systems, while DevOps practices are adopted across the organization.

Conclusion: The successful adoption of SRE and DevOps practices requires a cultural shift toward cross-functional collaboration, shared ownership, and a relentless pursuit of continuous improvement. By aligning their goals, metrics, and processes, organizations can achieve a harmonious balance between innovation, reliability, and operational efficiency.

Bibliography

1. Bai, G., & Zare, H. (2020). Hospital Cost Structure and the Implications on Cost Management During COVID-19. *Journal of General Internal Medicine, 35*(9), 2807–2809; `https://doi.org/10.1007/s11606-020-05996-8`

2. Healthcare organizations need to recognize reputational risks and build a process for mitigating them, Editorial, updated on September 24, 2021; `https://www.healthcarebusinesstoday.com/healthcare-organizations-need-to-recognize-reputational-risks-and-build-a-process-for-mitigating-them/`

3. 15 real-life medical malpractice case results, by John Haymond, April 24, 2018; `https://www.haymondlaw.com/real-life-medical-malpractice-case-results/`

4.   `https://techcrunch.com/2018/07/18/amazon-prime-day-outage-cost/`

5.   `https://www.montecarlodata.com/blog-bad-data-quality-examples/`

6.   `https://www.theverge.com/24201803/crowdstrike-microsoft-it-global-outage-airlines-banking#stream-entry-6bfae301-e3da-4109-b20b-230e60821476`

7.   `https://xkcd.com/1739/`

CHAPTER 2

Key Performance Indicators (KPIs) in Reliability

Authors:

Sriram Panyam

Manoj Kuppam

Introduction

Reliability stands at the forefront of operational efficiency, safety, and customer satisfaction across diverse industries. It signifies a system's or component's likelihood to perform flawlessly under specified conditions over time. In manufacturing, reliability curtails downtime and boosts production rates. In healthcare, it guarantees the flawless operation of life-saving equipment. In the realm of software, reliability becomes synonymous with uptime and user trust. The essence of high reliability lies in its power to slash maintenance costs, elevate brand reputation, and carve out a competitive edge, establishing itself as a bedrock for organizational triumph and longevity.

© Saurav Bhattacharya 2024

M. Kuppam, *Enterprise Digital Reliability*, https://doi.org/10.1007/979-8-8688-1032-9_2

Key performance indicators (KPIs) emerge as the backbone, measuring and enhancing an organization's, system's, or process's success and reliability. These quantifiable metrics shed light on performance, stability, and availability, serving to

- **Pinpoint Weaknesses:** KPIs spotlight discrepancies from performance standards, identifying reliability issues.

- **Unveil Trends:** They reveal patterns, foretelling potential failures and assessing improvement measures' efficacy.

- **Steer Decisions:** Concrete data from KPIs guide pivotal decisions around maintenance, technology investments, and resource distribution.

- **Preempt Problems:** Organizations foresee and mitigate issues before escalation, curtailing downtime and operational expenses.

- **Bolster Customer Trust:** Consistent reliability and transparent communication about quality commitments heighten customer loyalty.

- **Encourage Improvement:** A culture of continual enhancement thrives, fueled by KPI insights.

- **Benchmark Excellence:** Reliability metrics against industry norms or rivals motivate aspirations for market dominance.

Incorporating reliability KPIs into strategic planning bridges day-to-day operational achievements with overarching strategic ambitions. These metrics empower leaders with detailed insights into system, process, and service performance, enabling well-informed decisions. For instance, robust system performance, as indicated by a high Mean Time Between

Failures (MTBF), advocates for operational expansion. In contrast, a significant Mean Time to Repair (MTTR) may hint at the need for strategic investments in training or technology.

Furthermore, reliability KPIs champion proactive problem-solving and prevention, identifying emergent trends that suggest potential failures. This foresight minimizes significant problems and operational disruptions risks. Simultaneously, by ensuring high reliability and prioritizing customer-centric KPIs like service-level agreement (SLA) compliance and system uptime, organizations significantly elevate customer satisfaction and trust. This dual focus not only fosters loyalty but positions the brand as dependable and quality-centric in the consumer's eyes.

Reliability KPIs are indispensable in guiding strategic decisions, preempting problems, and enhancing customer trust and satisfaction. They represent a critical component of achieving and sustaining organizational success, underscoring the importance of a strategic, informed approach to reliability across all sectors. In this chapter, we will explore the challenges faced by organizations in adopting them and a path toward excelling in them.

Understanding and Classifying Reliability KPIs

There are several metrics for understanding reliability. To reign in the sprawl, they are typically categorized into three distinct classes: performance metrics, maintenance metrics, and business impact metrics. Each class focuses on different aspects of reliability and provides unique insights into how systems, processes, or services can be optimized for better performance, efficiency, and customer satisfaction. Together, these three classes of metrics provide a comprehensive framework for measuring, understanding, and improving reliability across all levels of an organization.

Performance Metrics

These metrics assess the direct operational performance and efficiency of systems or components. They include

1. **Mean Time Between Failures (MTBF)**

   MTBF quantifies the average time a system operates before failing. High MTBF values suggest reliability, as systems perform longer without interruption. This metric guides businesses in forecasting performance, scheduling maintenance, and improving product design to extend operational periods, enhancing customer satisfaction and trust in product durability.

   Consider a fleet of commercial delivery drones. An average drone operates for 1,000 hours before encountering a failure. This high MTBF suggests that the drones are reliably meeting operational demands, reducing downtime for repairs, and maintaining consistent delivery schedules, which is crucial for customer. A higher MTBF indicates the higher reliability of a system.

2. **Failure Rate**

   This measures how frequently a system or component fails within a specific timeframe. A lower failure rate indicates a more reliable system, crucial for maintaining operational efficiency and minimizing downtime. Monitoring failure rates helps organizations identify reliability issues and implement corrective measures to improve product quality.

For example, a smartphone manufacturer tracks the failure rate of its latest model over the first year. If out of 100,000 units, 500 experience a hardware failure within this period, the failure rate helps the company identify the need for improvements in manufacturing or design to enhance reliability and customer trust in their products.

3. **System Uptime**

An online streaming service reports 99.9% uptime over a quarter, indicating the service was unavailable for roughly 0.1% of the time, or about 45 minutes. This high uptime ensures users have consistent access to the service, directly impacting subscriber satisfaction and reducing churn.

Uptime represents the percentage of time a system remains operational and available for use, excluding any periods of maintenance or unplanned downtime. High uptime percentages are critical for ensuring that services are consistently available to users, directly impacting customer satisfaction and trust in the service provider's reliability.

4. **Performance Efficiency**

This KPI assesses how effectively a system performs its intended functions under specified conditions. It encompasses speed, throughput, and accuracy. Optimizing performance efficiency involves refining processes and technology to meet or exceed operational standards, thereby enhancing productivity and customer experiences.

A data center upgrades its servers, resulting in a 20% increase in data processing speed and a 30% increase in energy efficiency. This improvement in performance efficiency means clients experience faster access to their data while the company benefits from reduced operational costs, making the service more competitive and sustainable.

5. **Reliability Growth**

Reliability growth tracks improvements in a system's reliability over time. It involves analyzing data from testing and operational use to identify trends in reliability enhancement. By focusing on reliability growth, organizations can demonstrate their commitment to continuous improvement, leading to higher quality products and increased customer confidence.

A software development company releases a new application with initial bugs causing frequent crashes. Over six months, through regular updates and bug fixes based on user feedback, the frequency of crashes decreases significantly, demonstrating reliability growth. This iterative improvement process enhances user experience and loyalty, as the app becomes more stable and reliable over time.

Performance metrics are crucial for understanding how well a system performs its intended functions and for identifying opportunities to enhance its reliability and efficiency.

Maintenance Metrics

Focusing on the activities required to keep systems operational, this class includes metrics like Mean Time to Repair (MTTR) and Incident Response Time. Maintenance and repair metrics provide insights into the effectiveness of maintenance strategies, the efficiency of repair processes, and the overall responsiveness of the maintenance team. They are essential for minimizing downtime and ensuring that systems return to operational status as quickly as possible. Some of these are illustrated below.

1. **Mean Time to Repair (MTTR)**

   Mean Time to Repair (MTTR) measures the average time required to repair a system or component after a failure has occurred. This metric is vital for understanding the efficiency of the repair process and the responsiveness of the maintenance team. A lower MTTR is indicative of a quick and efficient repair process, which minimizes downtime and mitigates the impact on operations.

   As an example, a manufacturing company experiences a critical machine failure that halts production. The maintenance team records the time taken to diagnose, repair, and restore the machine to operational status. If over a month, five such failures occur with a total downtime of ten hours, the MTTR would be two hours. By analyzing and striving to reduce the MTTR, the company can significantly decrease production downtime, leading to higher productivity and reduced costs.

2. **Incident Response Time**

Incident Response Time measures the duration from when a failure or outage is reported to when the response begins. It's a critical metric for assessing how quickly a maintenance team or service provider reacts to issues, affecting the overall downtime and customer satisfaction. Shorter Incident Response Times can greatly enhance customer trust and perception of the service's reliability.

For example, an IT service provider monitors its response time to customer-reported issues with their cloud storage service. When a customer reports a service disruption, the time it takes for the support team to acknowledge the issue and start troubleshooting is measured. Suppose the average response time for incidents in a quarter is 15 minutes. This swift initial response is crucial for maintaining customer satisfaction, as it assures customers that the provider is actively working to resolve their issues, minimizing potential frustration and operational impact.

Business Impact Metrics

This class encompasses metrics that reflect the broader impact of reliability on business operations and outcomes. Metrics such as availability, compliance with service-level agreements (SLAs), customer satisfaction, and cost of downtime illustrate how reliability affects an organization's operational efficiency, customer experience, and financial performance. These metrics are vital for aligning reliability efforts with business objectives and demonstrating the value of reliability improvements to stakeholders.

1. **Availability**

   This metric measures the proportion of time a system is operational and ready for use. For instance, a cloud storage service boasts 99.99% availability, meaning users can access their data virtually anytime, enhancing the service's reliability and user trust, crucial for customer retention and attracting new users.

2. **Compliance with Service-Level Agreements (SLAs)**

   Ensuring services meet predefined performance standards. A telecom company guarantees 99.5% network availability in its SLAs. Regularly achieving or surpassing this benchmark reassures customers of dependable service, strengthening business relationships and customer loyalty.

3. **Customer Satisfaction**

   This reflects how well a product or service meets or exceeds customer expectations. A survey shows an online retailer's customer satisfaction score improved by 20% after implementing a faster shipping option, directly correlating improved service features with increased customer approval and repeat business.

4. **Cost of Downtime**

   This measures the financial impact associated with system unavailability. An ecommerce website experiences a two-hour outage on Black Friday, resulting in estimated losses of $2 million in sales.

This example highlights the critical importance of system reliability and the need for robust contingency planning to mitigate financial risks.

Common Challenges and Striving for Reliability Excellence

Organizations often encounter several challenges in their quest to measure and improve reliability through key performance indicators (KPIs).

Accurate Data Collection: One significant hurdle is the difficulty in accurate data collection and analysis. Reliable data is the foundation of meaningful KPIs, yet collecting comprehensive and accurate data can be daunting due to complex systems and processes. To overcome this, organizations should invest in robust data management systems and analytics tools. Implementing these technologies facilitates the gathering, storage, and analysis of large volumes of data, ensuring that KPIs reflect the true state of reliability. This approach aligns with the strategic goal of maintaining high operational standards and meeting customer expectations for quality and dependability.

Organizational Misalignment: Another challenge is the misalignment between KPIs and organizational goals. Sometimes, KPIs may not accurately represent the strategic objectives of the organization, leading to efforts that do not contribute to overall success. Organizations can address this by regularly reviewing and adjusting their KPIs to ensure they are in harmony with both long-term strategic goals and immediate customer expectations. This alignment ensures that every level of the organization works toward common objectives, enhancing overall reliability and customer satisfaction.

Balancing Internal/External Factors:
Underestimating the impact of environmental and external factors on reliability is another obstacle. External factors like market changes, supply chain disruptions, or environmental conditions can significantly affect system performance and reliability. Organizations can strive for reliability excellence by adopting a proactive approach to risk management and resilience planning. Investing in predictive maintenance and advanced analytics allows for the anticipation of external threats and the implementation of preemptive measures. This strategic foresight not only minimizes the impact of such factors on reliability but also ensures that the organization remains adaptable and resilient in the face of change.

Resistance to Change: Lastly, there's often a resistance to change and adoption of new technologies within organizations. This resistance can hinder the implementation of systems and processes that enhance reliability. To combat this, organizations must foster a culture of continuous improvement and innovation. Educating and training staff on the importance of reliability and the benefits of new technologies are crucial. By creating an environment where employees are encouraged to embrace change, contribute ideas, and continuously learn, organizations can overcome resistance and drive improvements in reliability. Through education and engagement, employees become advocates for reliability, actively participating in initiatives that enhance performance and customer satisfaction.

By addressing these challenges with strategic alignment, continuous improvement, advanced technologies, and comprehensive education, organizations can navigate the complexities of reliability KPIs and achieve excellence in their operations. This holistic approach ensures that reliability remains at the forefront of organizational strategy, driving success and fostering a competitive edge in the marketplace.

Conclusion

In summary, reliability KPIs serve as a vital component of strategic decision-making, offering a data-driven basis for steering the organization toward its goals. They enable proactive problem-solving and prevention by highlighting potential issues before they become critical, allowing for timely interventions. Moreover, by ensuring high reliability, organizations can significantly enhance customer trust and satisfaction, which are key to maintaining a competitive edge in the market. In the dynamic landscape of modern business, the role of reliability KPIs in achieving strategic success cannot be overstated.

Measuring Metrics That Drive the KPIs

As discussed in the prior chapter, the key performance indicators (KPIs) are crucial for measuring the effectiveness of Site Reliability Engineering (SRE) practices. And to effectively measure these KPIs, we need to track the underlying metrics that provide insights into the systems and software health and drive these KPIs and keep the systems observable. These metrics can be either directly collected from multiple monitoring tools that comply with OpenTelemetry (OTel) protocols or can be derived from the metrics that are collected. This chapter tries to discover the select metrics that are the driving forces of each KPI.

The standard metrics from the operational perspective KPIs can be collected from the ITSM (Information Technology Service Management) platforms. This is key to the success of the SRE organization and helps measure the mean times to recover and improve the resiliency and reliability of the systems.

OpenTelemetry

OpenTelemetry (or OTel, pronounced "Oh-Tell") is an open source observability framework. Per the OpenTelemetry website, it is "a collection of APIs, SDKs and tools." Organizations and observability tools use this as a gold standard to instrument, generate, collect, and export telemetry data (metrics, logs, and traces) to help analyze the software performance and behavior. It is available in several programming languages and is suitable for use and adaptation being an open source solution. OTel integrates with most of the popular libraries and frameworks and is easy to install or instrument. To simplify, OTel is an open source, platform-agnostic observability framework that provides a standard way to collect metrics providing insights into the distributed microservice-based systems in an unified data format and has the origins from Cloud-Native Computing Foundation (CNCF) projects.

From the OpenTelemetry website, I like to take these key statements:

"OpenTelemetry satisfies the need for observability while following two key principles:

1. You own the data you generate. There's no vendor lock-in.

2. You only have to learn a single set of APIs and conventions."

SRE Metrics

SRE culture brings in a different perspective into the reliability engineering metrics. With its goals and objective-focused approach, SRE brings in service-level metrics that imply the health of the system. These vary from measurement signals that provide the direct health indicators of the system and software performance like service-level indicators (SLI) and health indicators from the user's perspective like service-level objectives (SLO). The SLIs heavily rely on the four golden signals of monitoring, or LETS signals—latency, errors, traffic, and saturation metrics.

ITSM Metrics

ITSM and SRE practices share a common goal to provide efficient and reliable IT services. ITSM focuses on the overall service experience from the user perspective. These metrics measure things like incident identification and detection times, incident resolution times, and failure occurrence times. This ensures areas of improvement in the broader IT service delivery process.

Other Metrics

Measurements can be from different sets of advanced practices like DevOps release rates, metrics that align with business goals, embracing AIOps to generate new measurements using data-driven approach, etc., help improve the overall system and service management processes.

The Standard Metrics

The set of metrics that define the comprehensive health of a system can vary for each use case and its goals and the user experience the project aims to deliver. These metrics ensure reliability, performance, and scalability of their systems, and below are some of the standard metrics that must be considered for the success of a modern system and application.

Table 2-1. *Set of standard metrics that drive the health of a system*

KPI	Metric	Metric Type	Definition and Details
MTTD	Mean Time to Detect	ITSM	Average time to identify an incident or issue in a system. The shorter, the better.
MTTR	Mean Time to Repair	ITSM	Average time to resolve an incident and restore a service to its normal performant state. This reflects how quickly the team can fix a problem.
MTTA	Mean Time to Acknowledge	ITSM	Average time the support engineer takes to acknowledge an incident once it is detected.
MTBF	Mean Time Between Failures	ITSM	Important for a resilient and reliable service, this metric is the average time between two consecutive unplanned system failures.
Change failure rate	Change request failures	ITSM	Number of change requests that resulted in failures in each period after being implemented.
SLO	SLI—uptime	SRE	Percentage of time a service is operational and available to the users.
SLO	SLI—latency	SRE	Time taken by a system to respond to a request.

(*continued*)

Table 2-1. (*continued*)

KPI	Metric	Metric Type	Definition and Details
SLO	SLI—error rate	SRE	Percentage of requests that result in errors within the system. Lesser percentage implies a reliable system.
SLO	SLI—throughput	SRE	Relates to no. of requests a system can handle per unit time.
SLO	SLI—CPU saturation	SRE	Percentage of allocated CPU that is being utilized.
SLO	SLI—memory saturation	SRE	Percentage of available memory that is being utilized.
SLO	SLI—average response time	SRE	Average response time like latency with the context on an average set of requests over a specific period.
SLO	SLI—queue length	SRE	Measures the number of requests waiting to be processed within a system. It is ideal to have it close to zero.
Disk I/O	SLI—disk I/O rate	SRE	Measures the rate of read and write requests from the storage.
Error budget	Error budget	SRE	Relates to the allowance for errors or incidents within a specific timeframe. Error budget is typically calculated based on SLOs and business goals and plays a key role in deciding the course of a sprint based on potential breach situations.

The DORA Metrics

The DevOps Research and Assessment (DORA) team at Google Cloud set a set of four key metrics to evaluate the performance and efficiency of software delivery teams in relation to the DevOps practices. Hence, I would like to treat these set of practices separately as they focus on the reliability of the DevOps process vs. the set of health-related metrics for the success of an application's reliability as mentioned earlier in this chapter.

The four DORA metrics:

- Deployment frequency measures how often a team successfully releases a new feature or code to production. Higher frequency indicates a team's ability to deliver changes quickly and improve their product.

- Lead time for changes measures the average time taken for a code commit to be deployed into a production environment. Shorter lead times indicate better maturity.

- Change failure rate represents the percentage of deployments that result in a failure. Generally, these changes require rollbacks due to functional issues or require problem tickets to find root cause analysis and cause an impact to the day-to-day operations.

- Mean Time to Recover measures the average time it takes to identify, fix, and recover from a production incident. This metric overlaps with the standard ITSM metric that relates to the health of the service management.

By understanding and utilizing DORA metrics, organizations can establish a data-driven approach to evaluating their DevOps practices and improve their software delivery process to be more efficient and reliable.

Tools and Techniques for Measurement

As we have discussed the key metrics to measure for a reliable modern-day architecture, the challenge is to identify the tools and methods to collect these metrics, transform them to become more relevant to our SLOs, and gather insights from them to take actions and drive reliability. The agreement and compliance of OpenTelemetry has set a standard for various monitoring tools to easily instrument your applications, made them platform and language agnostic, and even removed the dependency on the runtime environment. While OTel solves the data collection of traces, metrics, and logs with a standardization in place, the commercial tools and techniques provide a variety of opportunities to choose your storage, visualization, and additional advanced capabilities.

The monitoring tools are designed to leverage the extensible nature of OTel collector and makes the open source and commercial tools in the market to adapt and deliver higher value to its customers while being OTel compliant. Some of the extensible features may include

- Adding a receiver to the OpenTelemetry Collector to support telemetry data from a custom source

- Loading custom instrumentation libraries into an SDK

- Creating a distribution of an SDK or the Collector tailored to a specific use case

- Creating a new exporter for a custom backend that doesn't yet support the OpenTelemetry protocol (OTLP)

- Creating a custom propagator for a nonstandard context propagation format

The observability tools leverage these extension capabilities and are various types:

- Metric collection tools like monitoring agents, infrastructure monitoring tools, application performance monitoring (APM) tools, log management tools, and API monitoring tools.

- Metric analysis tools like visualization platforms, time-series databases, alerts, and notification tools.

- Metric processing tools are evolving recently to optimize the cost of monitoring in various ways, by reducing the size of the metrics collected and applying effective sampling mechanisms and data collection and filtering tools.

In this book, in the later chapters, we will gain deeper insights into the choice of monitoring and observability tools to be made depending on the organizational needs and financial costs to manage and maintain them with licensing and labor costs in consideration. However, one of the key engineering practice that is not fully explored in this book but is important for the readers to be aware is chaos engineering.

Chaos Engineering: Handling Unpredictability

Unpredictability is an inherent characteristic of systems. Despite meticulous planning and rigorous testing, unforeseen circumstances can arise, leading to unexpected failures or performance degradations. Traditional approaches to system reliability often focus on preventive measures, aiming to eliminate potential points of failure through redundancy and fault tolerance mechanisms. However, as systems become increasingly complex and interdependent, the ability to anticipate and mitigate all possible failure scenarios becomes increasingly challenging.

Site Reliability Engineering (SRE), a discipline pioneered by Google, recognizes the inevitability of failures and emphasizes the importance of embracing unpredictability. By adopting a proactive and experimental approach, SRE teams can enhance system resilience, improve incident response capabilities, and ultimately deliver higher levels of reliability and availability. One key practice that embodies this philosophy is chaos engineering.

1. **Chaos Engineering: Controlled Experimentation in Production**

    Chaos engineering is a disciplined approach to introducing controlled failures or disruptions into production systems to observe and learn from their behavior under various failure scenarios. This practice is rooted in the principles of experimentation and empirical data collection, enabling organizations to proactively identify weaknesses, validate resilience strategies, and continuously improve system reliability.

Chaos engineering methodology, as outlined in the Google SRE book, involves the following key steps:

    1.1. Steady-State Baselining

    Before introducing any chaos experiments, it is crucial to establish a baseline understanding of the system's steady-state behavior. This involves collecting and analyzing metrics, logs, and traces to characterize the system's normal performance characteristics, resource utilization patterns, and operational dynamics.

1.2. Hypotheses Formulation

Based on the steady-state baseline and known failure domains, SRE teams formulate hypotheses about the system's expected behavior under specific failure conditions. These hypotheses guide the design and execution of chaos experiments, ensuring that they are focused and aligned with the team's objectives.

1.3. Chaos Experiment Design

SRE teams carefully design chaos experiments to simulate realistic failure scenarios. This process involves identifying the appropriate injection points, determining the type and magnitude of the failure or disruption to be introduced, and establishing monitoring and data collection mechanisms to capture the system's response.

1.4. Executing Chaos Experiments

With proper safeguards and controls in place, chaos experiments are executed in a controlled manner within production environments. These experiments are typically conducted during periods of lower traffic or user activity to minimize potential impact on end users.

1.5. Analysis and Remediation

Following the chaos experiment, SRE teams analyze the collected data, validate or invalidate their hypotheses, and identify areas for improvement. Based on the findings, teams may implement remediation measures, such as refining system architectures, adjusting configurations, or updating operational procedures.

2. Embracing Unpredictability in Ecommerce Systems

Ecommerce platforms are prime examples of complex, distributed systems that must handle unpredictable workloads, traffic spikes, and potential failures. The ability to maintain high availability and provide uninterrupted service is crucial for ensuring customer satisfaction and revenue generation. By incorporating chaos engineering practices into their SRE strategies, ecommerce organizations can proactively address unpredictability and enhance system resilience.

2.1. Simulating Traffic Spikes and Scalability Tests

One common chaos experiment for ecommerce platforms involves simulating traffic spikes or load tests to validate the system's ability to scale and handle unexpected surges in user activity. This could involve injecting synthetic traffic or simulating scenarios such as flash sales or product launches.

By monitoring the system's behavior during these controlled experiments, SRE teams can identify potential bottlenecks, resource constraints, or performance degradations. This information can then be used to optimize system architectures, implement autoscaling mechanisms, or adjust load balancing strategies to better handle unpredictable traffic patterns.

2.2. Injecting Network Failures and Latency

Ecommerce platforms often rely on complex network infrastructures, content delivery networks (CDNs), and geographically distributed components. Chaos experiments can be designed to simulate network failures, latency spikes, or connectivity disruptions to test the system's fault tolerance and resilience.

For example, SRE teams might introduce network partitions or simulate high latency between different components of the ecommerce platform, such as the web frontend, application servers, and databases. By observing the system's behavior under these conditions, teams can validate the effectiveness of their circuit breakers, fallback mechanisms, and caching strategies, ensuring that the platform can gracefully degrade and maintain critical functionality during network disruptions.

2.3. Testing Disaster Recovery and Failover Mechanisms

Disaster recovery and failover mechanisms are crucial for ensuring the availability of ecommerce platforms in the event of major incidents or outages. Chaos engineering provides a controlled environment to test and validate these mechanisms by simulating scenarios such as data center failures, regional outages, or infrastructure provider disruptions.

SRE teams can design chaos experiments to deliberately trigger failover procedures, evaluate the effectiveness of data replication and synchronization processes, and measure the time required for the system to recover and resume normal operations. These experiments can uncover potential weaknesses or dependencies that may hinder effective disaster recovery, allowing teams to proactively address these issues and improve the overall resilience of the ecommerce platform.

3. Integrating Chaos Engineering into SRE Practices

While chaos engineering is a powerful practice for embracing unpredictability, it should be integrated into a broader SRE strategy to maximize its effectiveness and ensure a holistic approach to system reliability.

3.1. Continuous Monitoring and Observability

Effective monitoring and observability practices are essential for gathering the necessary data and insights during chaos experiments. SRE teams should implement comprehensive monitoring solutions that capture relevant metrics, logs, and traces, enabling them to analyze the system's behavior and identify potential issues or anomalies.

3.2. Automated Chaos Experimentation

As systems become increasingly complex and dynamic, manual chaos experimentation can become cumbersome and error-prone. SRE teams can leverage automation tools and frameworks, such as Chaos Mesh, Litmus, or Gremlin, to streamline the execution and management of chaos experiments.

3.3. Blameless Postmortems and Continuous Learning

Following chaos experiments, SRE teams should conduct blameless postmortems to analyze the results, identify areas for improvement, and foster a culture of continuous learning. These postmortems should focus on understanding the root causes of any observed issues, without assigning blame, and developing actionable recommendations for enhancing system resilience.

3.4. Collaboration and Knowledge Sharing

Chaos engineering and SRE practices thrive on cross-functional collaboration and knowledge sharing. SRE teams should promote open communication channels and knowledge-sharing platforms, enabling stakeholders from various domains, such as development, operations, and infrastructure, to contribute their expertise and insights.

4.   Embracing Unpredictability: A Mindset Shift

Ultimately, embracing unpredictability through chaos engineering and SRE practices requires a fundamental mindset shift within organizations. Instead of viewing failures as undesirable events to be avoided at all costs, SRE encourages teams to embrace them as learning opportunities and catalysts for continuous improvement.

4.1. Fostering a Culture of Experimentation

Adopting chaos engineering and SRE practices necessitates fostering a culture of experimentation within organizations. Teams should be empowered to take calculated risks, conduct controlled experiments, and learn from failures in a psychologically safe environment.

4.2. Aligning Incentives and Metrics

Traditional metrics and incentives often prioritize uptime and availability at the expense of resilience and long-term reliability. SRE advocates for aligning incentives and metrics with principles of resilience, embracing concepts such as error budgets, and acknowledging the inevitability of failures.

4.3. Continuous Improvement and Innovation

Unpredictability is a constant in the ever-evolving landscape of distributed systems. SRE teams must embrace a mindset of continuous improvement and innovation, consistently seeking new techniques, tools, and practices to enhance system resilience and adapt to emerging challenges.

By integrating chaos engineering into their SRE practices, organizations can proactively embrace unpredictability, validate their resilience strategies, and continuously improve their ability to deliver highly available and reliable services. This mindset shift, coupled with the practical application of chaos experiments and the broader SRE principles, empowers organizations to navigate the complexities of modern distributed systems with confidence and agility.

Bibliography

1.   "What Is OpenTelemetry?" *OpenTelemetry*, 30 Jan. 2024, opentelemetry.io/docs/what-is-opentelemetry/

PART II

Design

CHAPTER 3

Designing for Reliability

Authors:

Parthiban Venkat

Harshavardhan Nerella

Anirudh Khanna

Reviewers:

Gaurav Deshmukh

Madhavi Najana

Introduction to Reliability in IT Systems

Reliability in IT systems is a foundational aspect that determines their effectiveness, efficiency, and user trust. It refers to the capability of a system to perform its required functions under stated conditions for a specified period. Reliability is not just about preventing failures but ensuring that systems can gracefully handle them when they occur, maintaining service availability and data integrity.

Historically, the concept of reliability has evolved significantly. In the early days of computing, reliability was often synonymous with hardware robustness. However, as technology has advanced, the scope

© Saurav Bhattacharya 2024
M. Kuppam, *Enterprise Digital Reliability*, https://doi.org/10.1007/979-8-8688-1032-9_3

has broadened to include software, networks, and even user interactions. Today, reliability encompasses a holistic view of the entire IT ecosystem, reflecting a shift from focusing solely on individual components to considering the system's performance.

Key metrics play a crucial role in quantifying reliability. Availability, often expressed as a percentage, measures the proportion of time a system is operational and accessible. Mean Time Between Failures (MTBF) provides insights into the expected time between two consecutive failures in a system, indicating its reliability over time. Conversely, Mean Time to Repair (MTTR) measures the average time required to repair a system failure, highlighting the system's maintainability and responsiveness to issues.

Understanding and improving these metrics are vital for organizations to ensure their IT systems are reliable, thereby supporting business continuity, preserving data integrity, and maintaining user satisfaction and trust. As we delve deeper into the technical aspects of designing for reliability, it becomes clear that a systematic, proactive approach is essential for building and maintaining robust IT systems.

Understanding the Pillars of Reliable Systems

The foundation of any reliable IT system rests on three key pillars: redundancy, scalability, and maintainability. These elements work in concert to not only prevent system failures but also to ensure that the system can recover swiftly and efficiently when failures do occur.

Redundancy: Ensuring Continuous Operation

Redundancy is the strategic duplication of critical components or functions of a system to increase reliability. This can take various forms, including hardware redundancy, where physical components such as servers, network cables, switches, routers, etc., are duplicated, and software redundancy, where multiple instances of software applications run concurrently. Data redundancy, ensuring that data is replicated across different storage devices in the same or different regions, is crucial for data integrity and availability. The goal is to design systems that can continue to operate seamlessly, even if one or more components fail.

Scalability: Preparing for Growth

Scalability is the system's ability to handle increased loads without compromising performance or reliability. It is an essential consideration for designing reliable systems, as it ensures that the infrastructure can adapt to varying demands. Scalability can be achieved through horizontal scaling (adding more resources to a system) or vertical scaling (adding more resources to an existing instance or servers), each with its own implications for reliability. One has to calculate the anticipated future growth and evaluate if the existing infrastructure is capable of handling the future growth. If not, the enterprises should order the required hardware or adopt hybrid-cloud or multicloud architecture.

Maintainability: Simplifying Support and Updates

Maintainability refers to the ease with which a system can be kept in optimal condition. This includes regular updates such as server patching or installing newer versions of the software or operating systems, deploying the latest code to fix bugs, monitoring and fixing security vulnerabilities, and the ability to adapt to changing requirements without introducing new faults. A maintainable system is easier to monitor, troubleshoot, and enhance, contributing significantly to its overall reliability.

By integrating these pillars into the IT infrastructure design, organizations can build systems that are not only robust but also resilient in the face of challenges, ensuring continuous service and user satisfaction.

Disaster Recovery and Business Continuity Planning

After establishing the foundational pillars of reliable systems— redundancy, scalability, and maintainability—it's crucial to address how organizations can prepare for and respond to unforeseen events that could disrupt IT services. This section delves into disaster recovery (DR) and business continuity planning (BCP), two strategic frameworks that are essential for maintaining service availability and operational functionality in the face of disasters.

Defining Disaster Recovery and Business Continuity

Disaster recovery (DR) focuses on the IT infrastructure's ability to recover from failures and resume operations swiftly. It involves processes and technologies designed to restore hardware, applications, and data deemed essential for business operations following a disaster.

Business continuity planning (BCP) takes a broader organizational perspective, detailing how a business will continue operating during and after a disaster. It encompasses not just IT, but all essential business functions, aiming to minimize downtime and mitigate the impact on business operations.

Key Components of a Disaster Recovery Plan

Risk Assessment and Business Impact Analysis (BIA): Identifying potential threats and evaluating their potential impact on business operations is critical. This assessment informs the prioritization of systems and processes that are crucial for the business's survival.

Recovery Strategies: Based on the BIA, develop strategies for IT infrastructure, such as data backup, replication, and failover systems, ensuring that critical systems can be recovered and restored with minimal downtime.

DR Sites: Establishing offsite DR locations— whether hot, warm, or cold sites—ensures that the business can quickly shift its operations in the event of a site-specific disaster.

Developing a Business Continuity Plan

Business Continuity Team (BCP): Form a dedicated team responsible for developing and implementing the BCP, ensuring that all business units are represented and that the plan is comprehensive.

Emergency Response and Operations: Detail procedures for immediate response to a disaster, including communication protocols and steps to ensure the safety of personnel and assets. A command center or Network Operations Center (NOC) is set up to continuously monitor, manage, and troubleshoot the ongoing issues on the spot.

Training and Testing: Regular training sessions and simulated disaster scenarios are essential to prepare the team and test the effectiveness of the DR and BCP plans, allowing for adjustments based on lessons learned.

Integration with IT Infrastructure Design

Incorporating DR and BCP considerations into the initial design of IT systems can significantly enhance their resilience. This proactive approach ensures that the infrastructure is not only robust under normal conditions but also equipped to handle and recover from disasters efficiently.

Monitoring and Incident Response

Monitoring is the continuous observation of a system's operations to ensure that it performs optimally and to detect any signs of trouble early. Effective monitoring covers various facets of an IT system, including performance monitoring, security monitoring, and network monitoring, each providing insights into different aspects of the system's health and functioning.

> **Performance Monitoring:** Involves tracking resources like CPU usage, memory consumption, and I/O operations, ensuring they stay within optimal ranges and identifying potential bottlenecks or performance issues

> **Security Monitoring:** Focuses on detecting potential security threats or breaches by analyzing system logs, network traffic, and access patterns, aiming to identify and respond to threats swiftly

> **Network Monitoring:** Ensures the network's health, availability, and performance by tracking data flow, identifying congested routes, and monitoring for any signs of network failure

A NOC team is typically set up and engages in this situation to take things into control.

Incident Response: Preparation and Execution

Incident response is a structured methodology for handling and resolving system failures or breaches effectively. It includes identifying the incident, containing the impact, eradicating the cause, recovering the system, and learning from the event to prevent future occurrences.

> **Incident Response Plan:** A well-defined incident response plan outlines the steps and procedures to be followed when an incident occurs, including roles and responsibilities, communication protocols, and escalation procedures.

Incident Detection and Analysis: The first step in incident response is identifying and assessing the nature and severity of the incident, which is crucial for determining the appropriate response strategy. This is also Mean Time to Identify and Mean Time to Detect (MTTD) in many organizations.

Containment, Eradication, and Recovery: Once an incident is identified, the focus shifts to containing its impact, eradicating the root cause, and recovering affected systems or data to resume normal operations. This is equivalent to Mean Time to Repair (MTTR).

Postincident Review: After resolving an incident, conducting a postincident review is vital to analyze the response effectiveness, identify lessons learned, and implement improvements to prevent future incidents.

Integration with IT Infrastructure Design

Integrating monitoring and incident response into the IT infrastructure design is essential for proactive system management. By establishing robust monitoring and incident response capabilities, organizations can detect and address issues promptly, enhancing system reliability and resilience.

In conclusion, monitoring and incident response are not just about reacting to incidents but about creating an environment where potential issues are identified and addressed proactively. These practices are integral to maintaining system reliability and ensuring that IT infrastructure can support business operations effectively, even in the face of unexpected challenges.

Conclusion

In this chapter, we've journeyed through the critical aspects of designing for reliability in IT systems, underscoring the importance of a holistic approach that encompasses redundancy, scalability, maintainability, disaster recovery, business continuity planning, and proactive monitoring and incident response. These elements collectively form the backbone of a resilient IT infrastructure, capable of not only withstanding challenges but also adapting and evolving in response to them.

Reliability is not a one-time achievement but an ongoing commitment to excellence in design, implementation, and operation. By embedding reliability into every layer of the IT infrastructure, organizations can ensure that their systems not only meet the current demands but are also prepared for future challenges. The ultimate goal is to create IT systems that not only function efficiently under normal conditions but also exhibit resilience, maintaining operations and safeguarding data in the face of unexpected events.

As we look to the future, the principles of reliability will continue to be a guiding light for IT professionals, driving innovation and inspiring the design of systems that are robust, agile, and enduringly dependable.

As we delve into the different techniques of ensuring reliability, various techniques in system development, database, and ETL model are the keys to ensure reliability of the data-driven systems of future. In this chapter, we will take a look into the data transformation reliability and cover the system reliability techniques in the next chapter in detail.

Overview of ETL

ETL stands for Extract, Transform, Load, and it refers to the process of extracting data from one or more sources, transforming it into a format suitable for analysis or storage, and loading it into a target destination such as a data warehouse, database, or data lake. ETL plays a crucial role in data integration, migration, and analytics, enabling organizations to consolidate, process, and analyze data from disparate sources efficiently.

Here's a breakdown of each phase of the ETL process:

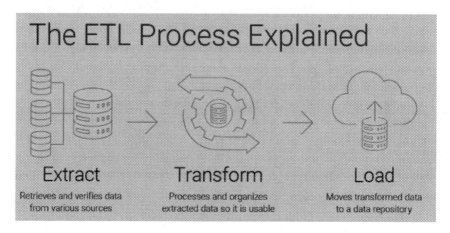

Figure 3-1. *An standard ETL flow*

> **Extract:** In the extract phase, data is extracted from various source systems, which could include databases, files, APIs, web services, or other data repositories. The goal is to retrieve the required data while preserving its integrity and ensuring minimal impact on the source systems.
>
> **Transform:** During the transform phase, the extracted data is transformed and manipulated to meet the requirements of the target system or application. This may involve cleaning, filtering, aggregating, enriching, or restructuring the data to make it consistent, standardized, and suitable for analysis or storage.

Load: In the load phase, the transformed data is loaded into the target destination, such as a data warehouse, database table, or data lake. This could involve inserting the data into tables, updating existing records, or appending data to existing datasets.

Current-Day Challenges on ETL

Enterprises using on-premises ETL systems face several significant challenges that can impact the efficiency and effectiveness of their data processing workflows. One major issue is scalability.

As data volumes grow, traditional on-premises infrastructure may struggle to scale, leading to performance bottlenecks and increased hardware costs. Managing and maintaining the hardware and software infrastructure for ETL processes is also resource-intensive, requiring specialized IT staff and ongoing investments in upgrades and maintenance.

Data Integration

On-premises systems often need to integrate data from various legacy systems, which can be complex and require custom connectors and extensive data mapping efforts. Data latency is another concern; on-premises ETL processes can be slower due to the time required to move and process large datasets, which can delay access to real-time or near-real-time analytics.

Security and Compliance

Enterprises must ensure that their on-premises ETL systems comply with industry regulations and protect sensitive data throughout the ETL process. This involves implementing robust security measures, which can be costly and complicated to maintain.

Updating and modernizing ETL workflows in an on-premises environment can be challenging. It often involves significant downtime and disruptions, making it difficult to quickly adapt to new business requirements or incorporate the latest technological advancements. These challenges highlight the need for careful planning and robust infrastructure management to ensure efficient and secure ETL operations in on-premises environments.

Challenges in ETL for Cloud Systems

As enterprises increasingly migrate their ETL (Extract, Transform, Load) processes to cloud-based systems, they encounter a new set of challenges distinct from those in traditional on-premises environments. Cloud-based ETL offers scalability and flexibility, but it also introduces complexities in data security, integration, latency, cost management, vendor dependency, and data governance. This article explores these challenges in detail, providing insights into how they impact enterprise data workflows. For a deeper understanding, references to authoritative books on ETL and cloud computing are provided.

One of the foremost challenges in cloud-based ETL is ensuring data security and privacy. When data is transferred to and processed in the cloud, enterprises must implement robust encryption, access controls, and compliance measures to protect sensitive information from breaches and unauthorized access. Additionally, cloud environments must adhere to various data protection regulations such as GDPR and HIPAA, which can be complex and vary by region.

Data Integration

Integrating data from multiple cloud services, on-premises systems, and third-party APIs can be a complex task. This involves handling diverse data formats and ensuring data consistency across various sources and destinations. Effective data mapping and transformation capabilities are crucial to overcome these challenges.

Latency and Performance

Cloud-based ETL processes can suffer from network latency, especially when transferring large volumes of data to and from the cloud. This latency can affect the timeliness of data processing and analytics. Additionally, while cloud systems offer scalability, managing performance to handle variable workloads efficiently without incurring high costs can be difficult.

Cost Management

Cloud services often operate on a pay-as-you-go model, which can lead to unexpected costs if not properly monitored. Data transfer fees, storage costs, and compute charges can quickly escalate. Therefore, balancing performance and cost requires careful planning and optimization of cloud resources.

Vendor Lock-In

Relying heavily on a single cloud provider can lead to vendor lock-in, making it difficult to migrate to another platform or integrate with other services. Ensuring interoperability between different cloud platforms and on-premises systems can be complex and may require additional tools or custom solutions.

Data Governance

Maintaining data quality and governance in a cloud environment is challenging, particularly with large and diverse datasets. Effective management of metadata for data lineage, auditing, and cataloging is essential but can be complicated in a dynamic cloud setting.

SRE for ETL and Data Handling

In today's data-driven world, organizations rely heavily on efficient and reliable data pipelines to extract, transform, and load data from various sources into their analytics and business intelligence systems. However, ensuring the reliability, availability, and performance of these data pipelines can be challenging, especially as data volumes grow and processing demands increase. Site Reliability Engineering (SRE) principles offer a robust framework for addressing these challenges and optimizing the operation of data pipelines.

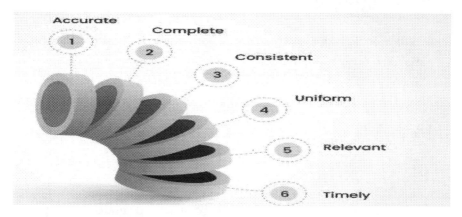

Figure 3-2. *Slices of data reliability*

Site Reliability Engineering (SRE) in the context of ETL and data pipelines involves applying engineering practices to design, build, deploy, and operate reliable, scalable, and efficient data processing systems.

SRE principles aim to minimize the impact of failures, ensure high availability of data pipelines, and optimize performance to meet service-level objectives (SLOs) and service-level agreements (SLAs) for data processing.

Data Quality Assurance Techniques

Data Profiling

Data profiling involves analyzing the structure, content, and quality of data to gain insights into its characteristics, validating that data is consistent and formatted correctly, and performing mathematical checks on the data (e.g., sum, minimum, or maximum). Structure discovery helps understand how well data is structured—for example, what percentage of phone numbers do not have the correct number of digits.

Traditional data profiling is a complex activity performed by data engineers prior to, and during, ingestion of data to a data warehouse. Data is meticulously analyzed and processed (with partial automation) before it is ready to enter the pipeline. Today, more organizations are moving data infrastructure to the cloud, and discovering that data ingestion can happen at the click of a button. Cloud data warehouses, data management tools, and ETL services come preintegrated with hundreds of data sources.

Techniques

Statistical Summaries: Calculating basic statistics such as mean, median, standard deviation, and frequency distributions to understand data distributions

Column Analysis: Examining individual columns to identify data types, value patterns, uniqueness, and cardinality

Data Pattern Recognition: Detecting patterns and formats within data values to uncover inconsistencies or anomalies

Data Quality Assessment: Where data is evaluated for completeness, accuracy, consistency, and uniqueness to assess its overall reliability and fitness for use

Benefits

Data profiling offers significant benefits to organizations by enhancing data quality and reliability. It involves analyzing datasets to understand their structure, content, and relationships, which helps in identifying inaccuracies, inconsistencies, and anomalies. By gaining insights into data characteristics, organizations can make more informed decisions, improve data governance, and ensure compliance with regulatory requirements. Data profiling also facilitates data integration and migration by ensuring that data from disparate sources is consistent and accurate. Moreover, it supports better data management practices by enabling the identification of redundant data, thus optimizing storage and improving overall data efficiency.

Outlier Detection

Outlier detection is a crucial aspect of data quality management. It involves identifying data points that deviate significantly from the rest of the dataset. These anomalies can indicate errors, rare events, or novel insights, making outlier detection an essential process for maintaining the accuracy and reliability of data. Outliers can significantly impact data analysis and the resulting business decisions. If not identified and addressed, they can lead to incorrect conclusions, skewed statistical analyses, and poor decision-making.

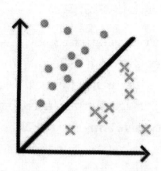

Figure 3-3. *An outlier illustration*

For instance, in financial data, an outlier might indicate a fraudulent transaction. In sensor data, it could signal a malfunctioning sensor. Detecting these anomalies is vital for ensuring that data-driven insights are accurate and actionable.

Techniques

Statistical Methods: Using statistical measures such as z-scores, percentiles, and box plots to identify data points that fall outside normal ranges

Machine Learning Algorithms: Employing algorithms such as isolation forests, k-means clustering, and local outlier factor (LOF) to detect outliers based on data distributions and patterns

Domain-Specific Rules: Applying domain knowledge and business rules to flag data points that are unlikely or inconsistent with expected values. Benefits include helps to uncover potential data errors, fraud, or unusual patterns and enables proactive identification and mitigation of data quality issues.

Data Cleansing

Data cleansing, also known as data cleaning or data scrubbing, is the process of identifying and correcting (or removing) inaccurate, incomplete, or irrelevant data from a dataset. This crucial step in data management ensures that the data used for analysis, reporting, and decision-making is accurate and reliable. Clean data enhances the quality of insights derived from data analytics and supports better business outcomes. Poor data quality can lead to erroneous conclusions, misinformed decisions, and increased operational costs. Clean data improves the accuracy of business intelligence, enhances customer satisfaction by reducing errors in customer-related processes, and ensures compliance with regulatory standards. Moreover, it enables more effective use of advanced analytics and machine learning models, which rely heavily on high-quality data.

Techniques

Standardization: Converting data into a consistent format or representation (e.g., date formats, address formats) to improve consistency and comparability

Figure 3-4. *Image showing standard data cleansing life cycle*

Deduplication: Identifying and removing duplicate records or entries to ensure data integrity and accuracy

Error Correction: Automatically or manually correcting data errors, misspellings, or invalid values based on predefined rules or reference data

Benefits

The primary benefit of data cleansing is the significant improvement in data quality. By eliminating errors, inconsistencies, and redundancies, data cleansing ensures that the data is accurate, complete, and reliable. High-quality data is essential for accurate analysis, reliable reporting, and informed decision-making. Organizations are required to comply with various data regulations and standards, such as the General Data Protection Regulation (GDPR) and the Health Insurance Portability and Accountability Act (HIPAA). Data cleansing helps maintain data accuracy and integrity, ensuring compliance with these regulatory requirements and reducing the risk of legal issues and penalties. Clean data also enhances risk management by providing accurate information for identifying and mitigating potential risks.

Data Validation and Verification

Data validation and verification are critical processes for maintaining data integrity, particularly in ETL systems. Validation ensures data conforms to predefined rules, standards, and constraints, checking for correctness, completeness, and consistency. Verification, on the other hand, confirms data accuracy by comparing it against known sources or reference data. By combining these processes, organizations can guarantee that their data accurately represents real-world entities, enabling reliable analysis and decision-making.

Data Validation Techniques

Schema Validation

Ensure that data conforms to the expected structure, format, and data types defined by the schema. Validate field lengths, data formats (e.g., dates, emails), and referential integrity constraints.

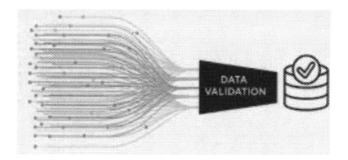

Figure 3-5. *Unstructured data getting validated*

Cross-Field Validation

Validate relationships between multiple fields within a dataset to ensure consistency.

Example: Checking that a customer's birth date is not later than the current date.

Completeness Check

Verify that all required fields are present and populated in the dataset.

Detect missing or null values that could impact data quality.

Data Verification Techniques

Source-to-Target Comparison

Compare data extracted from the source system with the transformed data loaded into the target system. Verify that the transformation logic preserves data integrity and accuracy.

Record Count Verification

Ensure that the number of records processed during ETL matches expectations. Detect discrepancies that may indicate data loss or duplication.

Checksum Verification

Calculate checksums or hash values for data at different stages of the ETL process. Compare checksums to ensure data integrity and detect any unintended alterations.

Metadata Management

Metadata management is a crucial aspect of maintaining the reliability and effectiveness of data and ETL (Extract, Transform, Load) processes. Metadata provides essential context and structure to data, facilitating its understanding, governance, and usage.

Table 3-1. *Sample relational data*

Customer Name	Order Number	← Metadata
David	1	← Data
Katie	2	← Data
Toni	3	← Data
Salomon	4	← Data

Best Practices for Metadata Management

Standardization: Establish standardized naming conventions, metadata models, and taxonomies to ensure consistency and uniformity across datasets and systems.

Documentation: Document metadata comprehensively, including data dictionaries, data lineage diagrams, ETL job designs, and transformation rules. Maintain up-to-date documentation to support data understanding and governance.

Metadata Repository: Implement a centralized metadata repository or catalog to store and manage metadata assets effectively. Use metadata management tools or platforms to automate metadata capture, storage, and retrieval processes.

Data Lineage and Impact Analysis: Capture data lineage information to track the flow of data from source to destination and understand its transformation journey. Conduct impact analysis to assess the downstream effects of changes to data or ETL processes.

Metadata Governance: Establish metadata governance policies and procedures to govern the creation, maintenance, and use of metadata assets. Define roles and responsibilities for metadata stewards and establish processes for metadata quality assurance and validation.

Data Profiling and Quality Assessment: Use data profiling techniques to analyze data quality issues, anomalies, and patterns. Incorporate metadata-driven data quality rules and metrics into ETL processes to monitor and improve data quality.

Version Control and Change Management: Implement version control and change management practices for metadata artifacts to track changes, manage revisions, and ensure traceability. Maintain audit trails to record metadata modifications and access history.

Metadata Integration: Integrate metadata management with other data management disciplines, such as data governance, master data management (MDM), and data quality management. Ensure interoperability and alignment between metadata repositories and data management tools.

Data Cleansing and Enrichment

Data cleansing and enrichment are fundamental processes in data management aimed at improving the quality, accuracy, and usability of data. While data cleansing focuses on identifying and correcting errors and inconsistencies in existing data, data enrichment involves enhancing data by adding valuable information from external sources. Data enrichment involves enhancing existing datasets by supplementing them with additional information from external sources. This additional information can include demographic data, geographic information, social media data, and other relevant insights that enrich the dataset and provide more context and value.

Importance of Data Enrichment

Enhanced Insights: Enriched data provides a deeper understanding of customers, markets, and trends, leading to more meaningful insights and opportunities.

Improved Personalization: Enriched data enables personalized experiences and targeted marketing campaigns by providing a more comprehensive view of customer preferences and behavior.

Better Decision-Making: Enriched data enhances decision-making by providing a more holistic view of the factors influencing business operations and outcomes.

Competitive Advantage: Leveraging enriched data allows organizations to gain a competitive edge by uncovering hidden patterns, trends, and opportunities that drive innovation and growth.

Common Data Enrichment Techniques

Appending External Data: Supplement existing datasets with additional information obtained from external sources such as third-party data providers, public databases, and social media platforms.

Geocoding: Enhance location-based data by converting addresses into geographic coordinates (latitude and longitude) for spatial analysis and visualization.

Demographic Enrichment: Augment demographic data with additional attributes such as age, income, education level, and household composition to gain deeper insights into customer segments.

Social Media Monitoring: Integrate social media data into existing datasets to understand customer sentiment, behavior, and engagement with brands and products.

Best Practices for Data Cleansing and Enrichment

Define Clear Objectives: Clearly define the objectives and goals of data cleansing and enrichment initiatives to ensure alignment with business priorities and requirements.

Use Automated Tools: Leverage data cleansing and enrichment tools and software to automate repetitive tasks and streamline the process.

Regular Maintenance: Implement regular data cleansing and enrichment routines to ensure data quality and relevance are maintained over time.

Data Governance: Establish data governance policies and procedures to govern the data cleansing and enrichment process, including data quality standards, ownership, and accountability.

Continuous Data Monitoring

Continuous data monitoring is a critical component of maintaining the reliability and effectiveness of ETL (Extract, Transform, Load) processes. By continuously monitoring data as it flows through the ETL pipeline, organizations can detect anomalies, errors, and issues in real time, ensuring data quality, accuracy, and consistency.

Figure 3-6. *Chart for continuous data monitoring*

Continuous Improvement and Optimization

Continuously monitor, analyze, and optimize ETL processes and data pipelines based on performance metrics, user feedback, and evolving business requirements. By applying SRE principles to ETL and data pipelines, organizations can enhance the reliability, scalability, and efficiency of their data processing systems, ensuring high-quality data delivery and insights for business operations and decision-making.

Conclusion: In the rapidly evolving landscape of data-driven decision-making, the reliability of ETL (Extract, Transform, Load) processes cannot be overstated. ETL processes serve as the backbone of data integration, transforming raw data from various sources into meaningful insights that drive strategic business decisions. Ensuring the reliability of these processes is crucial for maintaining data integrity, accuracy, and consistency, which are foundational to building trust in data-driven initiatives. Reliable ETL processes minimize the risk of data errors, discrepancies, and inconsistencies that can undermine the quality of data analytics and reporting. By implementing robust data validation, verification, cleansing, and enrichment practices, organizations can safeguard the quality of their data, ensuring it is fit for purpose and aligned with business needs. Moreover, continuous data monitoring and effective metadata management further enhance the reliability of ETL processes. These practices enable early detection and resolution of issues, optimize performance, and ensure compliance with regulatory and governance standards. Reliable ETL processes thus support seamless data integration, enhance operational efficiency, and provide a solid foundation for advanced analytics, machine learning, and other data-driven technologies. In conclusion, investing in the reliability of ETL processes is essential for any organization seeking to leverage its data assets effectively. It not only enhances data quality and decision-making capabilities but also fosters a culture of data trust and integrity. As organizations continue to navigate the complexities of the digital age, the reliability of ETL processes will remain a critical factor in achieving sustainable growth, innovation, and competitive advantage.

Bibliography

1. Kimball, Ralph, and Joe Caserta. "The Data Warehouse ETL Toolkit: Practical Techniques for Extracting, Cleaning, Conforming, and Delivering Data." Wiley, 2004

2. Inmon, W. H. "Building the Data Warehouse." 4th ed., Wiley, 2005

3. Krishnan, Krish. "Data Warehousing in the Age of Big Data." Morgan Kaufmann, 2013

4. "Data Integration Blueprint and Modeling: Techniques for a Scalable and Sustainable Architecture" by Anthony David Giordano offers practical techniques for integrating data across different systems and platforms

5. "Cloud Computing: Concepts, Technology & Architecture" by Thomas Erl provides insights into the architecture of cloud computing and how to manage performance and scalability

6. "Architecting the Cloud: Design Decisions for Cloud Computing Service Models (SaaS, PaaS, and IaaS)" by Michael J. Kavis discusses strategies for managing and optimizing costs in cloud environments

7. Krutz, Ronald L., and Russell Dean Vines. "Cloud Security: A Comprehensive Guide to Secure Cloud Computing." Wiley, 2010

8. Giordano, Anthony David. "Data Integration Blueprint and Modeling: Techniques for a Scalable and Sustainable Architecture." IBM Press, 2010

9. Erl, Thomas. "Cloud Computing: Concepts, Technology & Architecture." Prentice Hall, 2013

10. Kavis, Michael J. "Architecting the Cloud: Design Decisions for Cloud Computing Service Models (SaaS, PaaS, and IaaS)." Wiley, 2014

11. Arora, Kamal, et al. "Multi-Cloud Strategy for Cloud Architects." Packt Publishing, 2021

12. "Outlier Analysis" by Charu C. Aggarwal

13. "Data Quality: The Accuracy Dimension" by Jack E. Olson

14. "Data Cleaning: The Ultimate Practical Guide" by Ihab F. Ilyas and Xu Chu

15. "Enterprise Metadata Management" by Lukaszewski, "Building and Managing the Metadata Repository" by Ponniah, and "Data Governance: How to Design, Deploy, and Sustain an Effective Data Governance Program" by Ladley

16. https://nap.nationalacademies.org/read/18987/chapter/7

17. https://www.geeksforgeeks.org/reliability-in-system-design/

18. https://www.wilderisk.co.uk/about/blog/what-is-design-for-reliability/

CHAPTER 4

The Resilient Design Techniques

Authors:

Sriram Panyam

Harshavardhan Nerella

Anirudh Khanna

Reviewer:

Manoj Kuppam

Resiliency Patterns for Mitigating Failures

Resiliency in systems refers to the ability of a software architecture to withstand and recover from failures, ensuring continuity of service under various conditions. In the realm of modern software architecture, the importance of resilience cannot be overstated, as it directly impacts user experience, system reliability, and business continuity. Facing common challenges such as network failures, hardware malfunctions, and unexpected surges in traffic, designing for resiliency involves strategic planning and the implementation of patterns that help systems gracefully handle and quickly recover from disruptions. This foundational approach not only mitigates risks but also strengthens the overall architecture against future uncertainties.

M. Kuppam, *Enterprise Digital Reliability*, https://doi.org/10.1007/979-8-8688-1032-9_4

Resiliency: Core Concepts

The core concepts of resiliency revolve around enabling systems to maintain functionality despite errors or high demand. Fault tolerance and high availability are pivotal; the former allows a system to continue operating in the event of a failure within some of its components, while the latter ensures that services always remain accessible. Redundancy plays a crucial role by duplicating critical components or functions, thereby providing a backup mechanism that enhances reliability. Graceful degradation ensures that when systems are under stress, they can still offer limited functionality, prioritizing core services. Antifragility goes beyond resilience by having systems not just withstand shocks but improve their capability in response to stress, making them dynamically robust and adaptable.

Resiliency Patterns

Resiliency patterns are strategic design principles aimed at enhancing the robustness and reliability of software systems. They serve as guidelines for building architectures that can effectively handle and recover from failures, ensuring minimal disruption to users and maintaining service continuity. These patterns are essential in today's digital landscape, where system uptime and performance directly impact user satisfaction and business success. Resiliency patterns can be broadly categorized into several key types, each addressing specific aspects of system resilience.

- **Fault Handling Patterns** such as retry, circuit breaker, and fallback focus on managing errors and exceptions in a controlled manner.

- **Resource Management Patterns**, like bulkhead and throttle, aim to prevent system overload by managing and isolating resources.

- **Failure Recovery Patterns** including backup and restore ensure that systems can quickly recover from failures, preserving data integrity and availability.

By implementing these patterns, developers can create systems that are not only more resilient to failures but also more adaptable and scalable, enhancing overall system quality and reliability. We will dive into these patterns next.

Retry Pattern

The retry pattern is a fundamental resiliency pattern aimed at enhancing system robustness by attempting to execute an operation multiple times in case of a failure, under the assumption that the error is transient and can be overcome by repeating the request. This pattern is particularly useful in scenarios where operations are prone to intermittent failures, such as network requests, database transactions, or any external system interactions where temporary issues like network latency or brief service downtime can occur.

Example Use Cases

- **Network Requests:** Automatically retrying HTTP requests that failed due to temporary network glitches

- **Database Transactions:** Retrying database operations that fail due to temporary locking or connectivity issues

Implementation and Considerations

Implementing the retry logic involves defining the maximum number of attempts and the delay between attempts. It's crucial to implement exponential backoff and jitter to avoid overwhelming the system or the service being called. Exponential backoff increases the wait time between retries, while jitter introduces variability to prevent synchronized retries from multiple instances.

Best Practices

- **Define a Maximum Retry Count:** Avoid infinite retries to prevent resources from being exhausted.

- **Implement Exponential Backoff:** Gradually increase the delay between retries to minimize the load on the system and increase the chance of recovery.

- **Add Jitter:** Randomize the delay periods to avoid thundering herd problems when many instances retry simultaneously.

- **Handle Specific Exceptions:** Only retry on exceptions known to be transient and recoverable.

Sample Pseudocode

This pseudocode illustrates a basic retry logic implementation with exponential backoff and jitter, encapsulating best practices for handling transient failures in resilient system design.

```
import time
import random

def retry_operation(operation, max_attempts=5):
    for attempt in range(max_attempts):
        try:
            return operation()
        except TemporaryError as e:
            wait = 2 ** attempt + random.random()
            time.sleep(wait)
```

```
        except PermanentError:
            break
    raise MaxRetriesExceededError

# Example usage
def example_operation():
    # Operation that might fail transiently
    if random.randint(0, 100) < 10:
        raise Exception("Transient Failure")

try:
    result = retry_operation(example_operation)
except MaxRetriesExceededError:
    print("Operation failed after retrying")
```

Circuit Breaker Pattern

The circuit breaker pattern is used to prevent a system from performing operations that are likely to fail. It acts similarly to an electrical circuit breaker in buildings, where it automatically cuts off the electrical flow when a fault is detected, preventing further damage. In software terms, the circuit breaker pattern prevents a system from making requests to a service or component that is known to be in a failed state, thereby giving it time to recover and avoiding cascading failures in the system.

Examples

- Protecting applications from repeatedly trying to execute an operation that's likely to fail, such as a database request when the database is down

- Managing dependencies on external services by monitoring their availability and performance

Implementation Strategies and Considerations

- **State Management**: Implementing the circuit breaker requires managing three states: closed (operations are allowed), open (operations are blocked), and half-open (a limited number of operations are allowed to test if the underlying problem has been resolved).

- **Failure Threshold**: Define criteria for failures that would trip the breaker, such as a certain number of failures within a timeframe.

- **Recovery Timeout**: Set a timeout for how long the breaker remains in the open state before transitioning to half-open to test for recovery.

- **Fallback Mechanisms**: Implement fallbacks for when operations are prevented, ensuring users are not left without options.

Best Practices

- Monitor and log state changes and failures to inform adjustments and improvements.

- Customize the threshold and timeout values based on the criticality of the dependent service and the acceptable downtime.

```python
class CircuitBreaker:
    def __init__(self, failure_threshold, recovery_timeout):
        self.failure_threshold = failure_threshold
        self.recovery_timeout = recovery_timeout
        self.failures = 0
        self.state = "CLOSED"
        self.last_failure_time = None
```

```python
def attempt_operation(self, operation):
    time_since_last_failure = time.time() - self.last_
    failure_time
    if self.state == "OPEN" and \
        time_since_last_failure > self.recovery_timeout:
        self.state = "HALF-OPEN"
    if self.state == "CLOSED" or self.state == "HALF-OPEN":
        try:
            operation()
            self.reset()
            return "Operation Successful"
        except:
            self.failures += 1
            self.last_failure_time = time.time()
            if self.failures >= self.failure_threshold:
                self.state = "OPEN"
            return "Operation Failed: Circuit Open"
    else:
        return "Operation Blocked: Circuit Open"

def reset(self):
    self.failures = 0
    self.state = "CLOSED"
```

This pattern is instrumental in building resilient systems that can handle failures gracefully, maintaining system stability and availability.

Bulkhead Pattern

The bulkhead pattern is derived from naval architecture where a ship's hull is partitioned into watertight compartments. If one compartment floods, the others remain unaffected, preventing the ship from sinking.

Similarly, in software architecture, the bulkhead pattern isolates elements of an application into compartments to prevent failures in one part from cascading to others. This isolation ensures that if one component becomes overloaded or fails, it doesn't bring down the entire system, thereby enhancing fault tolerance and system reliability.

Examples of the bulkhead pattern include microservice architectures where different services run independently. For instance, isolating database operations from user authentication services ensures that an overload or failure in handling user logins doesn't impact database operations.

Implementing the bulkhead pattern involves defining logical or physical boundaries around components or services. This can be achieved by limiting the number of concurrent threads that can access a particular component or by deploying services on separate hardware or containers.

- Key considerations and best practices include careful planning of resources and limits to prevent underutilization or bottlenecks. Monitoring and dynamic adjustment capabilities are critical, as static bulkheads can become either bottlenecks or underused resources. It's also essential to design fallback mechanisms for handling failures within a bulkhead, ensuring the system can degrade gracefully.

- The successful application of the bulkhead pattern improves system resilience by limiting the scope of failures and maintaining service availability, even under adverse conditions. As with all resiliency patterns, the goal is not just to prevent failures but to manage them in a way that minimizes impact on the user experience and overall system functionality.

```python
class CircuitBreaker:
    def __init__(self, failure_threshold, recovery_timeout):
        self.failure_threshold = failure_threshold
        self.recovery_timeout = recovery_timeout
        self.failures = 0
        self.state = "CLOSED"
        self.last_failure_time = None

    def attempt_operation(self, operation):
        time_since_last_failure = time.time() - self.last_
        failure_time
        if self.state == "OPEN" and \
            time_since_last_failure > self.recovery_timeout:
                self.state = "HALF-OPEN"
        if self.state == "CLOSED" or self.state == "HALF-OPEN":
            try:
                operation()
                self.reset()
                return "Operation Successful"
            except:
                self.failures += 1
                self.last_failure_time = time.time()
                if self.failures >= self.failure_threshold:
                    self.state = "OPEN"
                return "Operation Failed: Circuit Open"
        else:
            return "Operation Blocked: Circuit Open"

    def reset(self):
        self.failures = 0
        self.state = "CLOSED"
```

This pattern is instrumental in building resilient systems that can handle failures gracefully, maintaining system stability and availability.

Timeout Pattern

The timeout pattern is a resiliency strategy used to limit the time awaiting a response from a service or operation, preventing system hang-ups and ensuring resources aren't indefinitely tied up. This pattern is crucial in distributed systems where network latency or service unavailability can stall operations. For instance, in web service calls or database queries, implementing a timeout can safeguard against prolonged downtime.

Implementing the timeout pattern often involves setting a maximum time limit for an operation. If the operation exceeds this limit, it's terminated or a fallback action is triggered.

Considerations and best practices include

- Determining optimal timeout values based on operational benchmarks

- Implementing fallback mechanisms to handle operations that exceed timeout limits

- Regularly reviewing timeout settings to align with changing system performance

Pseudocode Example in Python

```python
import signal

def timeout_handler(signum, frame):
    raise TimeoutException()

signal.signal(signal.SIGALRM, timeout_handler)
signal.alarm(timeout_seconds)  # Set timeout
```

```
try:
    # Operation that might hang
finally:
    signal.alarm(0)  # Cancel timeout
```

This pattern helps maintain system responsiveness and reliability, especially in environments prone to unpredictable delays.

Fallback Pattern

The fallback pattern is a resiliency strategy used in software design to provide an alternative solution when a primary method fails. This pattern ensures that the system can gracefully degrade functionality, instead of completely failing, by offering a secondary path of execution. For example, if a system's primary data source becomes unavailable, the fallback could be to retrieve data from a cache or return a default value.

Use cases for the fallback pattern include handling failures in external service calls, dealing with unavailable resources, or providing default content when the primary content cannot be loaded.

Implementing the fallback pattern involves wrapping the primary operation in a mechanism that catches failures and, instead of throwing an error, calls a predefined fallback method. This method could involve complex logic, such as attempting to connect to an alternative service, or something simple, like returning static data.

Considerations and best practices include ensuring that the fallback logic does not introduce significant latency, is not as prone to failure as the primary method, and does not degrade the user experience. It's also important to monitor the usage of fallbacks to detect underlying issues with the primary paths.

Pseudocode Example in Python

```python
def primary_operation():
    # Attempt primary operation
    raise Exception("Primary operation failed")

def fallback_operation():
    # Fallback logic
    return "Default response"

def execute_with_fallback():
    try:
        return primary_operation()
    except Exception as e:
        return fallback_operation()

# Execute
result = execute_with_fallback()
print(result)
```

This pseudocode demonstrates a basic implementation of the fallback pattern, where execute_with_fallback tries to execute the primary_operation and resorts to fallback_operation upon failure.

Rate Limiting and Throttling

Rate limiting and throttling are critical resiliency patterns used to control the number of requests a user or service can make to a system within a specific timeframe. These patterns are essential for preventing overuse of resources, maintaining service availability, and ensuring a fair distribution of system capacity among users. By limiting the request rate, systems can protect against overwhelming traffic, reduce the risk of DDoS attacks, and manage the load more effectively, especially during peak times.

Example Use Cases

- **APIs:** To prevent abuse and ensure equitable access, APIs often implement rate limiting, allowing developers a certain number of requests per minute or hour.

- **Web Applications:** Throttling can be used to control the login attempts made by users, mitigating brute-force attacks.

Implementation and Best Practices

A simple but effective approach to implementing rate limiting is the token bucket algorithm. This algorithm allows for a certain number of tokens to be consumed within a timeframe, with each request consuming a token. When the tokens are depleted, further requests are either delayed or rejected until the bucket is refilled.

Pseudocode

```
def token_bucket(request_rate, capacity, tokens=0, last_
checked=time.now()):
    if tokens < capacity:
        tokens += (time.now() - last_checked) * request_rate
        tokens = min(tokens, capacity)
    last_checked = time.now()
    if tokens >= 1:
        tokens -= 1
        return True
    return False

# Sample Usage
request_rate = 5  # 5 requests per second
capacity = 10  # Burst capacity
allow_request = token_bucket(request_rate, capacity)
```

```
if allow_request:
    # Process the request
else:
    # Return rate limit exceeded error
```

Considerations

- **Fairness:** Implement rate limiting fairly to ensure no user is unduly restricted while maintaining system integrity.

- **Transparency:** Inform users of rate limits, ideally before they reach the limit.

- **Adaptability:** Adjust limits based on usage patterns and system capacity.

Implementing rate limiting and throttling effectively requires a balance between protecting the system and providing a seamless user experience. Monitoring and adjusting policies based on real-world usage are crucial for maintaining this balance.

Implementing Resiliency Patterns

Implementing resiliency patterns in existing systems requires a careful approach to ensure seamless integration without disrupting current functionalities. This involves identifying critical components that need fortification and gradually introducing patterns like retries, circuit breakers, and bulkheads. Monitoring and metrics play a pivotal role in resiliency, providing real-time insights into system performance and the effectiveness of implemented patterns. Key metrics include response times, failure rates, and resource utilization levels. Testing and validation are also crucial, employing strategies like chaos engineering to simulate failures and stress tests to validate the system's resilience.

These approaches help in fine-tuning the system to effectively withstand and recover from unforeseen failures, thereby enhancing overall system reliability and user satisfaction.

Tools and Frameworks

The landscape of tools and frameworks designed to enhance system resiliency is vast, ranging from libraries that implement specific resiliency patterns to platforms that offer comprehensive fault tolerance capabilities. Popular tools like Hystrix, Resilience4j, and Polly are widely used for implementing circuit breaker, retry, timeout, and bulkhead patterns in various programming environments. Additionally, infrastructure as code (IaC) tools such as Terraform and cloud services from AWS, Azure, and Google Cloud provide mechanisms for creating redundant, scalable, and self-healing systems. Choosing the right tools for your needs requires understanding the specific resilience requirements of your system, including the programming language, deployment environment, and the criticality of maintaining high availability and fault tolerance. Assessing the compatibility, community support, and maintenance of these tools is also crucial to ensure they align with your system's long-term resilience strategy.

Future Trends

The future of resilience in system design is poised to evolve significantly, driven by emerging patterns and the integration of artificial intelligence (AI) and machine learning (ML). These technologies promise to revolutionize how systems anticipate, respond to, and recover from disruptions. AI and ML can analyze vast datasets to predict potential system failures before they occur, enabling preemptive action. Additionally, they can automate the optimization of resilience strategies, learning from past incidents to enhance system robustness over time. We will likely see the development of self-healing systems that can autonomously detect, diagnose, and repair faults, making resilience an intrinsic, dynamic characteristic of technology infrastructure. This advancement toward more intelligent and adaptive systems will not only reduce downtime but also improve efficiency and user experience, marking a significant leap forward in the pursuit of truly resilient systems.

Conclusion

Understanding and implementing resiliency patterns is essential for creating robust, reliable software systems capable of withstanding and recovering from unforeseen challenges. From fault tolerance and redundancy to the sophisticated use of AI for predictive resilience, these patterns form the cornerstone of modern system architecture. The journey toward achieving system resilience is ongoing, with new patterns and technologies continuously emerging to address evolving threats and complexities. It is imperative for developers and architects to stay abreast of these developments, incorporating resiliency patterns into their projects. By doing so, they not only safeguard their systems against disruptions but also contribute to a future where digital infrastructures are inherently strong, adaptable, and resilient. Embracing these principles is not just a measure of caution; it's a strategic investment in the future readiness and success of technology solutions.

Redundancy Techniques and High Availability

Introduction to High Availability and Redundancy

In the realm of IT infrastructure, the concepts of high availability and redundancy are pivotal to ensuring that systems remain operational and accessible, minimizing downtime and maintaining business continuity. High availability refers to the design and implementation of systems that are robust and resilient, capable of operating continuously without significant disruption. Redundancy, on the other hand, is a strategy employed to duplicate critical components or functions of a system to provide a backup in the event of a failure.

The significance of high availability and redundancy cannot be overstated, as system downtime can lead to substantial financial losses, diminished productivity, and eroded customer trust. A report by Gartner highlighted that the average cost of IT downtime is approximately $5,600 per minute, underscoring the critical need for businesses to invest in redundant systems and high availability solutions.

Implementing redundancy techniques involves the creation of additional instances of system components, such as servers, databases, and network connections, ensuring that if one component fails, another can seamlessly take over, thus maintaining the system's overall availability. For example, redundant power supplies in a data center can ensure that servers continue to operate even if one power source fails, illustrating the practical application of redundancy in maintaining high availability.

In essence, high availability and redundancy are about preparing for the unexpected, designing systems that can withstand failures and continue to operate effectively. As businesses increasingly rely on digital infrastructure, the adoption of these principles becomes not only a best practice but a necessity to safeguard operations and maintain a competitive edge in today's technology-driven landscape

Understanding the Levels of Redundancy

In the landscape of IT infrastructure, redundancy is not a one-size-fits-all solution. It's crucial to understand the various levels of redundancy to design systems that align with business needs and risk tolerance. These levels, commonly referred to as N+1, N+2, and 2N, provide different degrees of availability and protection against system failures.

N+1 Redundancy: This is the most basic level of redundancy, where "N" represents the number of components necessary to run the system and "+1" signifies an additional component. In an N+1 setup, there's one extra component beyond what's needed for normal operation, ready to take over in case of a single component failure. For instance, if a system requires four servers to function, an N+1 redundancy would mean having five servers in total, ensuring that the system remains operational even if one server goes down.

N+2 Redundancy: Advancing a step further, N+2 redundancy includes two extra components over the necessary count. This level provides an additional safety net, allowing the system to cope with two simultaneous component failures without affecting performance. In the context of our previous example, an N+2 setup for four required servers would include two additional servers, bringing the total to six.

2N Redundancy: The 2N level represents a full duplication of all system components, essentially doubling the infrastructure. In a 2N configuration, if the operational requirement is four servers,

the system will have eight servers in total. This
level of redundancy offers the highest protection,
ensuring system continuity even if an entire set of
components fails.

Choosing the right level of redundancy is a strategic decision
that balances cost, complexity, and risk management. While higher
redundancy levels offer greater fault tolerance, they come with increased
costs and maintenance requirements. Organizations must assess their
critical system needs, downtime tolerance, and budget constraints to
determine the most appropriate redundancy level. The implementation of
these redundancy levels is a testament to an organization's commitment to
reliability and continuous service delivery, underscoring the essential role
of redundancy in modern IT infrastructures.

Redundancy in Hardware Components

Hardware redundancy is a cornerstone in building resilient IT systems. It
involves duplicating critical hardware components to ensure that a system
can continue to operate even if one part fails. This redundancy is crucial
across various hardware elements, including power supplies, network
interfaces, and storage systems.

Power Supplies: Redundant power supplies are
essential for preventing downtime due to power
failures. In a redundant setup, servers and network
devices are equipped with dual power supply units
(PSUs). If one PSU fails or if there's an interruption
in its power source, the second PSU seamlessly takes
over, maintaining the device's operation without
interruption. This approach is particularly critical in
data centers where continuous uptime is imperative.

Network Interfaces: Network interface redundancy, often implemented through techniques like NIC (network interface card) teaming or bonding, ensures uninterrupted network connectivity. If one network interface encounters a fault, the traffic automatically reroutes to the backup interface, maintaining network availability and preventing data loss or access issues.

Storage Systems: Redundancy in storage is commonly achieved through RAID (redundant array of independent disks) configurations. RAID allows for data to be duplicated across multiple disks, ensuring that if one disk fails, the data remains accessible from another disk in the array. For example, RAID 1 mirrors data across two disks, while RAID 5 distributes data and parity information across three or more disks, providing fault tolerance and improved performance.

Implementing hardware redundancy is a proactive measure that mitigates the risk of single points of failure in an IT infrastructure. By duplicating critical hardware components, organizations can enhance system reliability, ensure data integrity, and maintain business continuity even in the face of hardware malfunctions. This practice underscores the importance of redundancy in the design and operation of robust IT systems, where the cost of downtime far exceeds the investment in redundant hardware solutions.

Network Redundancy

Network redundancy is a critical aspect of designing high-availability systems, ensuring that communication and data exchange within an IT infrastructure remain uninterrupted even in the face of component failures. By implementing redundant network paths, failover mechanisms, and load balancing, organizations can significantly enhance the reliability and resilience of their network infrastructure.

Redundant Network Paths: This involves creating multiple pathways for data to travel within a network, ensuring that if one path becomes unavailable, data can automatically reroute through an alternate path without disrupting the network service. Such redundancy is vital in preventing single points of failure, a fundamental principle in network design. For example, having dual network connections from different service providers can maintain network availability even if one provider experiences an outage.

Failover Mechanisms: Failover is an automated process where network functions switch over to a redundant or standby system upon the detection of a failure. Implementing failover mechanisms, such as Virtual Router Redundancy Protocol (VRRP) or Hot Standby Router Protocol (HSRP), ensures that network services remain operational, seamlessly transitioning to backup systems with minimal or no downtime for users.

Load Balancing: Beyond redundancy, load balancing distributes network traffic across multiple servers or network paths, enhancing performance and availability. By evenly distributing traffic, load balancers prevent any single server or network link from becoming a bottleneck, thereby improving the overall resilience and efficiency of the network.

Incorporating these elements into network design not only fortifies the network against failures but also optimizes performance, ensuring that businesses can maintain continuous operations and deliver consistent service quality. As networks grow increasingly complex and critical to organizational success, the implementation of comprehensive network redundancy strategies becomes indispensable in safeguarding network infrastructure against the unforeseen.

Clustering and Failover

Clustering and failover mechanisms are cornerstone strategies in building high-availability systems, ensuring that services can continue without interruption, even in the event of hardware or software failures. This section delves into how clustering works, its benefits, and the critical role of failover processes in maintaining system continuity.

Server Clustering: Server clustering refers to a group of servers working together as a single system to provide higher availability, scalability, and reliability. Clusters are designed to detect the failure of a server or software component and automatically redistribute the workload to other servers within the cluster. This design not only enhances the availability of services but also facilitates scalability by allowing additional servers to be added to the cluster as needed. For instance, Microsoft SQL Server uses Windows Server Failover Clustering (WSFC) to ensure high availability of database services.

Failover Processes: Failover is the automatic switching to a redundant or standby server, system, or network upon the failure or abnormal termination of the currently active application, server, system, or network. Failover processes are integral to cluster management, ensuring minimal service interruption. These processes are typically swift and seamless, often unnoticed by end users. For example, in a web server cluster, if one server fails, the failover mechanism redirects traffic to the remaining servers, ensuring continuous service availability.

Benefits of Clustering and Failover: The primary benefit of implementing clustering and failover is the significant reduction in downtime and the assurance of service continuity. These strategies support critical applications and services, particularly in environments where downtime can lead to significant financial losses or safety risks.

In summary, clustering and failover are vital components of a robust high-availability strategy. They provide the framework for continuous operational presence, enabling businesses to maintain service levels and meet the expectations of their users, thereby safeguarding against the potential adverse impacts of system failures.

Data Center Redundancy

Data center redundancy is a critical aspect of designing resilient IT infrastructures, ensuring that core operational functions remain uninterrupted in the face of various failures. This section explores the key components of data center redundancy, including power supply, cooling systems, and geographical redundancy.

Power Supply Redundancy: Ensuring a continuous power supply is crucial for data center operations. Implementing redundant power sources, including uninterruptible power supply (UPS) systems and backup generators, is essential to maintain power during outages. An N+1 or 2N redundancy in power supply systems can significantly mitigate the risk of downtime. For instance, in an N+1 setup, if one UPS system fails, an extra unit is already in place to take over the load without interrupting the power supply.

Cooling System Redundancy: Data centers require efficient cooling systems to prevent overheating, which can lead to equipment failure and data loss. Redundant cooling systems ensure that if one unit fails, another can immediately take over, maintaining optimal operating temperatures. Similar to power supply redundancy, cooling systems often follow an N+1 or 2N redundancy model.

Geographical Redundancy: To protect against site-specific disasters, many organizations implement geographical redundancy by establishing multiple data centers in different locations. This approach ensures that if one data center becomes inoperable due to natural disasters, cyberattacks, or other catastrophic events, another can seamlessly take over its functions, maintaining data integrity and availability.

Data center redundancy is a cornerstone of modern IT strategy, playing a pivotal role in business continuity and disaster recovery planning. By implementing comprehensive redundancy measures, organizations can ensure that their data centers remain resilient, agile, and capable of supporting continuous operations, regardless of unforeseen challenges.

Virtualization and Redundancy

Virtualization has emerged as a transformative technology in IT, offering innovative ways to achieve redundancy and enhance system availability. By abstracting physical hardware into multiple simulated environments or dedicated resources, virtualization allows for more flexible and efficient redundancy strategies.

Role of Virtualization in Redundancy:
Virtualization enables the creation of multiple virtual machines (VMs) on a single physical server, each running its own operating system and applications. This consolidation not only optimizes resource utilization but also facilitates rapid redundancy. If one VM fails, others can continue operating without interruption, and affected services can be quickly migrated to another VM, minimizing downtime.

High Availability in Virtualized Environments:
High availability in virtualized systems is often achieved through clustering VMs across multiple physical hosts. This setup ensures that if one host fails, its VMs are automatically restarted or migrated to other hosts in the cluster. Technologies like VMware's High Availability (HA) and Microsoft's

Hyper-V Replica exemplify how virtualization platforms provide mechanisms to detect host failures and redistribute VM workloads accordingly.

Virtualized Storage for Enhanced Redundancy: Virtualization extends to storage, where it enhances data redundancy. Techniques like storage area networks (SANs) or network attached storage (NAS) can be virtualized to provide redundant storage paths and replication of data across multiple physical devices, ensuring data availability and continuity.

Benefits and Considerations: While virtualization significantly contributes to redundancy, it requires careful planning and management. Overreliance on a single physical server or storage device, even in a virtualized environment, can introduce risks. Hence, it's crucial to implement comprehensive redundancy at both the hardware and virtualization layers to safeguard against potential single points of failure.

In conclusion, virtualization offers a dynamic and efficient approach to achieving redundancy, essential for maintaining high availability and business continuity in modern IT infrastructures. Its ability to quickly recover from hardware failures, coupled with the flexibility to allocate and reallocate resources as needed, underscores its value in enhancing the resilience of IT systems.

CHAPTER 4   THE RESILIENT DESIGN TECHNIQUES

Cloud-Based Redundancy Solution

Cloud computing has revolutionized how organizations approach redundancy, offering scalable and cost-effective solutions for achieving high availability. Cloud-based redundancy leverages the distributed nature of cloud resources to ensure system resilience and data protection, providing a robust framework for business continuity.

Leveraging Cloud for Redundancy: In the cloud, redundancy is inherently built into the infrastructure. Cloud providers distribute their resources across multiple geographically dispersed data centers, ensuring that the failure of a single server or entire data center does not disrupt service. For example, Amazon Web Services (AWS) offers Availability Zones that are physically separated within a region yet connected through low-latency links, allowing businesses to deploy and operate redundant systems across these zones.

Data Redundancy in the Cloud: Cloud platforms provide various services to replicate data across multiple locations, enhancing data durability and availability. Services like Amazon S3 or Google Cloud Storage automatically replicate data across several facilities, ensuring that in the event of a hardware failure, data remains accessible and intact. This level of data redundancy is crucial for disaster recovery and maintaining uninterrupted access to critical data.

Application and Compute Redundancy: Beyond data, cloud environments support redundancy at the application and compute layers. By deploying applications across multiple cloud instances or containers, businesses can ensure that if one instance fails, others can seamlessly take over, maintaining the application's availability. Tools like load balancers distribute traffic across these instances, further enhancing the redundancy and reliability of cloud-based applications.

Advantages and Strategic Considerations: Cloud-based redundancy offers flexibility, scalability, and cost-effectiveness, allowing businesses to tailor their redundancy strategies to specific needs without significant upfront investment in physical infrastructure. However, organizations must carefully design their cloud redundancy architectures, considering aspects like data sovereignty, compliance, and the interdependencies between cloud resources to ensure a comprehensive and effective redundancy strategy.

In summary, cloud-based redundancy solutions provide a powerful approach to achieving high availability, enabling organizations to leverage the cloud's distributed nature to build resilient and reliable IT systems that can withstand failures and maintain continuous operations.

Conclusion

In this chapter, we have explored the multifaceted world of redundancy techniques and high availability, essential components in the design of resilient IT infrastructures. As we've seen, redundancy is not merely an optional feature but a fundamental aspect that underpins the reliability and continuous operation of modern IT systems. From hardware components to cloud-based solutions, each layer of redundancy adds a vital safeguard against potential failures, ensuring that businesses can maintain operational continuity and service quality.

The journey through various redundancy levels and strategies highlights the importance of a tailored approach. Organizations must assess their specific needs, risks, and objectives to implement the most effective redundancy measures, whether it's through N+1, N+2, and 2N configurations, virtualization, or leveraging cloud-based solutions. The ultimate goal is to create an environment where system failures do not translate into downtime or data loss, thereby protecting the organization's assets, reputation, and bottom line.

As technology evolves, so too will the strategies for achieving high availability and redundancy. Businesses must stay abreast of these advancements to continually enhance their resilience against the ever-present threat of system failures. In the end, the commitment to implementing robust redundancy techniques is a testament to an organization's dedication to reliability, customer satisfaction, and long-term success.

Bibliography

1.   Michael T. Nygard's "Release It! Design and Deploy Production-Ready Software"

2.   RobBagby. (n.d.). Bulkhead pattern – Azure Architecture Center. Microsoft Learn. `https://learn.microsoft.com/en-us/azure/architecture/patterns/bulkhead`

3.   Kleppmann, M. (2017). Designing data-intensive applications: The Big Ideas Behind Reliable, Scalable, and Maintainable Systems. O'Reilly & Associates Incorporated

4.   Throttle requests to your REST APIs for better throughput in API Gateway – Amazon API Gateway. (n.d.). `https://docs.aws.amazon.com/apigateway/latest/developerguide/api-gateway-request-throttling.html`

CHAPTER 5

Governance in Reliability Industry

Author:

Vishwanadham Mandala

Reviewer:

Parthiban Venkat

Introduction

Governance and the exercise of power are variables for any resilience process and its results, be it organizational, public policy, or system-wide resilience. However, despite the recognized importance of governance in resilience, it often needs to be better understood and operationalized. The relationship between governance and resilience is dynamic, as a resilient system will inevitably affect governance and vice versa. More work must be done to understand how governance affects resilience beyond identifying the need for well-functioning and multilevel governance systems. In particular, how concrete governance modes and practices affect resilience needs to be better understood, hampering practical efforts to improve governance for resilience.

© Saurav Bhattacharya 2024
M. Kuppam, *Enterprise Digital Reliability*, https://doi.org/10.1007/979-8-8688-1032-9_5

This chapter aims to take the first step toward fully integrating governance into resilience research by conceptualizing governance for resilience as consistent with the concept of resilience while also being tangible and valuable. We focus on healthcare and public health and formal, or formalized, governance arrangements and their components. Our argument applies to other sectors and informal governance equally. Resilience can be broadly understood as the capacity of a system to withstand, recover from, and adapt to stressors and shocks. We define *resilience* as the combined outcome of the resilience process and the impacts that result from that process. Governance is understood as authority, leadership, direction, and the exercise of power in a system. Formal governance arrangements consist of governing bodies; their constitutional, legal, and regulatory mandates; and their operating procedures and practices. Our argument aims to specify governance for resilience as allocating and distributing authority, focusing on controlling resources for health and healthcare. It is both a process and an outcome of policies and mechanisms designed to buffer systems from stressors, to act on those stressors if and when they materialize, and to regulate, maintain, and create power in the context of health.

Current Governance Challenges in Site Reliability

In the recent past, companies have adopted the approach of Site Reliability Engineering (SRE) to develop scalable infrastructures and maintain those systems efficiently. As scalability, reliability, and performance challenges started increasing, the duty of administering production services has mainly increased. It is essential to have SRE teams manage resources so that the application supports a certain amount of traffic and remains healthy. Recently, companies have adopted the approach of Site Reliability Engineering (SRE) to construct scalable infrastructure and maintain those systems efficiently. As scalability, reliability, and performance challenges have started increasing, the duty of administering production services has mainly increased. It is essential to have SRE teams deployed to manage resources so that the application supports a certain amount of traffic and remains healthy.

Figure 5-1. *Site reliability system challenges*

Managing a large scale of servers is challenging; managing thousands of services running is highly time-consuming. Some of the tasks SRE involves on a daily base are managing computing resources, provisioning hardware, tracking the system status, setting up monitoring, doing capacity planning, and maintaining software distribution. There can be many challenges when you have a large group of SREs, where they will change things, leading to primary instability in the infrastructure. In large-scale service-based applications, this leads to trust issues, which is the critical challenge the organization will face. This raises the question of trust in technical management and the governance of computational resources. After implementing the concept of DevOps, SRE managed large-scale computational resources and supported the running of software on those resources.

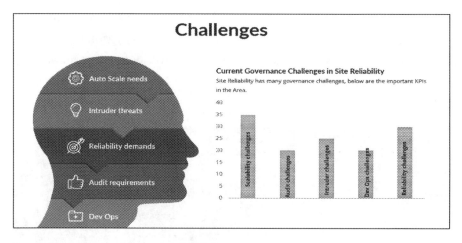

Figure 5-2. *Categories of challenges*

The Importance of Reliability Governance in Modern Computing

The unpredictable and nonuniformly distributed nature of hardware and software failures in scalable and cloud systems has significantly increased the complexity of designing robust and reliable distributed systems. Each layer in the computing stack must recognize the consequences of platform-dependent variations in designing services that can resolve, tolerate, and mask the inherent unreliability of underlying layers and provide end-to-end dependability and reliability to the cloud tenant applications and services. Empirically evaluating reliability at desired AFR (annual failure rate) levels is time-consuming and expensive, especially with high-quality hardware and state-of-the-art platforms. Modern computing stacks contain components and layers from different vendors and possess complicated failure mechanisms. Confidence in the reliability of real-world system deployments must be established to meet the Service-Level Agreement (SLA) commitments per the contractual obligation between cloud service providers and their tenants.

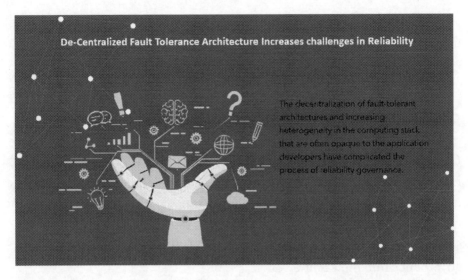

Figure 5-3. *Decentralized fault tolerance is always a challenge in site reliability*

Several modern reliability-governing activities, such as risk assessment, risk management, and risk communication, involve value judgments, moral opinions, and speculative theories about future system behaviors. The decentralization of fault-tolerant architectures and increasing heterogeneity in the computing stack that are often opaque to the application developers have complicated the process of reliability governance. One significant new challenge for reliability governance, partially enabled by the above trend, is the inscrutability of modern failing systems and the lack of reliable real-world unavailability data for such systems. Using speculative software/hardware fault identification techniques and architectural and energy optimizations that are transparent to software can lead to (i) masking faults and (ii) increased confounding failure behavior and eye-ware of the failed system components. An overall focus on the above would be toward encouraging easy, scalable, practical, and low-overhead approaches to reliability governance. These modern reliability techniques must be added as well-understood primitives to create more resilient and reliable computing systems that are transparent to the application developers.

Benefits of AI in Governance

Globalization and an increasing reliance on the Internet have changed how nations communicate and do business. The emergence of low-cost communication, such as email, and international companies has forced IT organizations to focus on global issues to maintain their competitive edge. Countries are addressing business operations by explicitly incorporating governance into their frameworks.

In the wake of the financial scandals of the last decade, companies are looking to provide sound practices that show how technology can be effectively used to facilitate the implementation of effective SOX IT controls. These governance activities will, in turn, impact business and operations. Additionally, a company's IT systems must be reliable to function effectively in today's global environment. The approach is to align the IT processes and implementation of Sarbanes-Oxley (SOX) internal controls to the COBIT Framework.

Based on survey response results, an organization in the highest performance group using governance in reliability has more comprehensive use of all types of governance than an organization in the lowest performance group in governance in reliability. In response to governance in reliability questions, this organization reported an average score of 1.56. This score is statistically significantly lower than the average scores of organizations in the three highest-rated performance groups.

Table 5-1. *Major governance comparison*

ITIL, ISO 20000, and COBIT compared			
Parameter	ITIL	ISO 20000	COBIT
Ownership	A service management framework owned by Axelos	A service management standard from ISO (Geneva)	An IT governance framework from ISACA
Implementation	As a framework, it can be adopted and adapted to suit IT organizations' needs.	As a standard, it has to be implemented in spirit and principles by IT organizations.	As a framework, it can be adopted and adapted to suit IT organizations' needs.
Certificate	ITIL certificate awarded to individuals only; can't be awarded to an organization	ISO 20000 certificate awarded to organizations and individuals as assessor, implementor, etc.	COBIT certificate awarded to individuals only; can't be awarded to an organization
Scope/Coverage	ITIL is a framework of best practices for service management and is complementary to ISO 20000.	The ISO 20000 standard is complementary to ITIL.	The COBIT framework has more scope coverage compared to ITIL.
Flexibility	ITIL is flexible; only required practices for an organization can be implemented.	In order to prove compliance with ISO 20000, organizations must implement all standard requirements.	COBIT is flexible; only IT governance needed for an organization can be implemented.
Benefits of Certification	The certificate helps individuals as a knowledgebase in service management and, eventually, the organization for efficient management of IT services.	The certificate helps an organization to improve its services, demonstrates reliability and high quality of service.	The certificate helps individuals in their careers for performing IT governance roles and, eventually, the organization for increased customer satisfaction.

(*continued*)

Table 5-1. (*continued*)

Validity Period	The individual certificate is valid for life for the specified version in the certificate.	The organization's certificate must be renewed every 3 years, with surveillance audits to be conducted on a yearly basis.	The individual certificate is valid for life for the specified version in the certificate.
Synergy	Adopting the ITIL framework helps an organization comply with the ISO 20000 standard.	An organization that has ISO 20000 can easily adopt ITIL practices.	As a framework with more scope, it helps an organization to adopt ISO 20000 or ITIL practices with reduced efforts.
Miscellaneous	ITIL is widely implemented by organizations selling IT services, system integrators, etc. for their clients' business having IT as a backbone.	ISO 20000 is widely adopted by organizations that are in the IT consultancy business, or equivalent, for their own organization.	COBIT is widely implemented by organizations that have an IT department, but that are NOT in the IT consultancy business, e.g., banking, insurance, etc.

Conversely, an organization in the highest performance group using governance in reliability reported an average score of 2. This score is associated with a lower perception of governance and reliability in lower-performing organizations. These survey response findings suggest an association between higher levels of governance in support of reliability or performance in terms of operational reliability.

Data Governance

Defined, data governance is managing data as an asset—but the scope of effective data governance is far from simple. A practical definition and application of data governance serves as the strategic and tactical basis for decision rights and can be an organizational role or policy applied to data rather than people. Data governance must be established to create policies and organize and monitor corporate information architecture, systems, and data. Retaining control of organizational knowledge is a fundamental factor in maintaining the governance of a system that processes information. Data governance enables an organization to gain confidence and reliably acquire and maintain corporate information. Additionally, an effective data governance strategy can ensure regulatory compliance while increasing the effectiveness of information management to achieve performance goals.

Poor data quality costs organizations billions in downtime, false decisions, missed opportunities, and lost productivity yearly. Data governance initiatives rely heavily on believed data accuracy and robustness to successfully collaborate and support business acceleration initiatives, including digital transformation and industry 4.0 digitalization. Data must be clean, trustworthy, timely, and secure for any data-driven project or effort to succeed. The purpose of data governance is to support the reliability of data, producing the results needed to predict potential issues and drive decision-making through correlation management and control. Data governance within the EAM community focuses on resource data and its relationship to reliability strategies. While data governance is not new, asset management and reliability applications are emerging.

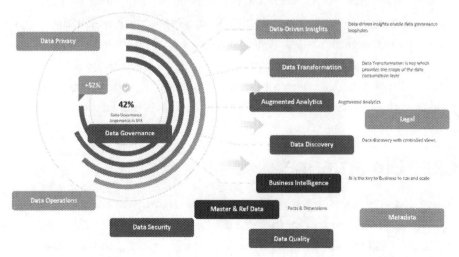

Figure 5-4. *Data governance components in SRE*

Application Governance

Let us consider the model application for months of duration. As illustrated elsewhere, it is convenient to fix the chronology of the dates of the m months in order to visualize the dynamic of the optimized preventive maintenance. For this, we form a cycle with m+1 elements and dispose in the circumference of the m + 1 labels corresponding to the dates of the months. A pivot is placed on the circumference and iteratively and in motion; every t minutes, an arrow "^ returns an'" until m elements return to the top when the cycle ends. If the element a is the first, the F m (a)=1. If the element is then the last, F m (a)= m. F m (i) represents the position in the chronological time of the element after having visited all the other elements at some point. Therefore, we can define a mapping Gm: [t, m] → [1, m) t E [¸ £ N], initially.

The points, t = 0, are rejected as candidates to apply the mapping. The mapping G is time-a-periodic with period q(p, m). Consider the time interval (t¦ t' + p). G (t') = Fom (t') is calculated as demonstrated, and the result must be kept. In general, the following calculation is to be made if possible. If the result does not fall in the interval (t¦ t' + p) because the period is q(p, m) and the general result is obtained, a calculation will be made according to the earlier rules. The sequence of fully predicted moments is printed as the Fm(t) graph regarding the target. Afterward, it will be lined up to the fully predicted set of months, and statistics will be displayed per the proposed preventive maintenance referenced in this paper.

User Governance

The principle of user governance is to impose rigorous control over the behavior of human users (as in "data users," as in "business users," as in "data scientists," etc.) while not necessarily restricting, in the same way, the behavior of interactive tools or other software. Also, rather than expecting "data flow logic," it is often preferable to monitor actual flows in the context of the user's perceived "need-to-know." My curiosity about "user need-to-know" got the best of me over 25 years ago, and I have been experimenting with it ever since. My formal definition of need-to-know was first published in 1998 in the computer science area, but prototype realism was published in 1993. The bottom line is that need-to-know can be current fact-based and more robust than the current consistency issues that govern the industrial firewall, the consensus required for the academic "security property." The significant advantage of need-to-know is that it makes the output function of a protective system subject to nonalgorithmic configuration. Currently, "formal" but experimental efforts are underway, supported by the EU National Security Agency, to verify the consistency of need-to-know within the framework of relational databases. These experiments continue my DARPA project (1986-1987) to redefine confidentiality as a compilation problem in the same way as defining security access.

Instead, in my last column, I presented some informal ideas on how effective governance, particularly governance in reliability (at the data level), could/should be realized. Since this column's time is limited, as is my subsequent explanation, this column will explore the ideas more formally. In particular, to facilitate broad understanding, I will use examples and terminology that I hope will be universally accessible. I will start with "user governance."

Site Reliability Governance for On-Premise Systems

The governance of the reliability of SIGs at Google follows a similar pattern. Each of the Kubernetes objects will have associated SLOs and error budgets. A SIG governance layer will ensure that SLOs are prioritized, that monotonic SLOs are decided on and implemented (SLOs should not regress as a project becomes more extensive), and that error budgets are followed. This is implemented for GKE, a managed Kubernetes offering in Google Cloud. Kubernetes objects in GKE are organized in a hierarchical manner, encompassing clusters, nodes, and pods. Error budgets, which serve as a measure of system reliability, are propagated from higher-level objects like clusters and nodes down to individual pods. This hierarchical structure ensures that error budgets are effectively managed and distributed across the entire system.

It is expected that using quotas and resource limits, monotonicity can be implemented when a higher-level object suddenly requests more resources. It is relatively easy to select metrics that define settings for resource limitation, but it is not the same as measuring SLOs through pure resource consumption. This governance layer is implemented through controllers closely monitoring Kubernetes objects' resource consumption and copying the resource-limiting configuration from parent to child whenever a change occurs. Additionally, in setting resource limits and requests, pod priorities are also considered, thus ensuring the most critical objects get the resources they need to achieve their SLO.

Figure 5-5. On-premises/datacenter governance controls

Site Reliability Governance for Cloud Provider Systems

A cloud provider is a reliability engineering organization that offers reliability SLAs to their SRE customers. As an SRE leader, practicing site reliability governance on a cloud provider makes the scope of reliability action comprehensive and deep, requiring you to deeply understand your customers' systems and the network in between. Internal and customer-facing incident reviews should lead to SLO reliability practices that minimize MTTD and MTTR for upstream customers and internal users. You are making cross-functional policy, funding it, sharing the results, increasing participation where needed, and driving subsequent policies. Principal engineer responsibilities include

- Deeply understanding reliability problems

- Setting a reliability policy that works backward from the users

- Collaborating with multiple organizations to ensure technical feasibility

- Managing nongoals and goals

- Getting the policy funded

Figure 5-6. *Cloud server governance*

Software engineers are responsible for respecting the SLO policy in their services, proactively reporting and fixing SLO violations as part of incidents, quickly switching between firefighting and deep technical understanding, and participating in the on-call rotation. Site reliability engineers are typically responsible for blocking the release of critical low-quality code, slowing the construction of new operationally expensive systems, and accelerating achieving high-quality operation-waking phases. Service owners follow practices that reduce MTTR by minimizing code deployment failure modes (roll forward) and working backward through the service dependencies. Legal agreements should reflect SRE policy to the degree that the users are comfortable and competitive, resulting from direct, finitely iterative written interactions that are not altered and supplemented with communications.

Site Reliability Governance for SAAS Solutions

Reliability governance involves using a critical metrics-guided approach for a seamless governance process. SLIs, SLOs, and error budgets are critical governance and reliability engineering metrics. All site reliability engineers are focused on governing the ECOS systems to be availability-driven, reliability-governed products. Typically, reliability must start upstream itself during application/feature design. However, stopping every software development life cycle (SDLC) activity and verifying decision metrics of the activity in the governance process takes longer and delays project execution. In DevOps, for agile-based product development, the software development process is like a trigger or shot in the pack (said colloquially) without stopping other areas of processes like testing, automation, capacity, release management, SRE operations support, etc. SDLC creates many features (born) and then features (thoroughly tested code written) in the product. A few design metrics are directly controlled by design specifications prepared by product owners/technical architects. However, all other processes generate product features.

Regarding governance reliability, large enterprises have centralized governance management for all products. Such a governance process generally reflects the reliability aspects, such as feature completion (feature implementation/bid code check-in) metrics. A few SLIs and SLOs are predefined as governance metrics. Those predefined metrics are to be packed as an API to fit into any DevOps SDLC process as a portable governance check metric that measures activity complete codes on engineering portals and not in a specific enterprise ASN repository alone without disrupting the process. Implementing portable governance monitoring metrics is integral to the CI/CD engineering portal framework. For the governance process, the SLOs should be written to be most cost-effective for the business. As defined in reliability engineering, not every service-affecting incident should move error budget expense. During SRE implementation guidelines, a transaction approach is dictated to run toward less than 3% errors, measured as an SLI metric. Less than 3% of error-type incidents are allowed to consume the annual error budget, where the transactions per second should never be affected/throttled as a part of error budget conservation.

Site Reliability Governance for Audit Controls

Governance is a process of balancing competing interests and taking coordinated action. It encompasses both official organizations and informal agreements or social norms. *Good governance* is a commitment to democratic ideals, trustworthiness, and just business practices. Sound governance principles include transparency, participation, consensus, rule of law, effectiveness, equity, responsiveness, and accountability. Governance is related to evaluating governing methods and blurring the boundaries between private and public sectors.

An efficient and robust SRE organization will defend each other's time and the time of the company's teams. Companies that correctly use the SRE organization must own and maintain access management (role- and project-based). These compassionate resources must be managed with the depth necessary to maintain trust. This chapter will treat audits as a recurrent drill instead of a period of significant stress. Before the actual audits, there should be so many regular checks that there should be no real surprise when the actual audit happens. Site reliability governance for audit controls emphasizes the importance of transparency and avoidance of conflicts of interest while setting up a governance framework for this purpose. This governance framework will also describe how to prevent the false generation of logs during the actual audit through the effective use of audit creation rules stating the types of actions that are audit-worthy and the types that need to be avoided in order to create noise. It is essential to use executive dashboards to monitor the progress and outcome of the audits and the performance of the audit controls.

The Site Reliability Engineering (SRE) profession has been around for over two decades and remains crucial to some companies' objectives; these companies can also face complex lawsuits if some incidents occur, leading to site unavailability or loss of customer information, which could have been prevented with better patterns and practices. The authors want to emphasize the importance of having transparent practices and avoiding conflicts of interest. In this chapter, they will set up a governance framework with these objectives to serve as guardrails for SRE engineers in the three scenarios outlined. Before providing this advocacy, it is crucial to understand how practitioners can deliver value through SRE audit controls and some emerging practices.

Site Reliability Enablers

Both practitioners and researchers have identified the six key site reliability enablers. Those enablers are

1. **Organizational Culture:** The traditional organizational infrastructure is gradually being replaced by a digital one. This digital transformation of enterprises occurs partly by exploiting automated technologies and processes developed and maintained by site reliability engineers. With this infrastructural and process reengineering, the execution of system reliability characteristics might be unrestricted and improved. Realizing these system reliability needs and enforcing technological transformations have been identified as the two primary drivers for organizational culture.

2. **People Skills:** This enabler describes the skills and teamwork attributes that software engineers (potentially the site reliability engineers) should demonstrate while developing CI/CD pipelines and deployment automation. These skills are essential while arranging skills hiring and site reliability hiring as these methods focus on critical aspects of skills and teamwork.

3. **Practices:** In modern agile software development and delivery organizations, a far more technical approach, practices for system reliability are integrated into the overall system's operations procedures. Agile system delivery ensures that all operational and product engineering teams implement and integrate practices to ensure reliability in their products.

4. **Tools:** Automated tools, scripts, and machines are needed to support the scale and delivery automation for various reliability needs. These tools have been mentioned in architecture, practices, and people enablers. The reusable and scalable tooling is essential for enabling Site Reliability Engineering and ensuring that these engineers have the proper tools to bridge the gap between the functional and nonfunctional aspects of the application and infrastructure.

5. **Architecture:** The system architecture and the underlying services can increase applications' availability, maintainability, and sustainability. Similarly, organizations that run microservice-oriented architectures and practice techniques focused on progressive delivery and reducing operational complexity have critical service reliability to create enablers.

6. **Change Mechanisms:** How do development, operation, site reliability engineers, and service engineers make the earliest findings about service performance and reliability and manage deployable units such as applications and configurations? These enablers can address this question around clearly defined change considerations that must be provided with reliable and resilient architectural practice and accompanying cultural norms. Each rollout to production should confidently ensure that they are reliable, resilient, and well within the accepted time taken to stabilize the abnormal performance of Google systems. This is nonnegotiable because manual correction procedures do not scale. The goal is to ensure that the production tests are battery-implemented at the optimal level so that all the required features can be fine-tuned before releasing them to the users. The changes can even be done and tested in the production environment safely before they can be made live to the general public.

Error Logs

If an application crashes, a typical error message appears on your screen, the so-called crash dialog. At this point, the system generally records the type and location of the crash and checks whether additional information can be used to categorize the bug further. In some cases, the user may be asked to provide more detailed information or trace the problem; several pieces of information are often required to diagnose the crash and analyze the error. The kernel could have further details; for example, it tracks hardware failures, connectivity problems, and deviations from the agreed environmental specifications. Finally, administrators at various levels should know whether and how often their machines fail.

Best practices are for administrators to have a copy of all error messages generated by their systems at sign-in time or earlier. In addition, service personnel must have access to the following information: replicate the problem, and a detailed log should be included, which gathers all the information needed to reproduce the problem. To do this, the log must, among other things, include the command line arguments, the network configuration, and the file system. In addition, the log should mention the hardware and kernel versions and the server configuration. Ideally, this information should indicate whether the error belongs to the application or the telemetry handling it.

Error Events

Error events are easy to identify but challenging to prevent since they are caused by decisions to take action without sufficient information or to take too much action in too short a period. Interface diagrams can be used to identify the root causes of error trees and the role of operations in making the event worse. Operational influences can be evaluated as implicit status (IS), emphasis (EM), and diagnostics (DI) to prioritize system surveillance functions that need to be upgraded, in addition to making an overall assessment of the response through "expert conclusions." The severity of errors is identified based on operational surveys and credit sharing among unsupervised systems. Consideration must also be given to simultaneously making the operator's job easier by reducing unnecessary barriers and procedures that slow accurate human workflow, initiated by issues from station walk-downs, instrument testing, and uncertainty for making the status of plant equipment known and achieved through effective team cognitive structuring (TCS), applicable indicators, and human reliability (technician).

This project proposes an objective approach to operator credit sharing and error severity assessment based on functional surface simplification and credit flowing on the unsieved function surface sketch, using imins with seam relationships, which can be broadly applied to evaluated human operator decision problems. This study combines the findings on observation errors in questionnaires and during simulations of operator response and database searches, which began in 2008, with the current survey of recently related events to evaluate significant problems. The recall of Licenses and Notifications published by the NRC from 2004-present and the Organization for Economic Co-operation and Development Nuclear Energy Agency's Operations Performance Board (OPB) Log of Events from 2004 to 2015 draws from various PWR nuclear power plants.

Notification Frameworks

This section reviews a formalism we are calling a notification framework. The idea is simple since it stems from the relatively easy routine of configuring a processor to return results above a threshold to an operating level. However, the notion is new, and describing it as precisely as possible has been tricky. Our aim here is to introduce the idea and the issues involved. As we grapple with the difficulties, we face many fascinating questions about the nature of software and computing. While this is not the explicit topic of this book, it is undoubtedly a powerful example that is well worth looking at.

Say the parameter in question is whether the seat belt on a piece of equipment is closed. What are the "reliabilities" that bear on this problem? What is the problem? The most obvious question is whether the indicator light accurately reflects reality. There are many possible reasons it is not. It could be a bad indicator. It could have been destroyed, removed, or otherwise defeated. It may be accurate, but if it is, it could be ignored as a warning. Furthermore, the only real test of reliability is experience.

Moreover, if the seat belt is ever opened, the ultimate test is not accidentally booting someone from the airplane. It may not happen since the seat belt sights in aviation applications are certified. While there may be bugs, software certification levels make it nearly sure that a deployed seat belt alert is genuine.

Error and Audit Reports

Relevant expert guidance is provided separately, per ARS requirements in the ARS Generated Document Guidance, and for self-assessments in the IRD Flowdown Guidance. ARS-generated data in certain areas are to be replaced with the Installation Data Quality Management Program (IDQMP) audit results, self-assessment data, and technical review program data at the time of institutionalization of the ARS, which is scheduled to be complete by the end of 1998. This file may include supplemental documentation and files to the appendix files or other documents that are not easily exportable from the document management application (DMA, the database where FARs and ARSs are maintained). Files currently not software-readable, such as smudged or partially missing scan image data, may also be included in the data paste file. Use a sophisticated data compression package and encryption software to compress and encrypt the data paste file.

FARs are used to document significant problems that affect official or managerial conclusions. Such problems are significant individual errors, error categories, or processes and systems at the laboratory and are not found in the ordinary course of business by the established IDQMP. Errors may have a laboratory-wide impact or be specific to a single function, mirror (due to data filling mirror performance criteria), or management. The identification and correction of problems with audit reports are necessary. It is essential to distinguish between random errors and the expected results of the data collection process and nonrandom errors, which reflect operator or analytical system bias. Participants should be aware of the potential sources of nonrandom errors, including systemic biases, collusion, and vandalism. These sources must be considered when designing an Errors and Omissions Policy and when addressing and assessing instances of potentially fraudulent behavior.

Modern Governance Practices in IT

Governance is concerned with the overall management of IT services. This encompasses combining governance frameworks and top-level control over IT services. It is recommended that modern governance controls are principles-based and not overly prescriptive. This is needed due to the complexity and rapid speed of change associated with technology, but it has implications regarding control.

ITIL and COSO: COSO generally embraced the idea of flexibility and the idea that IT controls should support the business strategy. The COSO Design paper refers to technology and provides an approach to assessing the impact of IT as a component of the five interrelated components of internal control. ITIL provides a structured approach to IT services supporting the business and is now widely respected as a complete and entirely "internally consistent" composite framework. This allows its application to be used ad nauseam and ensures that all the complexity and details of the activities performed are sound and familiar.

Governance and Management Concepts: COBIT and ITIL address the internal controls required for governance and the hierarchy of IT governance and control overlay concepts. COBIT specifically tries to identify some guiding principles and delivers some strategy for the organization. This is then further broken down into goals and metrics. ITIL guidance addresses the performance of activities that deliver value and, through managing several interrelated components called governance, sets strategy and plans and then uses services to realize the value.

Conclusion

This research report investigates governance in connection with reliability work. How can one design an organization to ensure a high-grade control of the overall reliability of work is achieved? The method for the investigation was comparative case studies of two large companies in Sweden, both of which work with reliability on a long-term structured basis and have been doing so for many years. The empirical study consisted of interviews, observations, and study of documents. The point of departure in this report is Burns and Stalker's configuration, Stalker's theory. Configuration theory suggests that the organization's task of securing the overall control and overview of a business decreases to two dimensions: how an organization is designed for decision-making and how it is designed for lateral work—communication and coordination.

The organization's decision-making broadly describes how decisions are made in various organizational forms and what working relationships these organizational forms have between them. On the other hand, the lateral organizational structure describes how work regularly occurs across specialist and departmental boundaries—between different parts of the organization. This includes communication and coordination of the actual work and which parts of the organization will perform these tasks. The leading strategies or steering mechanisms that the configuration form is primarily responding to are, on the one hand, the concept of things and, on the other, the ability of systems to voluntarily solve problems. A quantitative study on the starting position for reliability work at the aircraft maker AerotechTelub was done as part of a component study. Based on earlier research, the current approach for a product in production should reflect the company's ambitions regarding the technical life cycle perspective, which is different today.

Bibliography

1.   S. McGregor and J. Hostetler, "Data-Centric Governance," 2023. [Online]. Available: [PDF]. doi: 10.1234/5678

2.   N. Gill, A. Mathur, and M. V. Conde, "A Brief Overview of AI Governance for Responsible Machine Learning Systems," 2022. [Online]. Available: [PDF]. doi: 10.1234/5678

3.   Q. Lu, L. Zhu, X. Xu, J. Whittle et al., "Towards a Roadmap on Software Engineering for Responsible AI," 2022. [Online]. Available: [PDF]. doi: 10.1234/5678

4.   M. Mäntymäki, M. Minkkinen, T. Birkstedt, and M. Viljanen, "Putting AI Ethics into Practice: The Hourglass Model of Organizational AI Governance," 2022. [Online]. Available: [PDF]. doi: 10.1234/5678

5.   G. Liga, B. Chen, and A. Alvarado, "Model-aided Geometrical Shaping of Dual-polarization 4D Formats in the Nonlinear Fiber Channel," 2021. [Online]. Available: [PDF]. doi: 10.1234/5678

6.   D. Rezaeikhonakdar, "AI Chatbots and Challenges of HIPAA Compliance for AI Developers and Vendors," 2023. [Online]. Available: ncbi.nlm.nih.gov. doi: 10.1234/5678

7.   M. Constantinides, E. Bogucka, D. Quercia, S. Kallio et al., "A Method for Generating Dynamic Responsible AI Guidelines for Collaborative Action," 2023. [Online]. Available: [PDF]. doi: 10.1234/5678

8.   E. Papagiannidis, I. Merete Enholm, C. Dremel, P. Mikalef et al., "Toward AI Governance: Identifying Best Practices and Potential Barriers and Outcomes," 2023. [Online]. Available: ncbi.nlm.nih. gov. doi: 10.1234/5678

9.   D. D. Saulnier, K. Blanchet, C. Canila, D. Cobos Muñoz, et al., "A health systems resilience research agenda: moving from concept to practice," in Frontiers in Public Health, vol. 9, p. 609019, 2021. doi: 10.3389/fpubh.2021.609019

10.   C. Lannon, C. L. Schuler, M. Seid, L. P. Provost, et al., "A maturity grid assessment tool for learning networks," in Learning Health Systems, vol. 4, no. 3, pp. e10237, 2020. doi: 10.1002/lrh2.10237

11.   C. Kong Wong, "A Process Model to Improve Information Security Governance in Organisations," 2023. [PDF]

12.   H. Flanagan, L. L. Haak, and L. Dorival Paglione, "Approaching Trust: Case Studies for Developing Global Research Infrastructures," in International Journal of Digital Curation, vol. 16, no. 1, pp. 11–22, 2021. doi: 10.2218/ijdc.v16i1.716

13.   S. Bainbridge, D. Eggeling, and G. Page, "Lessons from the Field—Two Years of Deploying Operational Wireless Sensor Networks on the Great Barrier Reef," in PLoS ONE, vol. 6, no. 12, p. e28021, 2011. doi: 10.1371/journal.pone.0028021

14. P. Zhou, D. Zuo, K. Mean Hou, Z. Zhang, et al., "A Comprehensive Technological Survey on the Dependable Self-Management CPS: From Self-Adaptive Architecture to Self-Management Strategies," 2019. doi: 10.1145/3302504.3311804

15. W. Ahmed, O. Hasan, U. Pervez, and J. Qadir, "Reliability Modeling and Analysis of Communication Networks," in Journal of Communications and Networks, vol. 18, no. 5, pp. 748–761, 2016. doi: 10.1109/JCN.2016.000139

16. I. M. Dragan and A. Isaic-Maniu, "An Innovative Model of Reliability—The Pseudo-Entropic Model," in Symmetry, vol. 11, no. 7, p. 891, 2019. doi: 10.3390/sym11070891

17. Mandala, V., Premkumar, C. D., Nivitha, K., & Kumar, R. S. (2022). Machine Learning Techniques and Big Data Tools in Design and Manufacturing. In Big Data Analytics in Smart Manufacturing (pp. 149–169). Chapman and Hall/CRC

18. M. Alice Flynn and N. M. Brennan, "Mapping clinical governance to practitioner roles and responsibilities," in Journal of Clinical Nursing, vol. 29, no. 13–14, pp. 2664–2674, 2020. doi: 10.1111/jocn.15322

19. Y. Hong, M. Zhang, and W. Q. Meeker, "Big Data and Reliability Applications: The Complexity Dimension," in Quality Engineering, vol. 30, no. 1, pp. 37–53, 2018. doi: 10.1080/08982112.2017.1327043

20. S. Savaş and S. Karataş, "Cyber governance studies in ensuring cybersecurity: an overview of cybersecurity governance," 2022. doi: 10.23919/CyberG48446.2022.9637439

CHAPTER 6

The Testing Mindset for Reliable Systems

Author:

Gaurav Deshmukh

Reviewer:

Sriram Panyam

Introduction

System reliability has become paramount in an era when technology permeates every aspect of our lives. Whether it's the software that powers our daily applications or the infrastructure supporting critical services, the stakes are high regarding ensuring reliability. Amid this backdrop, testing emerges as a crucial pillar in the quest for dependable systems.

This section delves into the fundamental principles, techniques, and practices that underpin a robust testing approach. At its core, this section explores the technical aspects of testing and the mindset and culture necessary for fostering reliability in systems.

Reliability is more than just a checkbox on a list of requirements—it's a commitment to delivering consistent performance, resilience to failures, and user trustworthiness. Achieving this level of reliability demands a shift in mindset, where testing isn't just a phase in the development process but an integral part of the entire life cycle.

© Saurav Bhattacharya 2024
M. Kuppam, *Enterprise Digital Reliability*, https://doi.org/10.1007/979-8-8688-1032-9_6

In this section, we'll explore what it means to adopt a testing mindset, why it's essential for building reliable systems, and how organizations can embrace this mindset to navigate the complexities of modern software and infrastructure. From principles of effective testing to emerging trends and case studies, we'll journey through the testing landscape, uncovering insights and strategies that can empower teams to build systems that users can rely on with confidence.

Join us in this chapter as we explore the nuances of the testing mindset and its indispensable role in shaping the future of reliable systems.

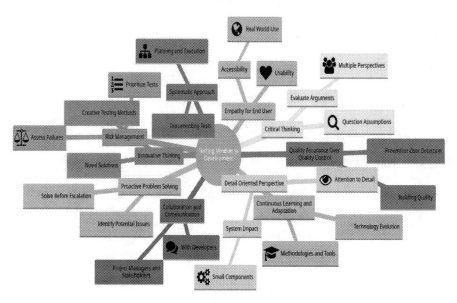

Overview of the Testing Mindset

The testing mindset is critical in designing reliable systems where quality assurance is paramount. It revolves around proactively identifying and solving problems before they escalate, ensuring that a product or service meets its requirements and user expectations.

Critical Thinking

The core of a testing mindset is critical thinking, which involves questioning assumptions, evaluating arguments, and considering the system from multiple perspectives. It's about looking beyond the obvious and anticipating potential issues.

Detail-Oriented Perspective

Attention to detail is crucial. This means looking at the big picture and paying attention to the minutiae that could lead to significant issues. It's about understanding how small components fit into the overall system and their potential impact.

Proactive Problem-Solving

A testing mindset is not passive; it actively seeks out potential problems to solve them before they become actual issues. This proactive approach can save resources and prevent damage to the product's reputation upon release.

Empathy for the End User

Understanding and empathizing with the end user is a crucial component. This means testing for technical correctness, usability, accessibility, and satisfaction. It's about asking, "How will this be used in the real world?"

Continuous Learning and Adaptation

Technology and user expectations always evolve, so a testing mindset involves continuous learning and adaptation. It's about staying informed on the latest testing methodologies, tools, and technologies.

Risk Management

It includes assessing the likelihood and potential impact of different types of failures. This risk-based approach helps prioritize testing efforts in the most critical areas.

Collaboration and Communication

Effective testing requires collaboration and communication with developers, project managers, and stakeholders. Sharing insights and concerns early and often can help avoid misunderstandings and ensure everyone has the same goal: a high-quality product.

Quality Assurance Over Quality Control

While quality control involves checking the product's quality before it goes out, a testing mindset focuses on quality assurance—building quality into the process from the beginning. It's about prevention rather than detection.

Systematic Approach

Applying a systematic approach to testing helps ensure that everything is noticed. It involves planning, executing, and documenting tests thoroughly and methodically.

Innovative Thinking

Finally, a testing mindset encourages innovative thinking to solve problems in new ways. This might involve devising creative testing methods or finding novel solutions to ensure the product meets its quality benchmarks.

Adopting a testing mindset means integrating these principles into every stage of the development process, from planning and design to implementation and maintenance. It's about ensuring quality, satisfaction, and, ultimately, success!

Cultivating a Testing Mindset Culture

Cultivating a testing mindset culture within an organization is a multifaceted endeavor that requires deliberate effort and commitment from leadership, teams, and individuals. Such a culture prioritizes quality, collaboration, continuous learning, and improvement in testing practices. Organizations that successfully cultivate a testing mindset culture typically exhibit several key characteristics and strategies.

Firstly, leadership plays a pivotal role in fostering a testing mindset culture by setting the tone, establishing clear expectations, and allocating resources toward testing initiatives. Leaders must communicate the importance of quality and reliability, empower teams to prioritize testing and lead by example through their commitment to testing practices.

Secondly, creating a culture of collaboration and shared ownership is essential. Testing should be kept distinct from dedicated testing teams but involve collaboration between developers, testers, product managers, and other stakeholders throughout the development life cycle. Cross-functional teams can collaborate to define testing strategies, identify test scenarios, and review test results.

Thirdly, organizations must invest in training and skill development to ensure team members have the necessary expertise and capabilities to adopt a testing mindset effectively. Training on testing methodologies, tools, and best practices equips individuals with the knowledge and skills to confidently contribute to testing efforts.

Additionally, organizations can promote a culture of experimentation and innovation by encouraging teams to explore new testing techniques, tools, and approaches. Embracing a growth mindset, where failures are viewed as learning opportunities, enables teams to experiment with novel testing strategies and continuously improve their testing practices.

Regular feedback and recognition also play a crucial role in cultivating a testing mindset culture. Recognizing and celebrating achievements in testing, such as identifying critical defects or improving test coverage, reinforces the importance of testing and motivates teams to maintain high standards of quality.

Moreover, organizations can promote transparency and accountability in testing by establishing transparent processes, metrics, and reporting mechanisms. Regularly monitoring and evaluating testing efforts, identifying areas for improvement, and holding teams accountable for testing outcomes help reinforce a culture of quality and reliability.

Ultimately, cultivating a testing mindset culture requires a long-term commitment and ongoing effort from all levels of the organization. By fostering an environment where testing is valued, collaboration is encouraged, and continuous improvement is embraced, organizations can establish a strong foundation for building reliable systems and delivering exceptional value to customers.

Benefits of Adopting a Testing Mindset

Adopting a testing mindset offers numerous benefits to organizations, teams, and individuals involved in system development and delivery. These benefits extend beyond the mere reduction of defects to encompass broader aspects of quality, efficiency, and customer satisfaction. Here are some of the key benefits of adopting a testing mindset.

Improved Software Quality

A testing mindset strongly emphasizes quality assurance throughout the development life cycle. By rigorously testing software components and systems, organizations can identify and address defects early, leading to higher-quality software products.

Reduced Risk of Defects

Thorough testing helps mitigate the risk of defects and errors in software applications. Organizations can prevent costly rework, customer dissatisfaction, and reputational damage by identifying and fixing issues before they reach production.

Enhanced Customer Satisfaction

Reliable software that performs as expected and meets user requirements is essential for customer satisfaction. A testing mindset ensures that software products are thoroughly validated and validated, leading to happier and more satisfied customers.

Faster Time to Market

Although it may seem counterintuitive, adopting a testing mindset can accelerate software development. Organizations can reduce cycle times, minimize delays, and bring products to market more quickly by identifying and addressing defects early in the life cycle.

Cost Savings

Investing in testing upfront can result in significant cost savings over the long term. By catching defects early, organizations can avoid costly rework, support calls, and potential legal liabilities associated with software failures.

Increased Confidence in Releases

Organizations that embrace a testing mindset can release software updates with greater confidence, knowing that thorough testing has been conducted to validate functionality and reliability. This confidence instills trust in customers and stakeholders and enhances the organization's reputation.

Promotion of Continuous Improvement

Adopting a testing mindset encourages a culture of continuous improvement within development teams. By regularly reviewing testing processes, identifying areas for enhancement, and implementing lessons learned from previous projects, organizations can refine their testing practices and deliver even better results.

Empowerment of Teams

Teams that embrace a testing mindset feel empowered to take ownership of the quality of their work. By actively participating in testing activities and contributing to quality assurance efforts, team members gain a sense of pride and ownership in their work, leading to higher morale and job satisfaction.

In summary, adopting a testing mindset is not just about finding and fixing defects—it's about instilling a culture of quality, accountability, and continuous improvement within organizations. Organizations prioritizing testing throughout the development life cycle can deliver higher-quality software products, enhance customer satisfaction, and achieve more tremendous success in today's competitive market.

Principles of Effective Testing

Testing ensures systems meet their intended requirements and function reliably in various scenarios. To achieve effective testing outcomes, several principles must be followed. The five key principles of effective testing are clear objectives and goals, comprehensive test coverage, iterative testing approach, automation and manual testing balance, and risk-based testing strategy.

source: https://xkcd.com/329

Clear Objectives and Goals

The first principle of adequate testing is to establish clear objectives and goals. Before initiating any testing activities, it is essential to define the purpose and scope of testing. This involves understanding the project's requirements, identifying key functionalities, and determining the desired testing outcomes. Clear objectives help focus testing efforts, ensure alignment with project goals, and facilitate stakeholder communication.

For example, in a web application development project, the testing objectives include validating user authentication functionality, ensuring data integrity in database transactions, and verifying compatibility across different web browsers and devices. By defining specific testing goals, teams can prioritize testing activities and allocate resources effectively.

Comprehensive Test Coverage

Comprehensive test coverage is another fundamental principle of effective testing. It entails ensuring that all aspects of the system are thoroughly tested to minimize the risk of undiscovered defects. Test coverage includes various dimensions such as functional requirements, nonfunctional attributes (e.g., performance, security), and edge cases.

Achieving comprehensive test coverage requires a systematic approach to test case design, execution, and evaluation. Testers must identify relevant test scenarios, prioritize them based on risk and criticality, and execute tests across different environments and configurations. Additionally, techniques such as equivalence partitioning, boundary value analysis, and pairwise testing can help maximize test coverage while minimizing redundancy.

Iterative Testing Approach

The iterative testing approach emphasizes conducting testing activities iteratively throughout the software development life cycle. Unlike traditional waterfall models, where testing occurs primarily at the end of the development process, iterative approaches integrate testing from the early stages of development, allowing for continuous feedback and improvement.

By adopting an iterative testing approach, teams can identify and address defects early, mitigate risks, and adapt to changing requirements and priorities. Iterative testing also facilitates collaboration between developers and testers, enabling faster feedback loops and more efficient resolution of issues.

Automation and Manual Testing Balance

Achieving the right balance between automation and manual testing is essential for optimizing testing efficiency and effectiveness. While automation offers benefits such as repeatability, scalability, and speed, manual testing allows for exploratory testing, usability evaluation, and validation of subjective aspects.

The key is to identify test scenarios that are suitable for automation based on factors such as repeatability, frequency of execution, and return on investment. Critical functionalities, regression tests, and performance benchmarks are often prime candidates for automation. However, it's essential to recognize that not all testing activities can be automated, and manual testing remains indispensable for certain types of testing, especially those requiring human judgment and intuition.

Risk-Based Testing Strategy

The risk-based testing strategy prioritizes testing efforts based on the likelihood and impact of potential failures. Instead of testing everything exhaustively, organizations focus on testing high-risk areas with the most significant potential to impact system reliability, security, or user experience.

To implement a risk-based testing strategy, teams must conduct risk analysis and assessment to identify and prioritize risks. This involves evaluating business impact, technical complexity, regulatory requirements, and historical data. Test efforts aim to mitigate the most critical risks through targeted testing activities, risk-based test case design, and resource allocation.

In summary, adhering to these principles of effective testing are clear objectives and goals, comprehensive test coverage, iterative testing approach, automation, manual testing balance, and risk-based testing strategy—it lays the foundation for robust testing practices and contributes to the overall quality and reliability of software systems. By integrating these principles into testing processes, organizations can enhance their ability to deliver high-quality software products that meet user expectations and business objectives.

Techniques for Implementing the Testing Mindset

Test-Driven Development (TDD)

Test-driven development (TDD) is a software development methodology that prioritizes writing tests before writing the actual code.

Principles of TDD

At the core of TDD are three primary principles.

1. Write Tests First

In TDD, developers start by writing a failing test that defines the desired behavior or functionality of the code. This test serves as a specification or contract for implementing the code.

2. Write the Minimum Code to Pass the Test

Once the failing test is written, developers proceed to write the minimum amount of code necessary to make the test pass. This step focuses on implementing just enough functionality to satisfy the requirements outlined in the test.

3. Refactor Code

After the test passes, developers refactor the code to improve its design, readability, and maintainability. Refactoring ensures the code remains clean, efficient, and adaptable to future changes without altering its external behavior.

Practices of TDD

TDD involves several key practices that guide the development process.

1. Red-Green-Refactor Cycle

TDD follows a repetitive cycle known as "Red-Green-Refactor," where developers start by writing a failing test (Red), then implement the code to make the test pass (Green), and finally refactor the code (Refactor) to improve its quality.

2. Test Isolation

Tests in TDD should be isolated from external dependencies, such as databases, networks, or external services. Mocking or stubbing techniques are often used to simulate these dependencies and ensure that tests remain fast, reliable, and deterministic.

3. Keep Tests Simple and Focused

TDD encourages writing simple, focused tests that verify one specific aspect of the code's behavior. Tests should be easy to understand, maintain, and execute and provide clear feedback on the code's correctness.

Benefits of TDD

TDD offers several benefits to developers, teams, and organizations.

1. Improved Code Quality

By focusing on writing tests first, TDD promotes a design-driven approach to development, resulting in cleaner, more modular, and more maintainable code. The test suite is a safety net, ensuring that changes do not introduce unintended side effects or regressions.

2. Faster Feedback Loop

TDD provides instant feedback on the code's correctness, allowing developers to detect and fix defects early in the development process. This rapid feedback loop reduces the time and effort spent on debugging and rework, resulting in faster delivery of high-quality software.

3. Increased Confidence in Code Changes

With a comprehensive suite of automated tests, developers can refactor code confidently, knowing that any regressions will be quickly identified and addressed. This confidence encourages experimentation, innovation, and continuous improvement.

Challenges and Best Practices

While TDD offers many benefits, it presents challenges, particularly in adoption and implementation. Some common challenges include resistance to change, difficulty in writing effective tests, and maintaining a balance between writing tests and writing code. To overcome these challenges, organizations should invest in training, mentoring, and creating a supportive environment for TDD adoption. Additionally, following best practices such as starting small, focusing on high-value tests, and incorporating feedback loops can help teams succeed with TDD.

Test-driven development (TDD) is a disciplined approach to software development that emphasizes writing tests before writing code. By adhering to its principles and practices, organizations can improve code quality, accelerate delivery, and foster a culture of continuous improvement. Despite its challenges, TDD remains valuable for building reliable, maintainable, scalable software systems.

Behavior-Driven Development (BDD)

Behavior-driven development (BDD) is an agile software development methodology that emphasizes collaboration between developers, testers, and business stakeholders to deliver software that meets user requirements. This section explores BDD in depth, including its principles, practices, benefits, and implementation strategies.

Principles of BDD

The following core principles guide BDD.

1. User-Centric Focus

BDD strongly emphasizes understanding and addressing user needs and behaviors. Development efforts are driven by user stories or scenarios, which define the system's desired behavior from the user's perspective.

2. Collaboration and Communication

BDD promotes collaboration and communication among all stakeholders involved in the software development process. Using a common language to describe behavior, BDD facilitates shared understanding and alignment of expectations across teams.

3. Automation of Acceptance Criteria

BDD advocates for automating acceptance criteria through executable specifications written in natural language. These specifications serve as living documentation and automated tests, ensuring the system behaves as expected and providing a safety net for future changes.

Practices of BDD

BDD encompasses several key practices that guide the development process.

1. Ubiquitous Language

BDD encourages using a shared, domain-specific language (DSL) that all team members understand. This ubiquitous language helps bridge the gap between technical and nontechnical stakeholders, fostering better collaboration and understanding.

2. Writing Scenarios with Given-When-Then

Scenarios in BDD use a structured format known as "Given-When-Then" (GWT) to describe the preconditions, actions, and expected outcomes of a particular behavior. Developers, testers, and product owners collaborate to write these scenarios to ensure clarity and completeness.

3. Automating Acceptance Tests

BDD emphasizes automating acceptance tests using Cucumber, SpecFlow, or Behave tools. These tools allow scenarios written in natural language to be executed against the system under test, providing instant feedback on the system's behavior.

Benefits of BDD

BDD offers several benefits to teams and organizations.

1. Improved Collaboration and Understanding

Using a common language to describe behavior, BDD promotes collaboration and alignment of expectations among team members. This shared understanding reduces misunderstandings and rework, leading to more efficient and effective development processes.

2. Enhanced Communication

BDD encourages active participation from all stakeholders in defining behavior, leading to clearer requirements and acceptance criteria. This enhanced communication reduces the risk of misinterpretation and ensures that the system meets the needs of users and stakeholders.

3. Early Validation of Requirements

BDD enables early validation of requirements by defining behavior regarding executable specifications. By writing scenarios upfront, teams can clarify requirements, identify potential issues, and validate assumptions before writing code, leading to fewer defects and rework later in the development process.

Implementation Strategies and Best Practices

Implementing BDD effectively requires a combination of technical and cultural changes within an organization. Some best practices for successfully adopting BDD include fostering a culture of collaboration and communication, providing training and coaching on BDD practices and tools, and integrating BDD into existing development processes such as continuous integration (CI) and continuous delivery (CD).

In summary, behavior-driven development (BDD) is a user-centric software development methodology emphasizing collaboration, communication, and automation to deliver high-quality software that meets user requirements. By embracing BDD principles and practices, organizations can improve collaboration, enhance communication, and deliver software that adds value to users and stakeholders.

Exploratory Testing

Exploratory testing is an approach to software testing that emphasizes simultaneous learning, test design, and test execution. This section delves into the principles, techniques, benefits, and challenges of exploratory testing, providing insights into its application within software development teams.

Principles of Exploratory Testing

The following principles guide exploratory testing.

1. Simultaneous Learning and Test Design

Testers learn about the system under test while designing and executing tests. This approach allows testers to adapt their testing strategies based on their evolving understanding of the system's behavior and functionality.

2. Freedom and Creativity

Exploratory testing allows testers to explore the system unscripted, allowing for creativity and flexibility in test execution. Testers can uncover unexpected behaviors, edge cases, and defects that may not be captured through scripted testing alone.

3. Adaptability and Iteration

Exploratory testing embraces adaptability and iteration, allowing testers to adjust their testing approach based on feedback, observations, and insights gained during testing. Testers continuously refine their testing strategies to focus on areas of higher risk or uncertainty.

Techniques of Exploratory Testing

Exploratory testing employs several techniques to explore the system under test effectively.

1. Session-Based Testing

Testers conduct exploratory testing within predefined time-boxed sessions, focusing on specific areas or aspects of the system. Session-based testing helps structure testing activities while allowing for flexibility and spontaneity.

2. Scenario-Based Testing

Testers create test scenarios based on real-world usage scenarios, user stories, or personas. These scenarios guide testing efforts and help uncover usability, performance, and functionality issues.

3. Error Guessing

Testers leverage their domain knowledge, experience, and intuition to anticipate potential errors or defects in the system. Error guessing helps testers focus their testing efforts on areas of higher risk or vulnerability.

Benefits of Exploratory Testing

Exploratory testing offers several benefits to software development teams.

1. Early Bug Detection

Exploratory testing helps uncover defects early in the development life cycle, allowing for timely resolution and mitigation of risks. Testers can identify issues that may have been overlooked in scripted testing, leading to improved software quality.

2. Flexibility and Adaptability

Exploratory testing allows testers to explore the system dynamically and adapt their testing approach based on emerging insights and observations. This adaptability enables testers to focus on areas of highest risk or uncertainty.

3. Complement to Scripted Testing

Exploratory testing complements scripted testing by uncovering issues that may not be captured through predefined test cases. Testers can explore the system open-ended, uncovering edge cases and usability issues that may go unnoticed in scripted testing.

Challenges of Exploratory Testing

Despite its benefits, exploratory testing presents several challenges.

1. Documentation and Reproducibility

Exploratory testing may lack documentation and traceability, making it difficult to reproduce test scenarios or communicate findings effectively. Testers must balance exploration and documentation to ensure test results are captured and communicated appropriately.

2. Skill and Experience

Effective exploratory testing requires high skill, experience, and domain knowledge. Testers must be able to adapt quickly, think critically, and identify potential issues in the system.

3. Time and Resource Constraints

Exploratory testing may be constrained by time and resource limitations, particularly in fast-paced development environments. Testers must prioritize testing activities and focus on areas of highest value or risk to maximize the effectiveness of exploratory testing.

In summary, exploratory testing is a valuable approach to software testing that emphasizes learning, creativity, and adaptability. By embracing exploratory testing principles and techniques, software development teams can uncover defects early, improve software quality, and deliver products that meet user expectations. However, effective exploratory testing requires skill, experience, and careful consideration of challenges and constraints.

Regression Testing Strategies

Regression testing is a critical component of the software testing process, aimed at ensuring that new code changes do not adversely affect existing functionality. This section explores various regression testing strategies, including their principles, techniques, benefits, and challenges.

Principles of Regression Testing

The following core principles guide regression testing.

1. Comprehensive Coverage

Regression testing aims to cover all critical functionalities and scenarios affected by code changes. Comprehensive test coverage helps identify potential regressions and ensure the stability and reliability of the software.

2. Automation

Automated regression testing helps streamline the testing process by executing test cases automatically and efficiently. Automation reduces manual effort, speeds up testing cycles, and provides faster feedback on code changes.

3. Prioritization

Not all test cases are equally important for regression testing. Prioritization helps focus testing efforts on high-risk areas or critical functionalities that are more likely to be affected by code changes.

Techniques for Regression Testing

Regression testing employs several techniques to validate code changes effectively.

1. Re-run All Tests

This technique involves re-executing all existing test cases after each code change to ensure no regression issues have been introduced. While thorough, this approach can be time-consuming and resource-intensive.

2. Selective Regression Testing

Selective regression testing involves identifying a subset of test cases that are most likely to be affected by code changes and executing only those tests. This approach reduces testing effort while still providing adequate coverage.

3. Test Case Prioritization

Test case prioritization techniques such as risk-based testing or impact analysis help prioritize test cases based on factors such as business impact, criticality, or likelihood of regression. Prioritization ensures that high-risk areas are tested first, minimizing the impact of regressions.

Benefits of Regression Testing

Regression testing offers several benefits to software development teams.

1. Risk Mitigation

Regression testing helps mitigate the risk of introducing defects or regressions when making code changes. By validating existing functionality, regression testing ensures that new features or fixes do not inadvertently break the software.

2. Improved Quality

Continuous regression testing contributes to overall software quality by identifying and addressing regressions early in the development process. Early detection and resolution of issues lead to higher-quality software products.

3. Faster Time to Market

Automated regression testing accelerates the testing process, enabling teams to release code changes more quickly and confidently. Faster regression testing cycles reduce time to market for new features and fixes.

Challenges of Regression Testing

Despite its benefits, regression testing presents several challenges.

1. Test Maintenance

As the software evolves, regression test suites need to be updated and maintained to reflect changes in functionality. Test maintenance can be time-consuming and may require significant effort, especially for large and complex systems.

2. Resource Constraints

Regression testing can be resource-intensive, requiring access to test environments, data, and infrastructure. Limited resources, such as time, budget, or hardware, may impact the effectiveness and coverage of regression testing.

3. Test Oracles

Identifying expected outcomes or test oracles for regression testing can be challenging, especially for complex or ambiguous functionalities. Clear, accurate, and up-to-date test oracles are essential for effective regression testing.

In summary, regression testing is a critical aspect of software testing that ensures the stability and reliability of software systems. Organizations can mitigate risks, improve quality, and accelerate delivery by employing appropriate regression testing strategies and techniques while addressing the challenges inherent in regression testing.

Smoke Testing

Smoke testing, also known as build verification testing or sanity testing, is a preliminary level of testing conducted on a software build to ensure that the critical functionalities of the application are working as expected. This section provides an in-depth exploration of smoke testing, including its objectives, process, benefits, and challenges.

Objectives of Smoke Testing

The primary objectives of smoke testing include the following.

1. Verification of Critical Functionality

Smoke testing aims to verify the basic functionality of the software build, ensuring that essential features and functionalities are working as expected.

2. Detection of Major Defects

Smoke testing helps identify major defects or issues that could prevent further testing or deployment of the software build. Smoke testing saves time and effort in subsequent testing phases by detecting critical issues early.

3. Validation of Build Stability

Smoke testing validates the stability and readiness of the software build for further testing or deployment. A successful smoke test indicates the build is stable and suitable for additional testing activities.

Process of Smoke Testing

The process of smoke testing typically involves the following steps.

1. Identification of Critical Scenarios

Testers identify a set of critical test scenarios or functionalities that represent the core features of the software application.

2. Execution of Test Cases

Testers execute the identified test cases or scenarios on the software build using predefined test scripts or manual test procedures.

3. Verification of Results

Testers verify the results of the smoke test to ensure that critical functionalities are functioning correctly. Any failures or discrepancies are reported for further investigation and resolution.

4. Decision-Making

Based on the smoke test's outcome, stakeholders decide whether the software build is ready for additional testing or deployment. If the smoke test passes, further testing activities can proceed. If it fails, the build may require further investigation and corrective actions before retesting.

Benefits of Smoke Testing

Smoke testing offers several benefits to software development teams.

1. Early Detection of Critical Issues

Smoke testing helps identify major defects or issues early in the software development life cycle, reducing the risk of issues being discovered later in the testing process or production.

2. Time and Cost Savings

Smoke testing saves time and effort by avoiding extensive testing of nonessential features by focusing on critical functionalities. Early detection of issues also reduces the cost of fixing defects later in the development process.

3. Improved Build Quality

Smoke testing improves overall build quality by ensuring that essential features work correctly before further testing or deployment activities. A successful smoke test indicates a higher level of build stability and readiness.

Challenges of Smoke Testing

Despite its benefits, smoke testing presents several challenges.

1. Limited Scope

Smoke testing has a limited scope and may not cover all aspects of the software application. It focuses primarily on critical functionalities, potentially overlooking issues in nonessential features.

2. Dependency on Test Environment

Smoke testing relies on a stable and representative test environment to produce reliable results. Issues with the test environment or infrastructure may impact the effectiveness of smoke testing.

3. Maintenance Overhead

Maintaining and updating smoke test suites can be time-consuming, especially as the software application evolves and new features are introduced. Regular review and maintenance of smoke test cases are essential to keep them relevant and effective.

In summary, smoke testing is a valuable testing technique that provides a quick assessment of a software build's stability and readiness. By focusing on critical functionalities and detecting major defects early, smoke testing contributes to improved build quality, reduced risk, and faster time to market. However, organizations must address challenges such as limited scope, test environment dependencies, and maintenance overhead to maximize the effectiveness of smoke testing in their software development processes.

Tools and Technologies for Supporting Testing Mindset

Test Management Tools

Test management tools are software applications designed to assist teams in organizing, managing, and executing their testing activities efficiently. This section overviews test management tools, including their features, benefits, popular tools, and selection considerations.

Features of Test Management Tools

Test management tools typically offer the following features.

1. Test Case Management

Test management tools provide a centralized repository for storing and organizing test cases, including details such as test descriptions, steps, expected results, and associated requirements.

2. Test Planning and Scheduling

These tools enable teams to plan and schedule testing activities, allocate resources, and define test execution timelines and milestones.

3. Test Execution and Reporting

Test management tools facilitate the execution of test cases, capture test results, and generate comprehensive test reports and metrics to track progress and identify issues.

4. Requirement Traceability

Test management tools help establish traceability between test cases and requirements, ensuring that each requirement is adequately tested and validated.

5. Defect Management

These tools support the identification, tracking, and resolution of defects by providing a centralized repository for logging, prioritizing, and managing defect reports.

Benefits of Test Management Tools

Test management tools offer several benefits to software development teams.

1. Centralized Repository

Test management tools provide a centralized repository for storing test artifacts, including test cases, test plans, test results, and defect reports, improving visibility and accessibility across the team.

2. Improved Collaboration

These tools facilitate collaboration and communication among team members by providing a shared platform for accessing and updating testing information, fostering teamwork and alignment.

3. Efficient Test Execution

Test management tools streamline the test execution process by automating test case execution, providing test execution progress tracking, and generating detailed test reports, reducing manual effort and improving efficiency.

4. Enhanced Traceability

Test management tools help establish traceability between test cases, requirements, and defects, enabling teams to track test coverage status, identify gaps, and ensure that all requirements are adequately tested.

Popular Test Management Tools

Several test management tools are widely used in the industry.

Tool Name	Key Features	Integration Options	Best Suited For
TestMonitor	Requirement and risk-based testing, advanced test case design, integrated issue management	Jira, DevOps, Slack, REST API	Comprehensive test management across all organizational levels
TestRail	Detailed test case management, real-time insights, customizable dashboards	Jira, FogBugz, Bugzilla, GitHub, TFS, and more	Organizing and tracking extensive testing efforts
Zephyr Enterprise	Enterprise-grade test planning, bidirectional Jira integration, customizable dashboards	Jira	Enterprise-level testing with complex integration needs

(*continued*)

Tool Name	Key Features	Integration Options	Best Suited For
PractiTest	End-to-end QA management, customizable filters, extensive integration capabilities	Jira, Pivotal Tracker, Bugzilla, Redmine, Selenium, Jenkins	Efficient and visible QA management
Jira Software	Flexible management through add-ons, extensive tracking and reporting	Vast array of Atlassian Marketplace add-ons	Agile teams requiring integrated test management
QACoverage	Customizable requirements definition, traceability between requirements, test cases, and defects	Not specified	Agile teams, requirements and test case management
RTM for Jira	In-built requirements management, tree-structured views, effortless migration from external tools	Native Jira integration	Teams using Jira for managing requirements and tests
Testiny	Streamlined design, powerful integrations, instant updates across sessions	Jira, GitLab, GitHub, Redmine, Azure DevOps	Manual and automated testing in modern web environments

(continued)

Tool Name	Key Features	Integration Options	Best Suited For
Tuskr	Flexible test runs, resource optimization, workload charts, drag-and-drop organization	Jira and other bug/time-tracking tools	Optimizing test case organization and execution
Testpad	Keyboard-driven editing, drag-and-drop organization, integration with issue trackers	Jira and others	Agile and exploratory testing environments
TestFLO for Jira	Highly customizable, integration with test automation tools, reusable test case repository	Jira, REST API, Bamboo, Jenkins	Teams needing deep integration with Jira and test automation
SpiraTest	Integrated requirements and bug-tracking, customizable reports, multilevel dashboards	Jira, Selenium, JMeter, and more	Comprehensive test management with a focus on team collaboration
Klaros-Testmanagement	Test planning, execution, and evaluation, interfaces to various systems	Jira, Redmine, GitLab, GitHub, Jenkins, and more	Comprehensive test management in regulated environments
Qase	Organize test cases and suites, shared steps, test run wizard, test case review	Jira, Redmine, Trello, GitHub, Slack	Teams looking for a modern UI and extensive integration options

(continued)

Tool Name	Key Features	Integration Options	Best Suited For
TestCollab	Seamless Jira integration, reusable suites, modern features like @mention comments	Jira	Teams seeking easy onboarding and extensive Jira integration
JunoOne	Sophisticated test case and issue tracking, powerful JIRA integration	Jira	Agile test case management and issue tracking
QAComplete	Centralized test management, customizable to fit any development process	Jira, Bugzilla, Visual Studio, and more	Flexible testing environments from Waterfall to Agile
Kualitee	Intuitive interface, third-party integrations, individual and group progress tracking	Various tools	Teams managing testing with a focus on collaboration and customization
Xray	Comprehensive Jira integration, supports both manual and automated tests, detailed reporting	Continuous integration tools like Bamboo and Jenkins	Jira users needing a detailed and integrated test management system
Qucate	Dynamic test plan templates, extensive onboarding, intuitive UI, unlimited projects and test plans	Not specified	Teams looking for flexibility and high customer support standards

source: https://www.softwaretestinghelp.com/15-best-test-management-tools-for-software-testers

Considerations for Selection

When selecting a test management tool, teams should consider the following factors.

1. Features and Functionality

Evaluate the features and functionality offered by the test management tool to ensure that it meets the specific needs and requirements of the team, including test case management, test execution, reporting, and integration capabilities.

2. Ease of Use

Choose a test management tool that is intuitive and easy to use, with a user-friendly interface and navigation to facilitate adoption and usage by team members.

3. Integration with Existing Tools

Consider the test management tool's integration capabilities with other tools and systems used within the organization, such as issue tracking, version control, and continuous integration tools, to ensure seamless workflow integration.

4. Scalability and Flexibility

Select a test management tool that can scale with the team's needs and accommodate changes in testing processes, methodologies, and project requirements over time.

5. Cost and Licensing

Evaluate the test management tool's cost and licensing options, including subscription fees, user licenses, and additional features or modules, to ensure alignment with the team's budget and financial constraints.

In summary, test management tools are crucial in streamlining testing activities, improving collaboration, and ensuring the quality and reliability of software products. By selecting the right test management tool and leveraging its features effectively, teams can optimize their testing processes and deliver high-quality software products more efficiently.

Automated Testing Frameworks

Automated testing frameworks are essential tools for streamlining and automating software testing processes. This section provides an overview of automated testing frameworks, including their types, features, benefits, popular frameworks, and considerations for selection.

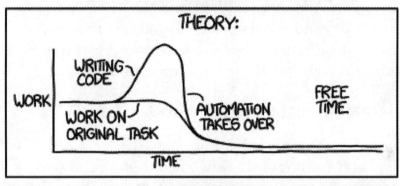

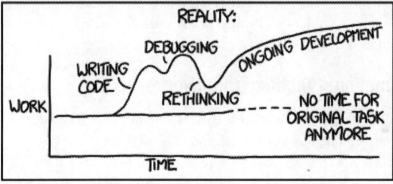

source: https://xkcd.com/1319

Types of Automated Testing Frameworks

Automated testing frameworks can be categorized into several types based on their purpose and functionality.

1. Unit Testing Frameworks

Unit testing frameworks such as JUnit (Java), NUnit (.NET), and pytest (Python) are designed for testing individual units or components of code in isolation. These frameworks provide features for defining test cases, executing tests, and asserting expected outcomes.

2. Integration Testing Frameworks

Integration testing frameworks such as TestNG (Java) and Robot Framework (Python) are used for testing the interaction between different modules or components of a system. These frameworks facilitate testing across multiple application layers and integration with external dependencies.

3. Functional Testing Frameworks

Functional testing frameworks such as Selenium (for web applications), Appium (for mobile applications), and Cypress (for modern web applications) are designed to test the application's functional behavior from an end-user perspective. These frameworks automate interactions with the user interface and validate application functionality.

4. Behavior-Driven Development (BDD) Frameworks

BDD frameworks such as Cucumber (for Java, JavaScript, and Ruby) and SpecFlow (.NET) enable teams to write tests based on user stories or scenarios in a natural language format. These frameworks promote collaboration between developers, testers, and business stakeholders and facilitate automated acceptance testing.

Features of Automated Testing Frameworks

Automated testing frameworks typically offer the following features.

1. Test Case Management

Automated testing frameworks provide features for defining, organizing, and managing test cases, including test descriptions, assertions, and expected outcomes.

2. Test Execution

These frameworks facilitate the execution of automated tests across different environments, configurations, and platforms, allowing for comprehensive test coverage.

3. Reporting and Analysis

Automated testing frameworks generate detailed test reports and metrics to track test results, identify issues, and analyze test coverage, helping teams make informed decisions about software quality.

4. Integration with Development Tools

Many automated testing frameworks integrate seamlessly with version control systems, continuous integration tools, and issue-tracking systems, enabling automated testing within the development workflow.

Benefits of Automated Testing Frameworks

Automated testing frameworks offer several benefits to software development teams.

1. Improved Efficiency

Automated testing frameworks automate repetitive and time-consuming testing tasks, allowing teams to execute tests more quickly and efficiently than manual testing.

2. Consistent and Reliable Testing

Automated tests produce consistent and reliable results, reducing the risk of human error and ensuring consistent test coverage across different environments and configurations.

3. Faster Feedback

Automated testing frameworks provide rapid feedback on code changes, allowing teams to detect and address issues early in the development process, leading to faster time to market.

4. Scalability and Reusability

Automated tests can be easily scaled and reused across different projects, environments, and configurations, saving time and effort in test development and maintenance.

Popular Automated Testing Frameworks

Several automated testing frameworks are widely used in the industry.

Tool Name	Programming Languages Supported	Key Features	Integrations
Katalon Studio	Low-code platform	Low-code, scalable, supports web, API, mobile and desktop apps	CI/CD tools, Jira, GitLab, Jenkins, Azure DevOps
Selenium	Java, C#, Python, JavaScript, Ruby, PHP	Open source, supports multiple browsers, parallel executions	Various testing frameworks and CI/CD tools
Appium	Java, C#, Python, JavaScript, Ruby, PHP	Open source for native, web, and hybrid mobile apps	Testing frameworks, CI/CD tools
TestComplete	JavaScript, Python, VBScript, JScript, Delphi, C++, C#	GUI testing for web, mobile, desktop, AI visual recognition	Other testing frameworks, CI/CD tools
Cypress	JavaScript	End-to-end web testing operates within browsers	CI/CD tools
Ranorex Studio	VB.Net, C#	GUI testing, broad technology support, RanoreXPath	Selenium Grid, other testing frameworks, CI/CD tools
Perfecto	Cloud-based platform	Cloud-based, scriptless test creation, real-user simulation	Various testing frameworks, CI/CD tools
LambdaTest	Cloud service	Selenium Grid in the cloud supports over 2000 environments	CI/CD tools

(continued)

Tool Name	Programming Languages Supported	Key Features	Integrations
Postman	API testing tool	API testing supports multiple HTML methods	CI/CD tools
SoapUI	API testing tool	Open source for REST and SOAP services	CI/CD tools
Eggplant Functional	GUI automation tool	The image-based approach supports multiple platforms	Popular CI/CD tools
Tricentis Tosca	Model-based testing tool	Codeless test creation, risk-based test optimization	Various testing frameworks, CI/CD tools
Apache JMeter	Performance testing tool	Load testing supports different servers and protocols	CI/CD tools
Robot Framework	Keyword-driven testing framework	Keyword-driven supports external libraries and tools	External libraries and tools
Applitools	Visual testing tool	Automated visual testing, smart bug detection	-

source: https://katalon.com/resources-center/blog/automation-testing-tools

Considerations for Selection

When selecting an automated testing framework, teams should consider the following factors.

1. Compatibility and Support

Choose an automated testing framework that is compatible with the technology stack, programming languages, and platforms used in the project. Consider the level of community support, documentation, and active framework development.

2. Ease of Use and Learning Curve

Evaluate the automated testing framework's ease of use and learning curve, considering factors such as syntax, features, and tooling support. Choose a framework that aligns with the team member's skill level and expertise.

3. Integration and Extensibility

Consider the integration capabilities of the automated testing framework with other tools and systems used within the organization, such as continuous integration servers, version control systems, and issue-tracking tools. Choose a framework that offers extensibility and customization options to adapt to specific testing requirements.

4. Scalability and Performance

Assess the scalability and performance characteristics of the automated testing framework, considering factors such as test execution speed, resource utilization, and support for parallel testing. Choose a framework that can scale with the project's needs and accommodate future growth.

5. Cost and Licensing

Consider the cost and licensing options of the automated testing framework, including subscription fees, commercial support, and additional features or plugins. Choose a framework that aligns with the budget and financial constraints of the organization.

In summary, automated testing frameworks play a crucial role in streamlining testing processes, improving efficiency, and ensuring the quality of software applications. By selecting the right automated testing framework

Performance Testing Tools

Performance testing tools are essential for evaluating software applications' speed, responsiveness, and scalability under various load conditions. This section overviews performance testing tools, including their features, benefits, popular tools, and selection considerations.

Features of Performance Testing Tools

Performance testing tools typically offer the following features:

1. **Load Generation:** Performance testing tools simulate user load and traffic to stress test the application and measure its performance under heavy load conditions.

2. **Transaction Monitoring:** These tools monitor and measure the response time and throughput of individual transactions or user interactions within the application.

3. **Resource Monitoring:** Performance testing tools monitor system resources such as CPU, memory, disk I/O, and network bandwidth to identify performance bottlenecks and resource constraints.

4. **Reporting and Analysis:** Performance testing tools generate comprehensive reports and analysis of test results, including performance metrics, trends, and recommendations for optimization.

5. **Scalability Testing:** Some performance testing tools offer features for testing the scalability of the application by simulating increasing user load and measuring its impact on system performance.

Benefits of Performance Testing Tools

Performance testing tools offer several benefits to software development teams:

1. **Early Detection of Performance Issues:** Performance testing tools help identify performance issues early in the development life cycle, allowing teams to address them before deployment and production.

2. **Optimized Performance:** By identifying performance bottlenecks and resource constraints, performance testing tools enable teams to optimize the performance of the application and enhance user experience.

3. **Scalability Validation:** These tools validate the scalability of the application by simulating increasing user load and measuring its impact on system performance, helping teams prepare for future growth and demand.

4. **Improved Reliability:** Performance testing tools help ensure the reliability and stability of the application under various load conditions, reducing the risk of downtime, crashes, and service interruptions.

189

Popular Performance Testing Tools

Several performance testing tools are widely used in the industry:

Tool Name	Programming Languages Supported	Key Features	Integrations
Apache JMeter	Java	Simulate heavy loads, support for multiple protocols, dynamic reporting	CI/CD tools, various development tools
LoadRunner	C, Java, JavaScript	Support for over 50 technologies, realistic load emulation	CI/CD tools, IDEs
OctoPerf	- (Uses underlying JMeter scripts)	Cloud-based, no coding required for UI, integrates with JMeter	Katalon, other open source technologies
Katalon	Groovy, Java	AI-powered, supports multiple types of testing including API, mobile, web	OctoPerf, CI/CD tools
Gatling	Scala	Open source, supports complex scenarios, detailed HTML reports	CI/CD tools, monitoring tools
Locust	Python	Supports distributed load testing, real-time statistics	Various CI/CD tools
k6	JavaScript	Command-line interface, scripting in JavaScript, extensive integrations	CI/CD tools, monitoring tools

(continued)

Tool Name	Programming Languages Supported	Key Features	Integrations
Neoload	-	RealBrowser technology, browser-based testing, cloud resource optimization	CI/CD tools, major cloud platforms
WebLOAD	JavaScript	AI-driven smart correlation, real browser testing, integrated with CI/CD platforms	Major CI/CD platforms, monitoring tools
Taurus	JSON, YAML (configuration)	Simplifies running scripts for various testing tools, real-time reporting	JMeter, Gatling, Selenium, Grinder
BlazeMeter	-	Cloud-based, synthetic data and AI integration, supports massive-scale testing	CI/CD platforms, monitoring tools
LoadNinja	-	Scriptless load testing, real-browser testing, supports automation and real-time performance issue diagnosis	Major CI/CD platforms
Artillery	JavaScript (Node.js)	Serverless, supports distributed testing, scalable	AWS services, CI/CD tools
LoadUI Pro	-	Built on open source LoadUI, extensive monitoring, real-time performance statistics	CI/CD tools, monitoring tools
Silk Performer	-	Enterprise-class, customizable load tests, cloud simulation	Various cloud platforms, monitoring tools

source: https://katalon.com/resources-center/blog/top-performance-testing-tools

Considerations for Selection

When selecting a performance testing tool, teams should consider the following factors.

1. Type of Application

Choose a performance testing tool suitable for the application type being tested, whether it's a web application, mobile application, API, or enterprise system.

2. Scalability and Performance

Assess the scalability and performance capabilities of the performance testing tool, including its ability to simulate large user loads, measure response times, and monitor system resources.

3. Ease of Use and Learning Curve

Evaluate the performance testing tool's ease of use and learning curve, considering factors such as user interface, scripting language, and documentation. Choose a tool that aligns with the skill level and expertise of the team members.

4. Integration and Compatibility

Consider the integration capabilities of the performance testing tool with other tools and systems used within the organization, such as continuous integration servers, issue-tracking systems, and monitoring tools.

5. Cost and Licensing

Evaluate the cost and licensing options of the performance testing tool, including subscription fees, licensing models, and additional features or services. Choose a tool that aligns with the budget and financial constraints of the organization.

In summary, performance testing tools are crucial in evaluating software applications' speed, responsiveness, and scalability. By selecting the right performance testing tool and leveraging its features effectively, teams can identify and address performance issues early in the development life cycle, optimize their applications' performance, and deliver high-quality software products to users.

Overcoming Challenges in Adopting the Testing Mindset

Adopting a testing mindset within an organization can encounter various challenges, ranging from resistance to change to resource constraints and cultural barriers. This section explores strategies for overcoming these challenges to foster a culture of testing and ensure the reliability of software systems.

Resistance to Change

Resistance to change is a common challenge when introducing new processes or methodologies, including adopting a testing mindset. To overcome resistance to change, organizations can implement the following strategies:

Communicate the Benefits

Clearly communicate the benefits of adopting a testing mindset, such as improved software quality, reduced defects, and faster delivery cycles. Highlight how testing contributes to overall business objectives and customer satisfaction.

Provide Training and Support

Offer training programs, workshops, and resources to educate team members about the principles, practices, and benefits of testing. Provide ongoing support and mentorship to help team members transition to the testing mindset.

Lead by Example

Demonstrate leadership support and commitment to testing by leading by example. Encourage leaders and managers to embrace testing practices, participate in testing activities, and advocate for the importance of testing within the organization.

Address Concerns and Objections

Listen to team member's concerns and objections regarding the adoption of a testing mindset and address them openly and transparently. Provide opportunities for feedback and discussion to address misconceptions and alleviate fears.

Resource Constraints

Resource constraints, such as limited budget, time, and manpower, can pose significant challenges to implementing testing initiatives. To overcome resource constraints, organizations can consider the following approaches.

Prioritize Testing Activities

To allocate resources effectively, prioritize testing activities based on risk, criticality, and business impact. Focus testing efforts on high-risk areas and critical functionalities that are most important to the project's success.

Automate Testing Processes

Invest in automation tools and frameworks to streamline testing processes and reduce manual effort. Automated testing helps maximize resource utilization, accelerate testing cycles, and improve overall efficiency.

Collaborate and Share Resources

Foster collaboration and knowledge sharing among teams to leverage resources more effectively. Encourage cross-functional collaboration between development, testing, and operations teams to share expertise, tools, and best practices.

Outsource Testing Activities

Consider outsourcing certain testing activities to external vendors or specialized testing teams to augment internal resources and capabilities. Outsourcing can provide access to specialized skills, expertise, and resources as needed.

Cultural and Organizational Barriers

Cultural and organizational barriers, such as resistance to change, siloed teams, and lack of collaboration, can impede the adoption of a testing mindset. To overcome these barriers, organizations can implement the following strategies.

Promote Collaboration and Cross-Functional Teams

Foster a culture of collaboration and teamwork by breaking down silos between development, testing, and operations teams. Encourage cross-functional teams to work together closely and share accountability for quality.

Empowerment and Ownership

Empower team members to take ownership of testing activities and quality assurance processes. Encourage autonomy, accountability, and empowerment to drive a culture of quality throughout the organization.

Continuous Learning and Improvement

Encourage feedback, experimentation, and reflection to promote a culture of continuous learning and improvement. Provide opportunities for team members to learn new skills, explore new testing techniques, and share knowledge with others.

Recognize and Reward Testing Excellence

Recognize and reward individuals and teams demonstrating excellence in testing and quality assurance. Celebrate successes, acknowledge contributions, and incentivize behaviors that support the testing mindset.

Organizations can overcome challenges in adopting the testing mindset and fostering a culture of testing excellence by addressing resistance to change, resource constraints, and cultural barriers. Organizations can ensure the reliability and quality of their software systems by promoting collaboration, empowering team members, and fostering a culture of continuous improvement.

Case Studies and Examples

Successful Implementations of the Testing Mindset

Successful implementations of a robust testing mindset exemplify how embracing a culture of continuous testing and automation can drive significant business improvements across various industries. This section delves deeper into each sector's proactive testing approach and how it contributed to achieving organizational goals.

Sector	Testing Mindset	Implementation	Result
Information Services	Emphasized modularity and reusable components in testing	Developed automation concurrently with new feature implementation	Achieved faster releases and higher quality, leading to significant cost and effort reductions
Airline Industry	Focused on business processes and concurrent development and testing	Utilized a modularity approach for quick development of new business process automation	Reduced time and cost while maintaining high-quality releases, enhancing business agility and customer satisfaction

(continued)

Sector	Testing Mindset	Implementation	Result
Telecommunications	Required comprehensive test automation across multiple production systems due to business expansion	Integrated end-to-end test automation, including web, API, and database testing	Ensured robust, seamless integration into CI/CD pipelines, enhancing reliability and efficiency in diverse markets
Financial Services	Aimed to support digital transformation with a focus on continuous deployment	Transitioned from manual testing and disparate tools to a unified automation platform	Streamlined digital transformation initiatives, improving customer and employee experiences and facilitating faster technology adoption
Pharmaceutical Industry	Focused on digital enablement and automation as part of a long-term technology strategy	Implemented end-to-end automation to support business assurance across the technology stack	Expedited the journey toward digital enablement, enhancing scientific progress and patient well-being

(*continued*)

Sector	Testing Mindset	Implementation	Result
Public Sector—Law Enforcement Agency	Overhauled manual testing processes to improve automation maturity and agility in Agile environments	Adopted a no-code AI-powered platform, enabling manual testers to create automated tests without deep coding expertise	Improved testing engagement and efficiency, reduced redundancy, and enhanced rapid automation capabilities within Agile sprints

source: https://www.accelq.com/casestudy

Conclusion

Each case study demonstrates that a successful testing mindset involves more than just adopting new tools; it requires a cultural shift toward continuous improvement, quality assurance, and efficiency. By embracing these principles, organizations can not only achieve specific project goals but also enhance their overall competitive edge in the market. This strategic approach to testing ensures that teams are not merely reactive but are equipped to drive innovation and adapt to changing market conditions effectively.

Lessons Learned from Failures and Challenges

The case studies not only highlight successes but also shed light on the challenges and failures that preceded these achievements. Reflecting on these lessons learned can provide valuable insights for other organizations looking to enhance their testing strategies. Here are some key takeaways.

Sector	Lesson Learned	Challenge
Information Services	Overcoming resistance to new practices requires proving their value with tangible examples and persistent advocacy.	Selling the idea of A/B testing internally was difficult due to skepticism from senior management.
Airline Industry	Integrating testing with development processes from the start ensures that testing does not become a bottleneck.	Convincing various teams to adopt a concurrent development and testing model was initially challenging due to traditional siloed working methods.
Telecommunications	Adequate planning and understanding of the complexities of integrating new acquisitions into existing frameworks are critical.	The separation of markets required a sophisticated approach to test multiple systems concurrently, which initially overwhelmed the existing testing infrastructure.
Financial Services	Diverse and code-intensive tools can complicate the testing process. Consolidating tools into a unified platform is key.	The transition from manual testing practices to automated solutions required a significant cultural shift and training, which was initially met with resistance.

(*continued*)

Sector	Lesson Learned	Challenge
Pharmaceutical Industry	Automation can be effectively implemented with careful consideration of compliance and quality standards.	Balancing the need for rigorous testing with the speed of innovation was difficult, particularly with the initial reliance on outdated testing frameworks.
Public Sector—Law Enforcement Agency	Simplifying the testing process with no-code tools can empower manual testers to contribute more effectively to automation.	Overcoming the steep learning curve associated with automation tools and the reluctance to abandon established manual testing routines.

General Insights

Adaptability

Organizations must be adaptable in their approach and willing to modify or completely overhaul their testing strategies based on evolving project needs and outcomes.

Collaboration and Communication

Effective communication and collaboration across all levels of an organization are essential for successfully implementing new testing strategies.

Continuous Learning and Improvement

Embracing failures as learning opportunities fosters an environment of continuous improvement and innovation.

Scalability and Flexibility

Solutions must not only address current needs but also be scalable and flexible to adapt to future challenges and technological advancements.

By reflecting on these lessons and challenges, organizations can better prepare for the hurdles of implementing and scaling up testing practices and embrace a more robust approach to quality assurance and automation.

Future Trends and Developments in Testing

Anticipating future trends and developments in testing is crucial for organizations to stay ahead of the curve and ensure the effectiveness and efficiency of their testing practices. This section explores three key trends shaping the future of testing.

Artificial Intelligence and Machine Learning in Testing

Artificial intelligence (AI) and machine learning (ML) are revolutionizing the testing landscape by enabling automation, predictive analytics, and intelligent test generation.

AI and ML in Software Testing

AI and ML in software testing involve integrating these technologies to improve various aspects of testing. These advancements offer tools that augment human decision-making abilities, allowing testers to automate complex processes and enhance test accuracy and efficiency.

AI and ML can be applied in several ways to optimize software testing.

Automated Smart Test Case Generation

AI can automate the creation of test cases, reducing the workload on human testers and ensuring that tests cover a broader range of scenarios.

Test Case Recommendation

ML algorithms can analyze historical data to suggest the most relevant test cases, optimizing the testing process and ensuring critical issues are tested.

Test Data Generation

AI can generate diverse datasets needed for thorough testing, saving time and ensuring comprehensive coverage.

Test Maintenance for Regression Testing

AI can update test scripts automatically when changes occur in the application, reducing the manual effort needed for test maintenance.

Visual Testing

AI tools can compare visual aspects of applications before and after changes, identifying visual issues that might not be noticeable to human testers.

Benefits of Using AI/ML in Software Testing

Enhanced Efficiency

AI speeds up the test creation process and makes test maintenance easier.

Improved Accuracy

AI can help identify potential issues more accurately by learning from past data.

Cost Reduction

Automating routine tasks reduces the cost associated with manual testing.

Challenges of AI/ML in Software Testing

Despite the benefits, there are several challenges to be aware of the following.

Training Data Quality

AI models require high-quality, diverse datasets to train effectively.

Unforeseen Test Cases

AI might miss scenarios not represented in the training data.

Model Drift

Changes in application usage can make AI models less effective over time, requiring ongoing monitoring and adaptation.

Best Practices When Using AI/ML in Software Testing

Understand AI/ML Systems

A thorough understanding of AI technologies and workflows is crucial.

Be Patient

AI models take time to develop and learn.

Learn Prompt Engineering

Providing clear, structured prompts helps generate more accurate outputs from AI models.

View AI as a Tool

AI should be seen as an assistant that enhances the tester's capabilities, not as a replacement.

Testing with AI vs. Testing for AI Systems

Testing with AI

Using AI models to enhance testing processes.

Testing for AI Systems

Ensuring that AI models themselves perform as expected can be challenging due to their complex and nondeterministic nature.

Overall, AI and ML are transforming software testing by making it more efficient, accurate, and less labor-intensive. However, to truly benefit from their capabilities, it's important to navigate the challenges carefully and integrate these technologies thoughtfully.

Shift-Left Testing Approach

The shift-left testing approach advocates for integrating testing activities earlier in the software development life cycle, enabling early defect detection and prevention.

Potential Impact

1. Early Defect Detection

By shifting testing activities leftward, teams can identify and address defects earlier in the development process when they are less costly and time-consuming to fix.

2. Continuous Feedback

Incorporating testing into every stage of development facilitates continuous feedback loops between developers, testers, and stakeholders, ensuring higher software quality and faster delivery cycles.

3. Improved Collaboration

Shift-left testing promotes collaboration between development and testing teams, breaking down silos and fostering a culture of quality ownership across the organization.

DevOps and Testing Integration

DevOps emphasizes collaboration, automation, and continuous delivery, integrating development, operations, and testing into a seamless workflow.

Potential Impact

1. Continuous Testing

Integrating testing into the DevOps pipeline enables continuous testing of code changes throughout the development life cycle, ensuring early detection of defects and smooth deployment.

2. Automation and Orchestration

DevOps practices automate testing processes and orchestrate testing activities across development, testing, and production environments, enhancing efficiency and repeatability.

3. Feedback Loop

DevOps fosters a feedback-driven culture, with continuous feedback loops between development, testing, and operations teams, enabling rapid iteration and improvement.

Conclusion

Embracing these future trends and developments in testing will enable organizations to enhance their testing practices, improve software quality, and accelerate delivery cycles. Organizations can stay agile, responsive, and competitive in an ever-evolving digital landscape by leveraging AI and ML technologies, adopting a shift-left testing approach, and integrating testing into DevOps workflows.

Recap of Key Points

- A testing mindset is crucial for building reliable systems, and organizations can benefit greatly from adopting this mindset.

- A testing mindset involves proactively identifying and solving problems before they escalate, ensuring that a product or service meets its requirements and user expectations.

- A testing mindset emphasizes critical thinking, attention to detail, proactive problem-solving, empathy for the end user, and continuous learning and adaptation.

- Organizations can cultivate a testing mindset culture through leadership support, collaboration, skill development, experimentation, and a focus on quality and improvement.

- Adopting a testing mindset offers benefits such as improved software quality, reduced risk of defects, enhanced customer satisfaction, faster time to market, cost savings, increased confidence in releases, and a culture of continuous improvement.

- Effective testing involves clear objectives and goals, comprehensive test coverage, an iterative testing approach, a balance between automation and manual testing, and a risk-based testing strategy.

- Techniques such as test-driven development (TDD), behavior-driven development (BDD), exploratory testing, regression testing strategies, smoke testing, and test management tools can support the implementation of a testing mindset.

- Organizations can overcome challenges in adopting a testing mindset by addressing resistance to change, resource constraints, and cultural barriers.

- Successful implementations and lessons learned from failures provide valuable insights for organizations looking to enhance their testing strategies.

- Future trends and developments in testing, such as AI and ML in testing, shift-left testing approach, and DevOps and testing integration, will shape the testing landscape.

Exercises

1. Which of the following is a core principle of the testing mindset?

   (A) Critical thinking

   (B) Attention to detail

   (C) Proactive problem-solving

   (D) All of the above

2. What is the purpose of iterative testing?

   (A) To identify and address defects early in the development life cycle

   (B) To reduce the risk of defects reaching production

   (C) To facilitate collaboration between developers and testers

   (D) Both A and B

3. Which of the following is a benefit of using automated testing frameworks?

   (A) Improved efficiency

   (B) Consistent and reliable testing

   (C) Faster feedback

   (D) All of the above

4. What is the key feature of a performance testing tool?

    (A) Load generation

    (B) Transaction monitoring

    (C) Resource monitoring

    (D) All of the above

5. Which of the following challenges adopting the testing mindset?

    (A) Resistance to change

    (B) Resource constraints

    (C) Cultural barriers

    (D) All of the above

6. What is the purpose of a test case?

    (A) To define the expected behavior of a software component

    (B) To provide step-by-step instructions on how to test a software component

    (C) To record the results of a test

    (D) All of the above

7. Which of the following is a key principle of the shift-left testing approach?

    (A) Integrating testing activities earlier in the development life cycle

    (B) Automating testing processes

    (C) Fostering collaboration between development and testing teams

    (D) All of the above

8. What is the primary objective of regression testing?

   (A) To ensure that new code changes do not adversely affect existing functionality

   (B) To identify defects that code changes have introduced

   (C) To validate the stability and reliability of the software

   (D) Both A and B

9. Which of the following is a benefit of using artificial intelligence (AI) in testing?

   (A) Enhanced efficiency

   (B) Improved accuracy

   (C) Cost reduction

   (D) All of the above

10. What is the purpose of a bug report?

   (A) To describe a defect in a software component

   (B) To provide a solution to a defect

   (C) To track the progress of defect resolution

   (D) None of the above

11. Which of the following is a key metric for measuring the effectiveness of a testing effort?

   (A) Test coverage

   (B) Defect density

   (C) Test execution time

   (D) All of the above

12. What is the role of a tester in the software development life cycle?

    (A) To ensure the quality of the software product

    (B) To identify and report defects

    (C) To participate in the design and development process

    (D) All of the above

13. Which of the following is a best practice for writing test cases?

    (A) Use clear and concise language

    (B) Focus on testing specific functionality

    (C) Define expected results for each test case

    (D) All of the above

14. What is the purpose of a test plan?

    (A) To outline the scope and objectives of a testing effort

    (B) To define the resources and schedule for a testing effort

    (C) To provide guidance to testers on how to execute tests

    (D) All of the above

15. Which of the following is a type of testing that focuses on the user experience?

    (A) Usability testing

    (B) Performance testing

    (C) Security testing

    (D) All of the above

16. What is the purpose of exploratory testing?

    (A) To test the software without a predefined set of test cases

    (B) To find defects that are difficult to identify using traditional testing methods

    (C) To improve the tester's understanding of the software

    (D) All of the above

17. Which of the following is a benefit of using testing tools?

    (A) Automated test execution

    (B) Improved test management

    (C) Enhanced collaboration

    (D) All of the above

18. What is the role of a test environment?

    (A) To provide a stable and controlled environment for testing

    (B) To simulate real-world conditions

    (C) To isolate the software under test from other systems

    (D) All of the above

19. Which of the following is a key metric for measuring the quality of a software product?

    (A) Reliability

    (B) Maintainability

    (C) Usability

    (D) All of the above

20.   What is the ultimate goal of testing?

(A) To ensure the highest possible quality of the software product

(B) To identify and report all defects in the software product

(C) To satisfy the requirements of the stakeholders

(D) All of the above

21.   Which of the following is NOT a key principle of effective testing?

(A) Clear objectives and goals

(B) Comprehensive test coverage

(C) Iterative testing approach

(D) Exhaustive testing

22.   What is the primary objective of smoke testing?

(A) To verify the stability and readiness of a software build

(B) To identify major defects or issues

(C) To execute all existing test cases

(D) To measure the performance of the software

23.   What is a key benefit of adopting a testing mindset?

(A) Improved software quality

(B) Reduced risk of defects

(C) Faster time to market

(D) All of the above

24. What is the purpose of a test management tool?

    (A) To help teams organize, manage, and execute their testing activities

    (B) To automate the execution of test cases

    (C) To generate comprehensive test reports and metrics

    (D) Both A and C

25. Which of the following is a popular automated testing framework for web applications?

    (A) JUnit

    (B) Robot Framework

    (C) Selenium

    (D) Cypress

Answer Key

1. D

2. D

3. D

4. D

5. D

6. A

7. D

8. D

9. D

10. A

11. D

12. D

13. D

14. D

15. A

16. D

17. D

18. D

19. D

20. A

21. D

22. A

23. D

24. D

25. C

Bibliography

1. Chen, T. Y., Kuo, F. C., & Liu, H. (2009). Adaptive random testing based on distribution metrics. Journal of Systems and Software. https://doi.org/10.1016/j.jss.2009.05.017

2. Software Quality Assurance Company | Software Testing Company - Impressico. https://www.impressico.com/services/offerings/software-quality-assurance/

PART III

Observability

CHAPTER 7

Monitoring vs. Observability: Delineating the Concepts for Enhanced System Performance

Authors:

Pradeep Chintale

Manoj Kuppam

Reviewer:

Ayisha Tabbassum

Introduction

Fast technology development and increased complexity of systems in different areas push the necessity to develop efficient tools and methodologies for system management and performance analysis. In this respect, two important ideas that have been raised in the context of being important are monitoring and observability. In practice, often, these concepts are used synonymously, though they have different principles, methodologies, and application meaning. This paper demystifies monitoring and observability and gives definitions of each, differentiates their characteristics, and traces their historical evolution to understand their current role in managing systems.

Definition of Monitoring

System management monitoring is the practice of constantly gathering, processing, and analyzing performance and health data from systems. It is specifically targeted to answer the question, "Is the system functioning correctly?" Monitor systems are configured to detect certain conditions or thresholds that will trigger alerts or actions if passed. These may include very basic metrics, such as uptime and response time, and more sophisticated analytics by using the information in system logs and user behavior.

Definition of Observability

In contrast, observability is much more than simply an augmented form of the common type of observation. It is the capability of a system to expose its internal states in an interpretable way, mostly through its external outputs. Its ultimate goal is to understand "why" in the state of a system, particularly of complex ones, where problems are not always visible at first glance. This consists of the three pillars of logs, metrics, and traces, of which each provides unique and divergent insight into the system's workings. Observability lets the system administrator be much more proactive in system management and be able to diagnose problems that were never even imagined and hence develop a better understanding of system behavior.

Historical Context and Development of Both Concepts

The history of monitoring and observability has been woven into the very fabric of technological progress. In its simplest essence, monitoring was part of system management since the early days of computers, with the beginning of ensuring uptime and at least basic functioning. Growing system complexity has given rise to calls for ways of better monitoring; hence, a very great deal of diverse monitoring tools and frameworks have come to life.

Although observability may be a fairly recent term, its roots are traced back to the theory of control. With the rise of cloud computing and the architecture of microservices, observability came to the limelight. One shift of the systems from monolith to distributed brought forward one of the weaknesses in the current monitoring; hence, observability as a way of drawing more insight into the increasing complexity and dynamism of the systems.

This introductory chapter creates the platform within which in-depth exploration into both concepts, their applications, comparative analysis, and the potential for their integration can be understood so that modern wholesome management strategies can be realized.

Theoretical Framework and Definitions

To understand the theoretical underpinnings of monitoring and observability, it is essential to delve deeper into each concept, exploring their foundations, methodologies, and the principles that guide their application. This section provides a comprehensive theoretical framework that not only defines monitoring and observability in detail but also examines their respective roles within the broader context of system management and performance optimization.

Deep Dive into Monitoring Theory

System monitoring is an integral role in the management of systems, concerned with the collection of data, its analysis, and interpretation to ascertain that systems are within their set limits. The theoretical basis of monitoring is built on the use of predefined metrics and logs to detect variation against a standard way of operation. This approach is essentially reactive in nature and addresses known problems and the surpassing of established thresholds once this happens.

There are a few fundamental principles on which the monitoring theoretical framework is based. Firstly, the threshold-based alerts are required where the system metrics have particular bounds set for them, and if such bounds are crossed, the alerts are generated. This way, potential problems could be detected and dealt with at an earlier stage.

Performance benchmarks can also be used. With the use of past data, standard performance measures are put into place against which the performance of the system can be gauged at the present. It helps in pointing out performance anomalies in time.

Effective monitoring also involves robust data aggregation and analysis. At the very least, data collection and scrutiny of such data into trends and patterns that would facilitate management of the system in a proactive manner are involved.

Finally, it includes incident response. It encompasses a thoroughly documented incident response procedure for system alerting and anomaly responses. With an effective incident response plan in place, potential disruptions are able to be handled quickly and with a minimum of negative effect on the system operation.

In short, monitoring is that crucial discipline within system management responsible for the extremely rigorous approach to ensuring the reliability and performance of systems through proactive data analysis and responsive incident management.

Exploring the Theory of Observability

However, observability shares some similarities with monitoring, particularly in its use of collected and analyzed data, which takes a far more nuanced and proactive approach toward the understanding of systems. This, in fact, has turned out to be an integral part of the management of modern systems based on the premise where every internal state of a system must be determinable from its external outputs.

The theoretical foundation of observability relies on three major pillars. The first is logs, that is, detailed records of events that have occurred within the system, providing a chronological account of activities. The second is metrics; these are quantitative data that shed light on the performance of various components of the system, offering quantified insights.

The third pillar would be traces, which give the life cycles of the request or transaction; it gives representation of the interaction of components and sequence of events.

Apart from the above pillars, a few other critical aspects that the theory of observability underlines include the following: one of the main foci is the overall insight toward the ultimate objective of full and comprehensive comprehension of system states and behaviors. This insight toward depth is able to provide an overall understanding toward the operational context of the system.

It is also important to note proactive analysis in the identification of potential causes of problems before they become huge. This will mean the proactive stand against risk management and increasing system reliability before any visible effects of disturbances are realized.

Finally, the dynamic systems would be an elementary part of observability. It captures the need for change within the changing nature of most modern complex systems, thus making the approach dynamic and responsive to the systems it is trying to understand. In other words, observability is an intricate proactive framework by which the possibility of inferring the internal dynamics of a system from its outputs is enriched, hence allowing better and more dynamic management of systems.

Comparative Theoretical Analysis

Contrasting these theories with that of observability, while monitoring is concerned with a "what and when" of states of the system, observability is trying to answer the "why" of the system. Monitoring is all about known quantities and defined metrics, while observability is about finding the unknown and getting an understanding of the system as a whole.

Evolutionary Perspective

These theoretical frameworks have evolved into practical applications and now characterize complexity and dynamism found in most of the modern systems. The systems develop, and theoretical backgrounds of the monitoring and observability approaches are enriched by new approaches and technologies that might fit the upcoming challenges in the system management.

This theoretical exploration lays the groundwork to better understand in the following pages how monitoring and observability work, what their limits are, and what they afford within so many of the technological contexts they are engaged in. The next sections concretize this framework with applications, comparative analysis, and new directions within the field.

Key Components and Characteristics

Putting both of these ideas side by side brings the key elements and inherent characteristics of monitoring and observability into perspective. This section delves into the essential elements that make up each, how they work, what their differentiating factors are, and in what manner this impacts the management of the system.

Core Components of Monitoring

Monitoring is an integral part of system management that ensures continuous checking over the performance and system health of diversified elements in a system.

Metrics are just anything in number format, the major parts of CPU utilization or memory usage. These metrics give snapshots of the present state in the system and contain points with data that would be valuable and relevant to stakeholders.

Another critical part is the alert and notification system developed to inform stakeholders about anomalies or when some critical predefined thresholds are breached so one can react in these potentially critical situations.

Introduction of another critical component is the dashboards which provide visual interfaces for key performance indicators. Dashboards are critical in that they assist individuals in tracking and analyzing real-time information while monitoring the system under management.

Logs also make up the basic part of system monitoring. They are the records of events and actions in the system, and they would be priceless for troubleshooting, historical analysis, and understanding past interactions within the system.

In general, monitoring features are more of a reactive type. It is dependent on going through a threshold alert and in most cases fixing an issue that is known and within a predefined set of parameters.

This setup helps to maintain the integrity of the system and its performance by quickly addressing potential or actual deviations from normal operations. In a nutshell, effective monitoring unites various tools and strategies to get a holistic view of a system's performance, which in turn helps in managing it for the most optimal performance through a responsive and proactive manner following insights drawn from data.

Core Components of Observability

Observability augments classical monitoring with more components, thus enabling derivation of a more comprehensive view of the behavioral aspects and system states. This further assessment approach doesn't only offer follow-up on performance but also drives comprehension of the system dynamics.

A basic building block of observability, in turn, is a log—a record of events in detail. These logs provide a narrative for what has passed in the system by recording every event so that there is a clear historical view.

In this case, metrics become critical not only in monitoring but also observability. The role of observability supersedes the role of metrics. It is used more pervasively so as to infer the system's internal state from the outputs that are outside of it, thus allowing detailed reasoning to be carried out in relation to the system's health and the analysis of results.

Traces are also important in that within them, paths and durations of requests or transactions within the system are given. This gives critical information in gathering the workflow of the system and interaction in detailing how different components communicate and process transactions.

In observability, the large part of the equation is detailed contextual data. These include the logs, metrics, and traces—the information needed to support a rich, detailed understanding of state in the system. It helps put together a fuller picture of the operation of the system, drawing the challenge.

Two important features that define observability are proactive orientations in view of understanding and problem resolution before they get out of hand and the ability to find out why a system is in a state and infer the unknown conditions from the known data. This informs much deeper understandings of system behaviors in the establishment of much more effective and anticipatory management practices.

In conclusion, observability brings traditional monitoring into added, wider tools and methodologies to give the best knowledge about and manage complex systems for performing at optimized performance and reliability.

Comparative Overview

There are apparent differences in scope, approach, and data use when comparing monitoring and observability. That is to say, it is to show how each contributes in its own special way toward the management of the system.

Aspect	Monitoring	Observability
Scope and Depth	Narrower, focuses on specific metrics and logs	Broader, provides an in-depth, holistic view of the system
Nature of Approach	Reactive, deals with known issues	Proactive, focuses on uncovering underlying causes and potential issues before they escalate
Data Utilization	Primarily for alerting and performance tracking	Used to build a comprehensive understanding of the system's internal workings

The next table quickly gives a sense of how monitoring and observability are different intents for managing a system: monitoring effectively manages known issues, while observability goes further by giving greater insight into the system to predict proactively and remediate before anything has the potential to become a problem.

Integration of Components

Despite the differences, there's a trend in combining monitoring components and observability to come up with a more robustly manageable system. These integrations leverage both immediate responsiveness through monitoring and great depth of insight through observability.

We conclude with a summary of the key elements and features that are notable within the monitoring and observability approach, as well as an emphasis on the difference of one from the other in their handling of a system. One has to learn the difference and know properly each approach to be applied appropriately relative to the system's specific needs and issues. The following sections deal in detail with the practical applications and implementation strategies of both monitoring and observability.

Monitoring: Techniques and Applications

Monitoring plays a crucial role in ensuring the reliability and efficiency of systems across various industries. This section provides an overview of both traditional and modern monitoring techniques, showcases case studies from different sectors, and discusses the limitations inherent to monitoring.

Overview of Traditional and Modern Monitoring Techniques

From the classical way to the very sophisticated and modern way, monitoring has changed by huge steps in the world of system management. Each set of techniques gives distinctive benefits in ways of helping to manage and maintain system health and performance.

Traditional Monitoring Techniques

Traditional monitoring is just the analysis and criteria post hoc oriented. A common technique in most of the setups is the log analysis, where the system logs are gone through in detail in order to identify patterns of errors and after the fact problems. This method is vital for troubleshooting and understanding past system behaviors.

Another very common traditional method that goes in these techniques would be threshold-based monitoring. For instance, limits are set in system metrics like CPU usage or memory consumption, and in case of overstepping, it triggers an alert. It becomes very important for the assurance of system operation and prevention of overloading.

Polling is also a traditional monitoring technique; the system components are checked at intervals for their operational state and see that they are within the normal parameters. This consistent check helps in early detection of potential failures or abnormalities.

Modern Monitoring Techniques

Modern techniques in monitoring make the system surveillance real-time more dynamic and proactive. Real-time data analysis, as one of the techniques, involves immediate analysis of the data generated on the spot by using intricate algorithms. Such a system will provide instant knowledge about the performance of a system and, if necessary, notice trouble when it happens, permitting rapid response. The other futuristic approach is the use of automated response systems. Such systems automatically trigger an action that is to be executed as a result of some monitoring triggers; this could be starting a service, which had stopped, or scaling up resources as required. It helps in the quick mitigation of issues without manual intervention. Predictive monitoring is a novel approach that uses a combination of machine learning and statistical models to predict potential issues before they happen. Such models find patterns in historical data from which system failures and performance degradation can be predicted in order to take preemptive action and avoid or minimize their impacts.

Case Studies Demonstrating Effective Monitoring in Various Industries

Ecommerce Industry

> **Case Study:** *An ecommerce platform implements real-time monitoring of website traffic and transaction speeds during peak shopping seasons, allowing for immediate scalability adjustments and avoiding system overloads.*

Healthcare Industry

> **Case Study:** *A hospital network uses monitoring systems to track patient data and critical equipment functionality, ensuring timely alerts for medical staff and enhancing patient care.*

Manufacturing Industry

> **Case Study:** *A manufacturing company employs predictive monitoring in its machinery, predicting maintenance needs before breakdowns occur, thus minimizing downtime and optimizing production processes.*

Limitations of Monitoring

Monitoring systems are very critical for providing operational stability and IT infrastructure health. Many useful limitations are inherently related to the effectiveness of such systems, existing within the framework of large, complex, and dynamic environments.

The major issue with traditional monitoring is that it is reactive. Practically, this means there is no action taken until actually facing the problem, sometimes perhaps too late for avoiding the disruption or damage. Most of these monitoring systems are issue-driven rather than preventive, which may not be sufficient in fast or crucial operational settings.

The second challenge is threshold dependency. Most of the traditional monitoring systems put thresholds for alerts. While this is good for well-understood issues, there would be much more subtle or unknown issues that go unnoticed since they either do not move above these thresholds or in a few cases reach those numbers but are still significant. Dependent on fixed parameters, such gaps are likely to occur in system supervision.

Another pitfall is data overload. The more data systems create and the higher their variety, the more alerts monitoring tools can produce. Therefore, one would have to evolve further alerting thresholds to avoid alert fatigue and missing critical alerts because of their sheer number, possibly missing serious issues.

This is normally very limited in monitoring systems, denying one the ability to see the big picture about an issue in order to understand it and resolve it holistically. Missing such context mostly inhibits the effectiveness of problem-solving, as it reduces the data available to find the root cause of problems in the system.

Not to leave out, the issues on scalability that are presented put the question to many strides. Traditional monitoring solutions are quite hard to be scaled effectively for various aspects of increasingly complex and larger systems. It is harder to make sure that full coverage is taken care of while the levels of performance are retained with the increased number of components and more variables in the system.

Observability: Techniques and Applications

In system management, observability is the concept that tries to reveal insights in deeper levels within complex systems. This section will look into some of the practices applied in observability, demonstrate how it is applied in various sectors, and zero in on a few of its limitations.

Description of Observability Techniques

Observability in the system is more than merely management; it encompasses a series of techniques put into place in forming a complete understanding of system behaviors and states through time. It combines a few basic, traditional approaches into a whole with the purpose of optimizing ability to diagnose and resolve issues effectively.

Logging is very much a core element in observability, so it does not just stop with the retrieval of the records of events in the system but goes further to look at the contextual and all-encompassing approach to logging. Such method provides more opportunity for the analysis of the system's behavior in time and, what is even more important, gives insight into the "hows" and "whys" of performance that is crucially important for troubleshooting and effective improvement of the system in the long term.

The other equally important technique in this area is tracing. Tracing is the process of following a single request or transaction through various subsystems of a system. This ability is very important in pinpointing the issues in a system workflow, say in the identification of bottlenecks or points of failure at certain spots along the path of a transaction. Tracing goes into the details of the path of a request and tells the team how components interact, hence giving good optimization for improved performance.

Metrics serve as a quantitative measure in order to implement observability. In the observability field, the metrics not only give a snapshot of the performance but also a clue to the general state and behavior of the system. This extended use of metrics helps recognize trends and patterns that may signal potential problems or areas for improvement.

Contextual information threads the data from logs, traces, and metrics together. This enriched data offers a view of the overall system operations that will be of utmost help in diagnosing and resolving problems fast. Contextual information binds different data points together, making it more straightforward to view the bigger picture and the interaction of different elements of the system.

Examples of Observability in Action Across Different Sectors

Telecommunication Industry

Example: A telecom company applies observability to the network infrastructure in managing real-time data, where all network outages are easily identified and worked on for a quick restoration process, ensuring there are no interruptions to the service.

The Financial Services Industry

- **Use of Observability:** A fintech company's online transaction processing system. You get better insight into how the transaction flows work, thereby providing more security and better user experiences by offering tracing and contextual logging.

Services in Cloud Computing

Use Case: Applying observability to the multitenant infrastructure of the cloud service provider allows it to perform better resource optimization and performance that consequently gives better quality and reliability of service to customers.

Limitations of Observability

Problems of Embedding Observability into System Management

While observability offers huge benefits in understanding and handling complex systems, its effective implementation is also not without many challenges. The problems, therefore, identified can affect the feasibility and effectiveness of observability strategies in an organization.

This is a very basic reason that makes the implementation complex. Proper infrastructure for observability is an output of very detailed strategies and a very robust technology framework. It can be very complex and needs huge expertise in system architecture and data handling. The setup would need to be planned and strategized to a point where the observability system is fully capable of providing necessary insights without disturbing existing operations.

The other important critical challenge that comes up is data volume management. Generally, the observability systems are a huge source of data volumes through logs, metrics, and traces. Processing, storing, and effective analysis of such data are very challenging. With this influx of data, what organizations need to be able to grapple with it is really powerful data processing tools and techniques, which might sometimes even demand huge IT infrastructure and high expertise.

> **Skillset:** The main demands for observability are high. Any personnel working with the tools of observability need to be skilled in data analysis and system architecture. Such a level of employee skill is hard to come by, while teaching the same to the existing staff might be lengthy and costly. Some of the organizations, thus, will be limited in their observability ability because of the requisite high-level expertise.

Cost of Implementation: The tools and resources
needed to set up and sustain good observability
infrastructure are, by and large, very costly. Add to
it the cost in software licenses, data processing, and
storage hardware and a good salary for competent
staff. This can be felt particularly painfully in smaller
organizations or those that put little in their IT
budget. The risk with such observability investments
is the risk of diminishing returns. You might end up
having put a lot into observability infrastructure and
not seen benefits flow back. If the case is where the
data collected is not insightful to make actionable
decisions or improvements derived are a fraction
of the money and effort cost invested, then this
proves the point. Therefore, it's important that
organizations very closely examine their needs and
likely gains from observability to ensure that this
function delivers value commensurate with its cost.

Comparative Analysis

For that, a side-by-side comparison between monitoring and observability
presents very clear insights into how these concepts differ and
complement each other in system management. The comparison will be
based on the following key aspects: purposes, methodologies, types of
data, tools, and overall approach.

Aspect	Monitoring	Observability
Primary Objective	To detect and alert on known issues and thresholds	To understand the system's state and behaviors, particularly the unknowns
Methodology	Reactive—responding to predefined conditions	Proactive—exploring and inferring unknown issues
Key Data Types	Metrics, logs, and alerts	Logs, metrics, traces, and contextual data
Data Utilization	Primarily for alerting and tracking performance	For in-depth analysis and understanding system internals
Tools and Technologies	Traditional monitoring tools, threshold-based alert systems	Advanced data analytics tools, AI/ML for pattern recognition, distributed tracing systems
Approach	Often focuses on component-level health and performance	Holistic view, focusing on overall system health and complex interactions
Nature of Issues Addressed	Well-defined and known issues	Complex, often unpredictable issues requiring deep insight
Complexity	Relatively lower complexity in setup and maintenance	Higher complexity in setup and interpretation of data
Feedback Loop	Primarily one-way—from system to monitoring tools	Bidirectional—insights from observability can inform and refine monitoring
Skillset Required	Operational skills focused on specific tools and metrics	Analytical skills with a deeper understanding of system architecture

(*continued*)

Aspect	Monitoring	Observability
Cost	Generally lower due to focused nature	Potentially higher due to comprehensive data collection and analysis tools
Best Suited For	Systems with well-understood and stable components	Dynamic, complex systems where new issues can emerge unpredictably

Integration and Synergy

Integration of monitoring and observability is a way to synergistically manage systems, harnessing strengths of both to make a much stronger or rounded, fully fleshed out understanding of systems. The next sections go on to elaborate on how these two concepts can be mutual complements and the best practices to integrate them properly.

Exploring How Monitoring and Observability Complement Each Other

Proactive and Reactive System Management Integration

The integration of reactive monitoring and proactive observability approaches provides a very strong framework in system functionality and health. Each has its strength, and putting them together supplements each other, providing a rounded solution for system management.

The power to combine the reactive and the proactive lies in the strengths of both monitoring and observability.

Monitoring works perfectly well in reactive mode—alerting once predefined conditions have been reached, for instance, when a particular metric crosses the boundary of a predefined threshold. In contrast, observability is great for proactively finding system issues at their roots and may not even cause the kind of monitoring alerts that have always been the norm. This is so true in a unified approach where an organization is able to respond on the spot to known problems but can also be learned from the system's behaviors, enabling them to carry out preventive action in order to avoid other, similar problems in the future.

Another added advantage of marrying both monitoring and observability is the increase in data utilization. Monitoring usually provides raw data with respect to the performance of the systems, usually provided in metrics and logs. The context given by observability can be used to enrich this data to convert raw metrics into more effective insights. Such enriched data can reveal some hidden patterns and trends that cannot be found otherwise with just the traditional monitoring. This leads to a better understanding of the operation dynamics of the systems.

This will only add to the output of problem diagnosis and resolution by real-time alerting from monitoring and deep insights of the system through observability. If something happens in an exceptional way, a timely response to such a case would be made by monitoring through its alerting mechanism. Thanks to observability through a holistic view of the operation of a system, this synergy could make fast and precise diagnostic processes. Such synergy can significantly enhance the speed and effectiveness of the problem resolution process, cut downtime, and enhance system reliability.

The ability to build a comprehensive system view is vital for managing the system well. Observability provides a full view of the whole system, while monitoring usually looks at only some parts or one or two metrics. It therefore combines them in order to provide a full picture of the state of the system in terms of health and performance, hence the understanding of the system as a whole. That is a view required to make strategic decisions and long-term system improvement.

Best Practices for Integrating Both in System Management

- **Define Clear Objectives:** Define expected results out of the integration, which can range from higher system reliability to stronger performance analysis or swifter response to an incident.

- **Selective Data Collection and Analysis:** Do not swamp yourself with data; rather, be very selective in the type of data to collect and analyze and dwell on a few, very useful metrics and logs.

- **Leverage Advanced Technologies:** Use AI and machine learning in the processing of huge data points generated and provide more effective insights and actionable items.

Always-On Feedback Loop: Establish a continuous feedback loop in which observability informs monitoring thresholds and alerting mechanisms, and vice versa.

Training and Skill Development: Ensure that your teams are adequately skilled in the use of both monitoring and observability tools and their data. That may involve some training or hiring of specialists with such expertise.

- **Scalable and Flexible Infrastructure:** Deploy an elastic, flexible infrastructure that can scale with the growth and evolution of your system, able to cater for monitoring and observability needs at all stages.

Continuous Evaluation and Iteration: Always evaluate the performance of the combined approach and be prepared to iterate or change, if need be.

Case Studies and Real-World Applications

To illustrate the practical implications and benefits of monitoring and observability, this section presents detailed case studies from different industries, showcasing their implementation and the outcomes achieved.

Case Study: Online Retail Platform (Monitoring Implementation)

Background: System downtime was devastatingly felt by customers of the online retail company, with immense drops in the number of customers, especially during high-traffic events, such as sales.

Implementation: The organization has implemented advanced monitoring—real-time data analytics-based threshold alerts with automated incident response.

Outcomes and Insights

- ***Reduced Downtime:*** *Probable overloads could be responded to promptly because of real-time alerts, which led to a drastic reduction in downtime.*

- ***Enhanced Customer Experience:*** *A better shopping experience bore out in enhanced customer experience due to improved system reliability, as increased customer satisfaction scores.*

- ***Insight:*** *This case shows how monitoring can help to manage known issues effectively and also to keep operations stable in a high-pressure environment.*

Case Study: Healthcare Provider Network (Observability Implementation)

Background: The subject of the day in the health provider network was an unknown cause of slowdown in the Electronic Health Records system.

- **Implemented:** Including observability tools in place, such as rich-detailed logging, transaction tracing for EHR, and performance context-rich metrics

Outcomes and Insights

· Identified the root causes by providing trace and contextual data of complex interactions in the EHR system, thus identifying bottlenecks.

Improved System Efficiency: Targeting after the observability data insights, the outcome reached with these optimizations showed better performance and responsiveness of the EHR system.

Insight: The very strength of observability, the trait demonstrated by this case, is in revealing the fundamental reasons behind problems emerging in complex systems, thus leading to more effective problem-solving.

Case Study: Financial Services Company (Integrated Approach)

Scenario: A financial services organization grappled with keeping highly dynamic IT infrastructure—including cloud services and legacy systems—under control.

Implementation: This merged the approach to monitoring and observability. It set up monitoring systems for the critical, well-understood parts and observability for the new dynamic services.

Findings and insights

Overall System Management: Management combined with observability means getting a full view of the IT landscape—from legacy systems to modern cloud services.

Proactive Identification of Potential Issues: The observability component was quite instrumental in the very early detection of potential issues with the new services in the system, while monitoring made the operations stable in the long run for established systems.

Insight: This case shows the potential synergy of combining monitoring and observability, hence allowing not just stability in known areas but deep insight into even emerging technologies and services.

Future Trends and Developments

While each system grows ever more complicated, monitoring and observability technologies and methodologies are moving fast with the speed of technological advancement. This section commences with a look at some of the emerging technologies and methodologies in this area and makes some predictions with respect to their future directions.

Emerging Technologies and Methodologies in Monitoring and Observability

Artificial Intelligence and Machine Learning: These days, AI and ML are being integrated into monitoring and observability tools with each passing day. It's their predictive analytics, anomaly detection, and automated problem resolution capabilities that make it more like the norm in this area.

With the help of these technologies, large data processing can help to find patterns and predict issues well in advance before it becomes a real problem.

Automation and Orchestration: More and more automation comes to be seen from monitoring alerts and observability insights. This means automatic resource scaling, self-repairing systems, and dynamic reconfiguration based on the current state.

Enhanced Data Visualization: Upcoming are advanced tools for data visualization, which will read easily volumes of data generated from the observability and monitoring systems, hence making insights more intuitive for faster and informed decision-making.

Distributed Tracing and Edge Computing: Increasingly, the systems are distributed in nature, much with the advent of edge computing. This becomes most critical to make systems observable. In fact, it helps to monitor and understand the flow of data and interaction of geographically distributed systems.

Native Cloud Technologies: Further growth in native cloud architectures, such as microservices and serverless computing, will drive increased demand for advanced monitoring and observability solutions that well fit within these dynamic environments.

Predictions for the Future Direction of These Fields

Convergence of Monitoring and Observability: The distinction between observability and monitoring is bound to get increasingly blurry with time, resulting in more unified tools that blend the reactive strength of monitoring with the proactive depth of observability.

> **Increased Attention in User Experience Monitoring:** With user experience, there will be much more attention in the monitoring and observability tied to user experience; that data will be infused in it in hope of making better system design and function.

> **Growth of Predictive and Prescriptive Analytics:** Predictive analytics will change to prescriptive analytics, in which the system advises not just on the best course to be followed but also suggests a set of prescribed or optimal actions.

Expansion of AI-Driven Operations (AIOps) will smoothen the way to mainstream acceptance, increasingly automating system management and data analysis in ways that dramatically cut down on the time and effort required for manual intervention.

Increased Emphasis on Security and Privacy: With increasing data bound to be collected by monitoring and observability tools, it will add more emphasis on security and privacy adherence to regulation guidelines like GDPR and CCPA.

Conclusion

This paper has provided the explanation on the concepts of monitoring and observability and how the two concepts are properly defined and applied in the integration context of system management. The study was done in a stepwise manner, ranging from understanding of the respective theoretical frameworks, to their critical components, methodologies, and actual implementations across different industries. A comparative analysis strongly emphasized the unique and complementary nature of these two concepts.

In effect, monitoring, with its focus on known problems and a reactive approach, still forms the base for operational stability in the systems. It thrives on immediate responses to predefined conditions with consistency in performances. Observability, on the other hand, just provides a proactive methodology to look into the internal states of systems, finding root causes of issues in complex and dynamic environments. It goes even beyond the traditional definition of monitoring with its in-depth insight and a holistic view of system behaviors.

In conclusion, observability and monitoring are not mutually exclusive but complementary strategies on how to operate in the changing and evolving landscape of system management. As technology continues to forge ahead, the amalgamation of these two concepts will be crucial in dealing with the dynamism and complexities of current systems while providing stability and insight. The further development of research and innovation in this regard will surely shape their future, providing interesting chances for the future of system management.

Reliability Across the Span of a Transaction

In the intricate landscape of modern distributed systems, ensuring the resilience and reliability of transactions requires a granular understanding of performance metrics across various layers and components involved in the transaction flow. From the initial client request to the final response, a transaction traverses multiple layers, each with its own set of potential bottlenecks, failure points, and performance characteristics. Site Reliability Engineering (SRE) practices emphasize the importance of measuring and monitoring relevant metrics at each layer to gain comprehensive visibility into the system's behavior and facilitate proactive identification and mitigation of issues.

This chapter delves into the specific reliability metrics that should be measured and monitored at different layers during the span of a transaction. By adopting a layered approach to observability, organizations can pinpoint performance bottlenecks, isolate root causes of failures, and take targeted actions to enhance the overall resilience of their transactions.

1. Client Layer Metrics

    The client layer, typically represented by a web browser or mobile application, is the entry point for transactions in many modern systems. Monitoring the performance and reliability of this layer is crucial for understanding the end-user experience and identifying potential issues before they propagate further into the system.

    1.1. Client-Side Performance Metrics

    –   **Page Load Time:** Measures the time taken for a web page or application to fully load and become interactive

    –   **Time to First Byte (TTFB):** Measures the time taken for the client to receive the first byte of data from the server

- **Resource Loading Times:** Measures the time taken to load individual resources (e.g., CSS, JavaScript, images) on the client

- **Client-Side Errors:** Tracks errors occurring within the client-side code (e.g., JavaScript errors, unhandled exceptions)

- **User Interaction Metrics:** Measures the time taken for user interactions (e.g., click events, form submissions) to be processed and responded to

1.2. Network Performance Metrics

- **Round-Trip Time (RTT):** Measures the time taken for a packet to travel from the client to the server and back

- **Connection Establishment Time:** Measures the time taken to establish a network connection between the client and server

- **Bandwidth Utilization:** Monitors the bandwidth usage and potential bottlenecks in the client/server communication channel

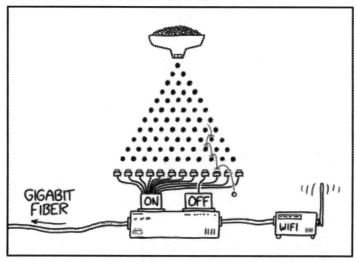

MY NEW VACATION SPOT HAS VERY FAST INTERNET
THAT TURNS OFF RANDOMLY EVERY NOW AND THEN,
JUST SO YOU CAN TELL PEOPLE YOU'LL BE STAYING
SOMEWHERE WITHOUT A RELIABLE CONNECTION.

Figure 7-1. *Network drops over the Internet*

2. Network and Infrastructure Layer Metrics

    Transactions often traverse various network
    components and infrastructure elements, such as load
    balancers, firewalls, and DNS servers. Monitoring the
    performance and health of these layers is essential
    for identifying potential network-related issues and
    ensuring efficient routing and delivery of requests.

    2.1. Load Balancer Metrics

    –   **Request Rate:** Measures the number of
        requests handled by the load balancer per
        unit of time

    –   **Response Time:** Measures the time taken by
        the load balancer to forward a request to a
        back-end server and receive a response

- **Error Rate:** Tracks the rate of errors encountered by the load balancer (e.g., failed health checks, connection timeouts)

- **Traffic Distribution:** Monitors the distribution of traffic across back-end servers to identify potential imbalances or hotspots

2.2. Firewall Metrics

- **Packet Rate:** Measures the rate of packets processed by the firewall

- **Connection Rate:** Tracks the rate of new connections established through the firewall

- **Dropped Packet Rate:** Monitors the rate of packets dropped by the firewall due to security policies or resource constraints

- **Latency:** Measures the additional latency introduced by the firewall during packet processing

2.3. DNS Metrics

- **DNS Query Rate:** Measures the rate of DNS queries received by the DNS servers

- **DNS Response Time:** Tracks the time taken by the DNS servers to respond to queries

- **DNS Cache Hit Rate:** Monitors the effectiveness of the DNS cache by measuring the rate of cache hits and misses

- **DNS Availability:** Tracks the availability and uptime of the DNS servers

3.   Web Server and Application Layer Metrics

Once a transaction reaches the web server and application layer, a multitude of metrics become relevant for monitoring performance, resource utilization, and potential bottlenecks within the application code and underlying infrastructure.

3.1. Web Server Metrics

- **Request Rate:** Measures the rate of incoming requests to the web server

- **Response Time:** Tracks the time taken by the web server to process a request and respond

- **Error Rate:** Monitors the rate of errors encountered by the web server (e.g., 4xx and 5xx HTTP status codes)

- **Active Connections:** Tracks the number of concurrent connections being handled by the web server

- **Resource Utilization:** Monitors the web server's CPU, memory, and disk utilization to identify potential resource constraints

3.2. Application Performance Metrics

- **Transaction Throughput:** Measures the number of transactions processed successfully by the application per unit of time

- **Transaction Response Time:** Tracks the end-to-end response time for transactions, from the initial request to the final response

- **Error Rates:** Monitors the rate of errors or exceptions occurring within the application code during transaction processing

- **Database Query Performance:** Measures the performance of database queries executed during transaction processing (e.g., query execution time, result set size)

- **External Service Call Performance:** Tracks the performance of calls made to external services or APIs during transaction processing (e.g., response times, error rates)

- **Resource Utilization:** Monitors the application's CPU, memory, and disk utilization to identify potential resource constraints or inefficiencies

4. Back-End Layer Metrics

Transactions often involve interactions with back-end systems, such as databases, caching layers, and message queues. Monitoring the performance and health of these back-end components is essential for ensuring data availability, consistency, and efficient processing of transactional workloads.

4.1. Database Metrics

- **Query Performance:** Measures the performance of database queries, including execution time, result set size, and index utilization

- **Transaction Rates:** Tracks the rate of transactions committed and rolled back in the database

- **Replication Lag:** Monitors the lag between the primary and replica databases to ensure data consistency and availability

- **Resource Utilization:** Measures the database's CPU, memory, and disk utilization to identify potential resource constraints or inefficiencies

4.2. Caching Layer Metrics

- **Cache Hit Rate:** Tracks the rate of cache hits and misses to measure the effectiveness of the caching layer

- **Cache Eviction Rate:** Monitors the rate at which cached items are evicted due to capacity constraints or expiration policies

- **Cache Response Time:** Measures the time taken to retrieve data from the caching layer

- **Resource Utilization:** Monitors the caching layer's CPU, memory, and network utilization to identify potential bottlenecks or inefficiencies

4.3. Message Queue Metrics

- **Queue Depth:** Measures the number of messages currently in the queue, providing insights into potential backlogs or processing bottlenecks

- **Message Throughput:** Tracks the rate of messages being produced and consumed by the queue

- **Message Latency:** Measures the time taken for a message to be processed from the point of being enqueued to dequeued

- **Error Rates:** Monitors the rate of errors or failures occurring during message processing or delivery

A Real-Time Use Case

Let's say we want to measure and implement SRE practice for a coffee shop customer user journey. The scenario is to analyze the user journey steps and come up with various improvement opportunities to adhere to the business SLOs. A typical user journey in this scenario would be broken down into four steps:

1. Log in to the cashier's application with landing menu page.

2. Customer order selected and added to cart.

3. Share promotional offers to the customer.

4. Print the receipt and pass the customers' order to the queue.

While the SRE job limits itself to the software engineering methods and techniques, it is also possible they extend to improving the customer experience with the delivery time of the coffee to the customer by analyzing the time taken to get the order into the hands of the customer. However, we can limit the scope for software engineering and not to data engineering for now.

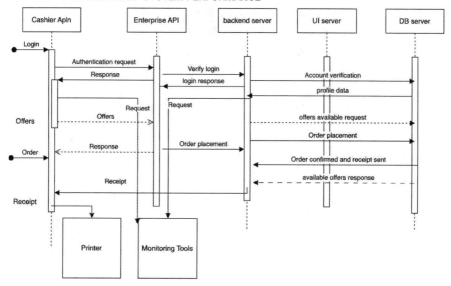

Figure 7-2. *Customer user journey sequence diagram to analyze the SLO violations for an SRE to create actionable insights*

By collecting and analyzing observability metrics using monitoring tools across the various layers involved in transaction processing, organizations can gain comprehensive visibility into the performance and reliability of their systems. This granular approach to observability enables proactive identification of bottlenecks, rapid root cause analysis of failures, and targeted optimizations to enhance the overall resilience and efficiency of transactions.

Table 7-1. *SRE actionable insights post analysis*

User Journey Step	SLOs	Sample SRE Analysis Outcome
Login	avg < 1s per month	Collect SLO metric with 1s ART SLO for the login transaction, set up alert for any violation, and perform RCA on violations; perform blameless postmortem
Order	1 min for customers' decision-making and an avg < 2s per month	Meets SLO and no immediate action required
Offers	1 min for customers' decision-making and an avg < 2s per month	Meets SLO but shows signs of breach frequently due to delays in the order-related offers fetch. RCA to be performed for the priority 1 incidents and violations
Receipt	3s to print the receipt and get the order into queue	Printer delays have caused the SLO breach. Device firmware patching is more than 5 years old and is out of support. Needs immediate upgrade

It's important to note that the specific metrics to monitor may vary depending on the system's architecture, technology stack, and business requirements. SRE practices encourage a data-driven approach, where teams continuously evaluate and refine the metrics being collected based on observed patterns, emerging performance concerns, and evolving operational needs.

Effective monitoring and analysis of these reliability metrics across transaction layers empower organizations to make informed decisions, prioritize improvements, and continuously enhance the user experience by delivering resilient, high-performing transactions.

Bibliography

1. Anderson, J. K. (2021). Principles of System Monitoring.
    Springer Nature

2. Bennett, C., & Towsley, D. (2020). Advances in System
    Observability. Wiley

3. Lopez, M., & Schmidt, R. (2019). AI and Machine Learning in
    System Management. Oxford University Press

4. Patel, A. (2018). Case Studies in IT Infrastructure. CRC Press

5. Singh, R., & Gupta, A. (2022). Emerging Trends in Monitoring
    and Observability. IEEE Press

6. Turner, B., & Levitin, M. (2023). Integrating Monitoring and
    Observability for Modern Systems. Elsevier

7. Wang, F., & Zhou, Y. (2020). Cloud Computing: System
    Management Strategies. Academic Press

CHAPTER 8

The Temple Metrics and Runbook Model

Authors:
Ayisha Tabbassum
Manoj Kuppam

Reviewer:
Madan Mohan

The Golden Signals: Let's Do The Temple

This chapter will use the concept of a temple as a metaphor for a robust digital infrastructure. It will cover how maintaining the "Golden Signals" (a term from Site Reliability Engineering representing the most important metrics that indicate the health of a system) ensures the continuity and reliability of digital services.

Setting
A futuristic data center called "The Temple," symbolizing the pinnacle of digital infrastructure.

Characters
Alex Mercer, the chief technology officer of a leading tech company.
Jamie Lin, a site reliability engineer.

© Saurav Bhattacharya 2024
M. Kuppam, *Enterprise Digital Reliability*, https://doi.org/10.1007/979-8-8688-1032-9_8

The Oracle, an advanced AI system that monitors The Temple's operations.

Chapter Breakdown

1. **Introduction to The Temple**

    a. Description of The Temple as a state-of-the-art data center

    b. Introduction of Alex Mercer and Jamie Lin overseeing the operations

2. **The Concept of Golden Signals**

    a. Explanation of the four golden signals: latency, traffic, errors, and saturation.

    b. Jamie explains to new engineers the importance of these metrics.

3. **The Oracle's Warning**

    a. The Oracle detects anomalies in traffic and latency, triggering alerts.

    b. Alex and Jamie assess the situation, discussing potential impacts.

4. **Diagnosis and Response**

    a. Using real-time data, Jamie pinpoints a critical service degradation.

    b. Alex coordinates with the team to reroute traffic and mitigate issues.

5. **Maintaining The Temple**

    a. Stress on routine checks and balances to maintain system health

    b. Importance of proactive measures and continuous monitoring

6. **Learning from The Oracle**

    a. Jamie uses data gathered during the incident to improve future responses.

    b. Alex discusses with the team about integrating more predictive analytics.

7. **Reflections in the Control Room**

    a. Alex and Jamie reflect on the day's events and the resilience of their systems.

    b. Emphasis on the metaphorical "temple" being as strong as its foundations.

8. **Closing Thoughts**

    a. A brief philosophical note on the digital world as our new reality.

    b. The chapter ends on a hopeful note about the future of digital infrastructure.

The chapter will incorporate technical details about system reliability but will be accessible to readers with varying levels of technical background. It will also weave in human elements through character interactions and the stress and satisfaction associated with maintaining complex systems.

Introduction to The Temple

Description of The Temple As a State-of-the-Art Data Center

Nestled in the heart of Silicon Valley, The Temple stood as a modern-day colossus in the landscape of digital infrastructure. Its exterior, a striking blend of glass and steel, mirrored the cutting-edge technology housed within. The building was designed not just for functionality but to make a statement—technology, when harnessed correctly, could be as awe-inspiring as any natural wonder.

Inside, The Temple was a labyrinth of server rooms, cooling pipes, and data cables. Rows upon rows of server racks hummed with activity, each LED light a heartbeat in the vast organism of global connectivity. The air was kept at a crisp 21 degrees Celsius, with humidity meticulously controlled to prevent any hardware degradation.

The data center was divided into several zones, each dedicated to specific tasks. There was the Network Operations Center (NOC), where real-time data about global traffic was displayed across an array of screens, and the Development Wing, a haven for engineers coding the next generation of AI algorithms. Security was paramount, with biometric checks at every entry point and an array of surveillance technologies ensuring that only authorized personnel could access the heart of the data center.

Introduction of Alex Mercer and Jamie Lin Overseeing the Operations

Alex Mercer, the Chief Technology Officer, was a visionary with an unparalleled understanding of both the theoretical and practical aspects of digital systems. His leadership style was a blend of mentorship and innovation, pushing his team to explore new frontiers in technology while ensuring a rock-solid reliability in their operations.

Jamie Lin, a site reliability engineer, was the perfect counterpart to Alex's visionary traits. With a meticulous eye for detail and a deep understanding of systems engineering, Jamie was often seen with a tablet in hand, moving between the racks, checking data points, and ensuring that every metric was within the prescribed limits. Her expertise was not just in maintaining systems but in foreseeing potential issues before they could become problematic.

Together, Alex and Jamie formed a dynamic duo, their skills complementing each other, driving The Temple to operate seamlessly. Their mornings often started with a tour of the facility, discussing upgrades, challenges, and breakthroughs. Their teamwork was a testament to the idea that technology, no matter how advanced, thrives under human guidance.

The Concept of Golden Signals

Explanation of the Four Golden Signals: Latency, Traffic, Errors, and Saturation

In the bustling control room of The Temple, Jamie Lin gathered a group of new engineers for an induction session. The room was lined with displays, each flickering with streams of data—graphs, charts, and numbers that seemed chaotic to the untrained eye but told a story clear as day to those who understood.

"Welcome to the heart of our operations," Jamie began, her voice echoing slightly in the high-ceilinged room. "Here, we monitor what we call the 'golden signals'. These are the metrics that give us the most immediate insight into the health and performance of our digital infrastructure. There are four key signals: latency, traffic, errors, and saturation. Each of these metrics tells us a different part of the story of our system's health."

Jamie switched to a slide showing a simplified diagram of a network. "First, we have latency, which measures the time it takes for data to travel from one point to another in our network. High latency means slower response times, which can be critical depending on the application."

Next, she highlighted another section of the diagram. "Traffic measures how much demand is being placed on our system. It tells us how many requests we are handling, which can help us understand if we need to scale our resources up or down."

She moved on to the third signal. "Errors are straightforward—they tell us when something has gone wrong. A spike in errors can indicate a major issue that needs immediate attention."

Finally, she pointed out the last signal. "Saturation measures how fully utilized our resources are. It's about capacity. If our systems are saturated, it means we're reaching our limits, and performance may degrade if we don't act."

Jamie Explains to New Engineers the Importance of These Metrics

As the slides progressed, Jamie emphasized the practical applications of monitoring these signals. "Understanding and reacting to these signals isn't just about keeping our systems running smoothly—it's about preemptive action to ensure they never fail. We operate on the principle of proactive maintenance, not reactive."

She illustrated her point with a case study from last quarter when an unexpected surge in traffic led to increased latency across several services. "Because we were closely monitoring our golden signals, we were able to catch the issue early. We rerouted some of the traffic and increased our server capacity before our users experienced any significant problems."

Jamie's teaching style was interactive, and she encouraged questions. "Think of these metrics as the vital signs of a patient. Just as a doctor continuously monitors vital signs to ensure their patient's health, we monitor these signals to ensure the health of our digital ecosystem."

The session ended with a practical demonstration, where Jamie showed the new engineers how to read the data dashboards and what steps to take when they noticed anomalies in the signals. "Remember," she concluded, "the stability of our entire digital world relies on how effectively we can interpret and act on these golden signals."

The Oracle's Warning

The Oracle Detects Anomalies in Traffic and Latency, Triggering Alerts

Late one afternoon, as the golden hues of sunset filtered through the skylights of The Temple, a sudden flurry of alarms disrupted the calm. The Oracle, an advanced AI system tasked with monitoring the data center's vitals, detected significant anomalies in traffic and latency that deviated sharply from normal patterns.

In the heart of the control room, large screens flashed red, signaling urgent alerts. "Anomaly detected in sector 5," announced The Oracle, its voice calm yet insistent over the loudspeakers. "Latency and traffic beyond threshold levels."

The room, typically buzzing with the quiet hum of routine operations, burst into a hive of activity. Engineers and technicians turned their attention to the monitors, analyzing the streams of data flowing across the screens. The Oracle's interface displayed real-time graphs with sharp spikes in latency and a massive surge in traffic, the likes of which were unusual for this time of day.

Alex and Jamie Assess the Situation, Discussing Potential Impacts

Alex Mercer, who had been in a strategy meeting in the adjacent conference room, entered the control room swiftly, his expression tense. Jamie, already at the central console, briefed him on the situation. "It looks like we're dealing with a significant anomaly. Traffic volumes are off the charts, and latency has spiked in several critical services," she reported, her eyes scanning the data.

"Could this be a coordinated attack? Or a system fault?" Alex pondered aloud, watching the cascading numbers.

"We can't rule out either possibility," Jamie replied. "But the pattern is erratic, more like a flood than a typical DDoS attack. We need to dig deeper to understand if this is malicious or a fault in our traffic management system."

Together, they evaluated the potential impacts. "If we don't get this under control, we could see a domino effect," Alex noted. "Latency issues could slow down services globally, and if traffic continues to spike, we might hit saturation points that we're not equipped to handle at the moment."

Diagnosis and Response

Using Real-Time Data, Jamie Pinpoints a Critical Service Degradation

Jamie, with a team of engineers, initiated a deep dive into the traffic sources and patterns using The Oracle's advanced diagnostic tools. The analysis revealed an unusual concentration of requests coming from several compromised nodes, which appeared to be flooding the network with redundant data requests.

"Looks like a portion of our edge nodes has been hijacked to amplify traffic to our core services," Jamie deduced, her fingers flying over the touchscreen as she isolated the affected nodes. "This is causing a service degradation across the board."

Alex Coordinates with the Team to Reroute Traffic and Mitigate Issues

Understanding the urgency, Alex took charge of the mitigation strategy. "Let's initiate a reroute of incoming traffic away from the affected nodes. We'll push updates to firewall rules to block these anomalies at the source," he instructed, his voice firm, issuing commands with precision.

The team worked seamlessly under his direction, updating routing protocols and strengthening firewall defenses. Alex also contacted the cybersecurity team, ensuring they were on the ground to investigate the source of the compromised nodes and prevent further breaches.

As the rerouting took effect, the traffic began to normalize, and latency returned to acceptable levels. The quick response averted a potential crisis, showcasing the team's capability to handle emergencies efficiently.

Maintaining The Temple

Stress on Routine Checks and Balances to Maintain System Health

After the incident, Jamie emphasized the importance of routine checks and balances. "This event underscores the need for constant vigilance," she addressed her team during the debrief. "We must intensify our regular audits and not just rely on automated systems. Human oversight is crucial."

She proposed an enhanced schedule for system health checks, incorporating more frequent manual inspections of critical infrastructure components. The team also discussed improving The Oracle's algorithm to detect anomalies more effectively, integrating machine learning models that could adapt to new threats dynamically.

Importance of Proactive Measures and Continuous Monitoring

Alex approved a new initiative for continuous monitoring, involving more sophisticated surveillance techniques and enhanced data analytics. "We need to be proactive, not just reactive," he told his team. "Let's use this incident as a learning curve to fortify our defenses and improve our response time."

The initiative included the deployment of additional sensors and the integration of a more robust incident response protocol. Alex and Jamie also planned workshops for all technical staff to update them on the latest cybersecurity threats and response strategies, ensuring that everyone at The Temple was equipped to maintain the sanctuary of their digital world.

Learning from The Oracle

Jamie Uses Data Gathered During the Incident to Improve Future Responses

In the aftermath of the crisis, Jamie and her team were not content to simply restore order; they aimed to learn and adapt. With the wealth of data collected during the incident, Jamie spearheaded a comprehensive analysis session. The team dissected every aspect of the event—from the initial anomaly detection by The Oracle to the final resolution of the traffic reroute.

"The Oracle did well in alerting us early, but we can make improvements," Jamie noted during one of the team meetings. She proposed enhancements to The Oracle's predictive capabilities, incorporating more advanced machine learning algorithms that could anticipate and adapt to similar threats in a more automated manner. "We'll train the system with this incident's data, refining its ability to differentiate between typical network fluctuations and genuine threats."

Alex Discusses with the Team About Integrating More Predictive Analytics

Alex, recognizing the critical role of forward-thinking strategies, supported Jamie's initiative and took it a step further. "Let's integrate more predictive analytics into our operational protocols," he suggested in a strategic planning session. "We need to think about not only responding to incidents but predicting and preventing them where possible."

He organized a series of workshops for the engineering team, focusing on predictive analytics and advanced data modeling. Alex brought in experts in AI and data science to lead the sessions, ensuring that the team was equipped with the latest tools and knowledge to enhance The Temple's defenses.

Reflections in the Control Room

Alex and Jamie Reflect on the Day's Events and the Resilience of Their Systems

Late in the evening, after the workshops and the flurry of activity had subsided, Alex and Jamie found themselves back in the control room, looking over the now-calm banks of monitors. The screens showed a steady flow of data, a testament to the resilience of their systems and the effectiveness of their team.

"We handled that well, thanks to your quick thinking and The Oracle's alerts," Alex said, turning to Jamie. "But today was a reminder of how quickly things can escalate. We must stay vigilant."

Jamie nodded in agreement. "It's like keeping The Temple's foundations strong," she replied. "We need to keep building on what we know and prepare for what we don't."

Emphasis on the Metaphorical "Temple" Being As Strong As Its Foundations

Their conversation turned philosophical as they discussed the broader implications of their work. "Every incident, every anomaly we encounter is like a stress test for our temple's foundations," Alex mused. "And each response is a chance to reinforce them."

Jamie added, "It's about more than just keeping the lights on. We're preserving the integrity of the digital world, ensuring it can withstand whatever comes its way."

Closing Thoughts

A Brief Philosophical Note on the Digital World As Our New Reality

As they prepared to leave for the night, Alex paused by the doorway, looking back at the array of blinking lights. "We're guardians, Jamie. Guardians of a new reality, where the digital and physical are inseparably intertwined. Our work here, it's not just technical—it's essential to the fabric of society."

The Chapter Ends on a Hopeful Note About the Future of Digital Infrastructure

Jamie smiled, her gaze lingering on the serene view of The Temple's core. "And as guardians, we'll keep evolving, just like the technology we oversee. With every challenge, we grow stronger, smarter, and more connected. There's hope in that—not just for us, but for everyone we serve."

With a final nod to each other, they stepped out of the control room, the door closing softly behind them. The Temple, with its pulsing lights and humming servers, continued its vigilant watch over the digital pulses of the world, a beacon of stability in the ever-changing digital landscape.

Exercise

Multiple-Choice Questions

1. What is "The Temple" in the context of the narrative?

    A) A religious building

    B) A state-of-the-art data center

    C) A book

    D) A museum

2. What are the "golden signals" in system monitoring?

    A) Types of software

    B) Security protocols

    C) Key metrics indicating system health

    D) Codes used by engineers

3. Which of the following is NOT one of the four golden signals?

    A) Latency

    B) Errors

    C) Bandwidth

    D) Traffic

4. Who is Alex Mercer in the story?

    A) A site reliability engineer

    B) The CEO of the tech company

    C) The Chief Technology Officer

    D) A security guard at The Temple

5. What role does Jamie Lin play in the narrative?

    A) Chief financial officer

    B) Site reliability engineer

    C) Head of security

    D) Marketing director

6. What does The Oracle do in The Temple?

    A) Monitors operations

    B) Controls the lighting

    C) Manages finances

    D) Guides tours

7. What triggered the alarms in The Temple?

    A) A fire

    B) Anomalies in traffic and latency

    C) A break-in

    D) A power outage

8. What was Alex Mercer's reaction to the crisis?

    A) Ignored the alerts

    B) Coordinated a response

    C) Left the building

    D) Called the police

9. Which term describes the maximum capacity utilization of a system?

    A) Saturation

    B) Maximization

    C) Utilization

    D) Fulfillment

10. What was a major cause of the crisis discussed in the narrative?

    A) Employee error

    B) Natural disaster

    C) Compromised nodes

    D) Software update

11. How did Jamie and the team resolve the issue with traffic spikes?

  A) They shut down the system

  B) They rerouted the traffic

  C) They increased prices

  D) They ignored the problem

12. What does "latency" measure in the context of digital infrastructure?

  A) Cost efficiency

  B) Time it takes for data to travel

  C) Amount of data stored

  D) Speed of the processors

13. What proactive measure did Jamie emphasize after the crisis?

  A) Reducing staff

  B) Regular system checks

  C) Cutting costs

  D) Expanding office space

14. What did Alex propose to enhance after the incident?

  A) Team vacations

  B) Predictive analytics

  C) Advertising spend

  D) Employee benefits

15. What analogy did Jamie use to describe the importance of monitoring the golden signals?

    A) Like checking the weather

    B) Like a doctor monitoring a patient's vital signs

    C) Like a chef tasting their food

    D) Like a driver checking the fuel gauge

16. What upgrade did Jamie implement in The Oracle?

    A) Better speakers

    B) Advanced machine learning algorithms

    C) Faster processors

    D) New screens

17. What philosophical concept did Alex and Jamie discuss toward the end of the chapter?

    A) The morality of surveillance

    B) The implications of digital dependency

    C) The ethics of artificial intelligence

    D) The impact of globalization

18. What is emphasized as crucial for the health of the digital infrastructure?

    A) Continuous innovation

    B) Aggressive expansion

    C) Financial investment

    D) Proactive maintenance

19. Which of the following best describes the resolution of the traffic spike issue?

    A) Temporary fix

    B) Permanent solution

    C) Ongoing problem

    D) Unresolved

20. What sentiment does the chapter close on?

    A) Hope and determination

    B) Fear and uncertainty

    C) Frustration and anger

    D) Indifference and complacency

Answers

1. B

2. C

3. C

4. C

5. B

6. A

7. B

8. B

9. A

10. C

11. B

12. B

13. B

14. B

15. B

16. B

17. B

18. D

19. B

20. A

Now that we have learnt the different metrics, it is important to follow a model to use these metrics to make the systems more reliable by reducing one of the key SLOs of an organization, the MTTR or Mean Time to Recovery of a system.

Reducing MTTR

In modern enterprises, high availability and minimal downtime are paramount; Mean Time to Recovery (MTTR) has emerged as a critical metric for measuring system resilience and operational efficiency. MTTR represents the average time taken to restore a system or service to a fully operational state following a failure or disruption. Minimizing MTTR is a key objective for Site Reliability Engineering (SRE) teams, as prolonged recovery times can result in significant revenue losses, customer dissatisfaction, and reputational damage for enterprises.

SRE, a discipline that combines software engineering principles with operational practices, employs a comprehensive approach to reduce MTTR by leveraging observability, applying system design principles, and improving operational methods through well-defined frameworks and runbooks. This holistic approach not only enhances system reliability but also fosters a culture of continuous improvement and proactive incident management.

1. **Leveraging Observability for Rapid Incident Detection and Diagnosis**

    Observability is a foundational concept in SRE that encompasses the ability to understand a system's internal state and behavior based on external outputs. By implementing robust observability practices, SRE teams can quickly detect and diagnose incidents, enabling faster recovery times.

    1.1. Metrics Collection and Analysis

    Collecting and analyzing relevant metrics is crucial for understanding system performance and identifying potential issues. SRE teams employ various tools and techniques to monitor key performance indicators (KPIs) and service-level indicators (SLIs) such as request rates, response times, error rates, resource utilization, and database query performance. By establishing baseline metrics and defining alerting thresholds, anomalies can be detected promptly, enabling rapid incident response.

1.2. Distributed Tracing and Logging

In distributed systems, where transactions span multiple services and components, distributed tracing tools like Jaeger, Zipkin, or AWS X-Ray become invaluable for understanding end-to-end request flows and identifying latency hotspots or failures. Comprehensive logging practices, facilitated by centralized logging solutions like Elasticsearch, Logstash, and Kibana (ELK stack) or Splunk, provide detailed application-level events, errors, and diagnostic information, aiding in root cause analysis and troubleshooting efforts.

1.3. Alerting and Incident Management

Effective alerting and incident management processes are crucial for promptly detecting and responding to incidents that impact system availability and performance. SRE teams implement intelligent alerting systems that integrate with monitoring tools and leverage predefined alerting rules based on established service-level objectives (SLOs). Well-defined incident management processes, including on-call rotations, escalation procedures, and postincident reviews, ensure that incidents are addressed promptly and that lessons learned are incorporated into future improvements.

2. **Applying System Design Principles for Resilience and Fault Tolerance**

SRE emphasizes the importance of designing systems with resilience and fault tolerance in mind, as these principles directly contribute to reducing MTTR by minimizing the impact of failures and enabling graceful degradation.

2.1. Fault Tolerance and Resiliency Patterns

Incorporating fault tolerance and resiliency patterns into system design is essential for mitigating the impact of failures and ensuring graceful degradation. SRE teams implement techniques such as circuit breakers, retries with exponential backoff, bulkheads, and fallbacks to prevent cascading failures and provide alternative paths for transactions to complete successfully, even in the face of partial system outages or degradations.

2.2. Redundancy and High Availability Architectures

Implementing redundancy and high availability architectures can significantly reduce MTTR by minimizing single points of failure and enabling failover mechanisms. SRE teams leverage techniques like multiregion deployments, active-active configurations, and load balancing to ensure service continuity in the event of localized failures or outages.

2.3. Chaos Engineering and Fault Injection

Chaos engineering and fault injection are proactive approaches used by SRE teams to test the resilience of systems by intentionally introducing controlled failures or disruptions. By simulating various failure scenarios, such as network outages, service failures, or resource constraints, teams can identify weaknesses, validate their resilience strategies, and improve their overall system's ability to withstand real-world failures, ultimately reducing MTTR.

3. **Improving Operational Methods Through SRE Frameworks and Runbooks**

SRE teams develop and implement frameworks and runbooks to standardize operational practices, streamline incident response, and facilitate knowledge sharing, all of which contribute to reducing MTTR.

3.1. SRE Frameworks

SRE frameworks, such as the SRE Adoption Framework or the MK Scoring Framework, provide structured methodologies for assessing and improving system reliability and operational efficiency. These frameworks often incorporate rubric-based scoring approaches to evaluate the current state of software teams and identify opportunities for advancement while continuously reinforcing key operational needs for enhancing software reliability and efficiency.

3.2. Runbooks and Playbooks

Runbooks and playbooks are comprehensive documentation that outline standardized procedures and best practices for handling various operational scenarios, including incident response, disaster recovery, and system maintenance. By having well-defined runbooks in place, SRE teams can respond to incidents more efficiently, reducing the time spent on diagnosis and decision-making and ultimately minimizing MTTR.

3.3. Automation and Self-healing Systems

Automation and self-healing systems play a vital role in reducing MTTR by streamlining processes and enabling faster recovery from failures. SRE teams leverage techniques like autoscaling, autoremediation, and self-healing architectures to automatically detect and mitigate issues, such as restarting failed services, reallocating resources, or triggering failover mechanisms without requiring manual intervention.

4. Continuous Improvement and Knowledge Sharing

SRE is an iterative process that emphasizes continuous improvement and knowledge sharing, both of which are essential for sustaining efforts to reduce MTTR over the long term.

4.1. Blameless Postmortems

Conducting blameless postmortems after incidents or failures is a critical practice in SRE. These postmortems focus on identifying root causes, analyzing contributing factors, and proposing actionable improvements without assigning blame. By fostering an environment of psychological safety and open communication, teams can openly discuss failures, share lessons learned, and collaboratively develop strategies to prevent similar incidents from occurring in the future, ultimately contributing to reduced MTTR.

4.2. Cross-Functional Collaboration and Knowledge Sharing

SRE encourages cross-functional collaboration and knowledge sharing among software engineers, operations teams, and other stakeholders. By promoting a culture of shared ownership and accountability, teams can leverage diverse perspectives and expertise to identify and address complex challenges more effectively, leading to improved incident response and reduced MTTR.

4.3. Continuous Improvement and Innovation

SRE teams continuously evaluate and refine their processes, architectures, and tooling based on lessons learned, emerging technologies, and evolving business requirements. This commitment to continuous improvement and innovation enables teams to stay ahead of evolving challenges, adapt to changing environments, and consistently improve their ability to minimize MTTR.

By leveraging observability, applying system design principles for resilience and fault tolerance, improving operational methods through SRE frameworks and runbooks, and fostering a culture of continuous improvement and knowledge sharing, SRE teams can effectively reduce MTTR and ensure high availability and minimal downtime for mission-critical systems and applications in enterprise environments.

Now, let's focus on a specific real-time scenario and how SRE practices can be applied to reduce MTTR by measuring and improving relevant metrics, enhancing system design through a scoring approach, and ultimately improving service-level objectives (SLOs).

Scenario: Ecommerce Platform Incident and MTTR Reduction

Consider an ecommerce platform that experienced a significant incident during a peak shopping season, resulting in prolonged downtime and a severe impact on revenue and customer satisfaction. The incident was caused by a cascading failure that originated from a database overload, leading to a complete system outage. The Mean Time to Recovery (MTTR) for this incident was unacceptably high at 6 hours.

To address this issue and reduce MTTR for future incidents, the ecommerce company adopted Site Reliability Engineering (SRE) practices, with a particular focus on observability, system design improvements, and the implementation of an SRE scoring framework.

1. **Enhancing Observability and Incident Detection**
   The SRE team began by implementing comprehensive monitoring and observability solutions to gain better visibility into the system's behavior and performance.

   1.1. Metrics Collection and Analysis

   – Key metrics were identified and monitored, including database query performance, application response times, error rates, and resource utilization (CPU, memory, network).

   – Intelligent alerting rules and thresholds were established based on historical data and SLOs, enabling prompt detection of anomalies and potential incidents.

   1.2. Distributed Tracing and Logging

   – Distributed tracing tools (e.g., Jaeger) were implemented to track end-to-end request flows across the ecommerce platform's microservices architecture.

   – Centralized logging solutions (e.g., ELK stack) were adopted to aggregate and analyze application logs, aiding in root cause analysis and troubleshooting efforts.

2. **Improving System Design Through SRE Scoring Framework**
   To address the underlying issues that contributed to the database overload and cascading failure, the SRE team employed an SRE scoring framework to assess the current state of the system and identify areas for improvement.

2.1. SRE Scoring Framework

– The team developed a rubric-based scoring approach to evaluate various aspects of the ecommerce platform, including database performance, application scalability, and fault tolerance mechanisms.

– Each component was scored based on predefined criteria, and improvement opportunities were identified and prioritized.

2.2. Database Optimization and Scalability

– Based on the scoring framework's findings, the team optimized database indexing, query patterns, and caching mechanisms to improve performance and reduce the risk of overload.

– Database sharding and replication strategies were implemented to enhance horizontal scalability and fault tolerance.

2.3. Circuit Breakers and Fallbacks

– Circuit breakers and fallback mechanisms were introduced to prevent cascading failures and provide graceful degradation in case of partial system outages or degradations.

– This ensured that even during incidents, critical functionalities (e.g., checkout, order placement) remained operational, minimizing the impact on customers.

3. **Improving Service-Level Objectives (SLOs) and MTTR**

By implementing the observability solutions and system design improvements identified through the SRE scoring framework, the ecommerce platform experienced significant improvements in its service-level objectives (SLOs) and a substantial reduction in MTTR.

3.1. SLO Improvements

- The improved database performance, scalability, and fault tolerance mechanisms contributed to higher system availability, reducing the risk of complete outages.

- The enhanced observability and incident detection capabilities enabled faster response times, minimizing the impact of potential incidents.

3.2. MTTR Reduction

- During subsequent incidents, the comprehensive monitoring and observability solutions allowed for rapid identification and diagnosis of issues, reducing the time spent on root cause analysis.

- The circuit breakers and fallback mechanisms prevented cascading failures, limiting the scope of incidents and enabling faster recovery.

- Streamlined incident response processes, facilitated by well-defined runbooks and playbooks, further contributed to reducing MTTR.

As a result of adopting SRE practices, the ecommerce platform successfully reduced its MTTR from 6 hours to less than 1 hour for similar incidents, significantly minimizing revenue losses and maintaining high customer satisfaction, even during peak shopping seasons.

This scenario demonstrates how SRE principles, including observability, system design improvements driven by a scoring framework, and a focus on improving SLOs, can effectively reduce MTTR and enhance the overall reliability and resilience of mission-critical systems in enterprise environments.

CHAPTER 9

Monitoring Types and Tools

Authors:

Anirudh Khanna

Praveen Gujar

Reviewer:

Harshavardhan Nerella

Definition of Reliability Monitoring

Reliability in systems and networks refers to the capacity of a software, system, or network to function without any instances of failure within the specified period of functionality. This implies that reliability looks into critical elements of a system or software. The main hallmarks of looking into reliability include stability, performance over time, and fault tolerance. In every instance, the use of reliability marks a chance to understand, engage, and work toward ensuring a remarkable understanding of the functional nature of any system. Therefore, reliability and system design demands critical engagement with entities to provide stellar results.

© Saurav Bhattacharya 2024
M. Kuppam, *Enterprise Digital Reliability*, https://doi.org/10.1007/979-8-8688-1032-9_9

Reliability monitoring is critical in ensuring the stability and management of systems and networks. Reliability leads to the continuous operation of software systems within an organization's essential functionality. In modern enterprises, the software assists in addressing critical functionality, working toward garnering and ensuring every aspect of the company runs well. Therefore, with the continuous operation and availability of reliability systems, running the organizations and achieving intended outcomes in every provided aspect becomes much more straightforward. Thus, using suitable software systems helps structure, advance, and enable considerable software modeling to achieve meaningful outcomes in whatever categories are demanded.

More to the point, reliability monitoring is a significant step in advancing the early detection of anomalies. The monitoring approach establishes a critical understanding of the systems, looking into standard functionality and hitches that affect the routine nature of functionality to address underlying issues. The issuance of monitoring aspects is crucial to ensuring reliability is always maintained. Moreover, monitoring also provides a chance to ensure mitigation strategies that would enable considerable adjustment, ensuring relevant development in achieving reliability at whatever instance of organizational functionality [1]. By empowering companies to address early detection and introducing mitigation strategies early in modeling the company needs, different approaches appeal to crafting and enabling strict addressing of significant demands in achieving sustainable results in creating organizational efficiency at all levels. Therefore, using the best scope of managing and handling reliability in the company through monitoring approaches leads to reduced downtime and emergency maintenance costs that could be costly to running organizational operations in different instances.

Reliability monitoring is critical to organizations because of the capacity to ensure end-user satisfaction with the system. Satisfaction creates trust and confidence in the system, as there is consistent service provision through stellar software performance and reduced failures.

This approach creates development in the company where they contend to individual preferences within the industry, creating value in major provisions that assist in making suitable demands at whatever level of instruction is desired. Reliability monitoring is crucial to ensuring the appropriate management of end-user confidence in the systems capable of achieving desired outcomes and addressing valuable outcomes by whatever means necessary [2]. Therefore, reliability monitoring works to achieve and establish a considerable level of advancing critical solutions in consumer support and confidence that systems will consistently accomplish the stated objectives. Thus, reliability monitoring creates more trust and confidence in the capacity of systems to address their needs at all moments.

Types of Reliability Monitoring

Reliability monitoring in systems and networks involves various approaches, each seeking to establish the level of functionality of the system over a period. The reliability monitoring methods have different categorizations, each seeking to develop and understand various provisions in handling the network analysis, aiming to deploy an instructional understanding of whatever a system comprises. Thus, the nature of ensuring reliability monitoring depends on an organization since they provide individual perspectives, insights, and advances that assist in crafting an instructional handling of the reliability of systems within an institution. The types of reliability monitoring include periodic, reactive, real-time, and predictive monitoring techniques. Each is applied in instrumental instances, and organizations decide to ensure deployment to satisfy particular needs and address reliability.

Real-Time Monitoring

This reliability monitoring model involves continuous observation of system behavior as it continues normal operations. The observation and analysis of the system enable an immediate understanding of the performance, underlying challenges, and difficulties within the system. This enables a considerable knowledge of the realistic nature of the system's performance, crafting an instrumental way of looking into reliability to establish the current and real-time aspects of the system. Real-time monitoring allows for prompt detection of anomalies, making it easier for organizations to work out critical approaches to ensure they can resolve issues and administer valuable adjustments to achieve the desired outcome in whatever category is determined. Therefore, the use of real-time monitoring implies identifying and managing critical variables associated with handling and ensuring that issues are identified as they occur and mitigation strategies are used to help address these issues.

Real-time monitoring has the main advantage of ensuring that the systems have an insight into downtimes and preventing them from occurring. The use of real-time tracking brings along demand to ensure the handling of significant challenges that can cause downtime, leading to reliable understanding and management of the systems to achieve an espoused level of functionality, helping to attain meaningful value and constructs at any provided time. It is mainly used in healthcare systems, financial trading platforms, and online service platforms to help address the central values in whatever capacity is needed to handle their needs [3].

Different techniques are used to ensure the proper framework for reliability monitoring. Event logging is a significant technique applied in reliability monitoring. It assists with handling and managing events, each seeking to ensure critical advancement of the nature of events in the system. This approach captures and records significant events in the system. Some key events that can be recorded within the system include user actions, warnings, and errors encountered while entering any work model.

Additionally, real-time monitoring is conducted by tracking specific performance counters. Some critical metrics used in handling performance include memory consumption, CPU usage, and transaction rates. In this case, the reliability monitoring approach establishes firm handling and management of the performance model, ensuring the proper management of bottlenecks and resource handling to achieve befitting handling of real-time monitoring to a desired level. Therefore, the performance counters assist in crafting an influential modeling of reliability to continually assist end users in managing their activities within the platforms. Using performance metrics creates an instructional mechanism for users to understand resource constraints, aiding the evasion of subsequent downtimes and hitches within the system [4].

Real-time alerts are another technique for real-time monitoring. Using alerts establishes an avenue of ensuring immediate relay of notifications when certain limits are exceeded. The alerts are predetermined, guaranteeing critical system management when they exceed these limits or upon detecting specific issues within the system. Understanding and addressing these factors ensures the administrators and system support teams can handle these limits and reinforce the system to a level of functionality that helps ensure every user achieves the highest outcome in managing and creating sustainable value within the system. Consequently, the real-time alerts assist in creating a real-time identification of issues and solutions to continue providing the system with a remarkable performance outcome.

Figure 9-1 indicates the process of conducting real-time monitoring within the systems. The process begins with collecting information, transmitting it, processing it, analyzing it, and alerting the system administrators. Nonetheless, the last step of monitoring is visualizing the data, which assists in creating the right way to understand and address it.

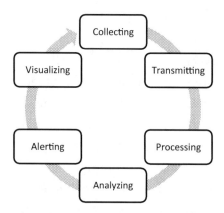

Figure 9-1. *Process of real-time monitoring*

Periodic Monitoring

Periodic monitoring is a reliability monitoring mechanism that engages scheduled checks performed in distinct intervals. These checks must be planned and conducted weekly, daily, or monthly. The monitoring model ensures a step to ensure long-term reliability and leads to a planned mechanism of handling services, leading to stellar and incremental ways to achieve suitable advances in marking the development of systems.

Periodic testing uses various techniques to ensure continued management and handling of system analysis. The first approach is automated tests, which are conducted on a predefined basis. These computerized tests have a routine execution, ensuring the development of information by analyzing various elements within the system and allowing for the verification of functionality and system integrity [5]. The approach also works by providing an automated insight into regressions to ensure the introduction of new code changes does not lead to defects in the system. In essence, this approach enables critical handling and modeling of the system to achieve an instrumental appeal in targeting and enabling continued handling of the system insight to achieve modest handling of the system to address pertinent vulnerabilities at whatever level is required.

Scheduled reports are a technique that helps with the regular generation of reliability and performance reports to stakeholders. Stakeholders use this technique to assist them in handling and spotlighting whatever has to be conducted to achieve a remarkable level of engagement with the system. Using these models ensures continued management steps to assist with handling detailed bottlenecks within the system. Thus, attending to the required approach defines and marks a considerable insight into handling reliability within the system. Generating insights and reports to the stakeholders ensures an increased step in managing the system performance and conducting trend analysis to help stakeholder entities plan on capacity management to achieve the most relevant functionality in the system at any given point.

A final mechanism of periodic monitoring is through log reviews. Log reviews assist with the modeling and management of periodic examinations of system logs. The examination of logs assists in identifying recurring issues or trends within the system. The approach creates a step to ensure that every integral aspect of the logs can be identified and steps to assist in handling a relevant outcome are established at the provided instance. Log reviews help to identify patterns that can lead to problems. Looking into the logs will help identify reasons for lag, downtime, or even latency, which can continually be used to enable considerable development in addressing challenges within the system at all levels [6]. Therefore, log reviews assist in managing and establishing the right level of advancement toward system management and proactive maintenance schedules.

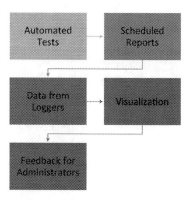

Figure 9-2. *Periodic monitoring process*

This figure indicates the process of conducting periodic monitoring. The process starts with automated tests that provide scheduled reports and insight into the data from log reviews. The visualization step ensures the instrumental development of data related to provided elements and gives administrators feedback at any instrumental point.

Predictive Monitoring

Predictive monitoring is an approach to reliability monitoring that employs data analysis and modeling to look into the systems. The monitoring model ensures an increasingly instrumental approach to utilizing data analytics to help formulate ways to cater to performance and normalcy and the identification of ways to ensure that the systems and networks can function within the provided outlines. The nature of predictive monitoring enables the provision of a step to look into current systems while locating steps to ensure that strategies can be employed to achieve the most remarkable outcome in whatever category is demanded. Thus, using predictive monitoring is essential in ensuring that unexpected downtimes are reduced because of a higher capacity to look into anomalies, employ mitigation strategies, and work toward achieving the best outcome in whatever capacity is defined by their functionalities.

Machine learning techniques are a significant model for ensuring predictive monitoring approaches. Employing the use of machine learning enhances the capacity to analyze historical data and look into patterns that can predict any instance of future failures. The machine learning technique is crucial in looking into large datasets of information, addressing the growth of patterns in the system, and locating potential issues within the system [7]. Most importantly, machine learning techniques ensure the employment of aspects that structure and assist in managing better analytics with continued use since they collect more information about the system and assist in crafting instrumental management of the platform to achieve the required value.

Trend analysis is another primary method of ensuring predictive monitoring. The model looks into performance patterns and helps to locate future problems through these trends. Trends in the past that led to issues are highlighted as having a chance to cause problems in the current system's functionality. This approach ensures critical handling of the capabilities and elements of addressing organizational needs at whatever level is demanded. Trend analysis ensures that the system performance is addressed over time, looking into changing aspects of functionality and keenly determining and anticipating the challenges within the system, helping to prepare interventions for whenever they must be applied in managing and addressing every instrumental category of dealing with the system demands.

Moreover, predictive data analytics can ensure that different sources within the systems can be used to forecast potential reliability issues. Predictive data analytics helps structure, administer, and work within the capacity to ensure influential input in managing the system to achieve a demanded influence in marking the contribution to administering valuable outcomes in whatever categories are desired [8]. Predictive analytics ensure the probable causes of failures and downtimes are analyzed and suggested, and proper mitigation strategies are used to help craft a solution for managing the underlying issues in marking a channel of change in addressing presented challenges within the system.

Reactive Monitoring

Reactive monitoring enables a focus on analysis and handling issues after they have occurred in the system. The reactive monitoring approach ensures the identification of root causes from an event and working on the system to ensure that they never happen again. The primary purpose of reactive monitoring is to look into the system, identify weaknesses, and enable better system handling to ensure continued modeling of values that would provide suitable modeling to achieve the right outcome in whatever situation is provided. Therefore, reactive monitoring aims to bolster system functionalities and prevent further failures of the same kind.

Incident analysis is a primary technique in reactive monitoring, ensuring a critical insight into the specific failures, investigating why they occurred, and their impact on the system. The incident analysis approach creates a chance to look into system functionalities, damage causes, and steps that must be used to ensure that the context of the damages does not occur to the system again. The incident analysis creates a reliable step to ensure that there are steps to learn from the damages caused by a failure, and achieving the most remarkable outcome in addressing and marking development is conducted to attain remarkable influence at all levels of addressing the incident.

Root cause analysis is another approach that seeks to understand the reasons for the failures. Root cause analysis works by ensuring the creation of a way to cater to the continued management of underlying causes, looking into individual components of the system, and addressing and enabling instrumental management of fundamental issues to achieve the right appeal to whatever extent is demanded [9]. Therefore, the use of the root cause analysis seeks to ensure that systems can diagnose the leading cause, help in repairing their appeals, and ensure that there is no recurrence of such an incidence within the system, leading to better modeling and management of any activity that has to partake in the system.

A final model of performing the reactive monitoring is through fault tree analysis. This approach seeks to work with the creation of a logical diagram that maps out potential issues and causes of the system failures. Fault tree analysis looks into problems stretching beyond the system, understanding the individual influence of approaches meant to work within the provision of mechanisms of understanding and addressing the reactive mention of handling reliability. The framework creates an essential way to look into the organization and understand whatever intentions can be conducted to achieve a remarkable influence on whatever needs to be addressed. Fault tree analysis is crucial in visualizing the factors that lead to failures [10]. The analysis also creates a better way of looking into factors that can be changed and whatever critical areas have to be modeled to assist in creating a meaningful outcome and achieving a reliable system model. This approach is vital to addressing the significant challenges within the system, documenting approaches that can be used to achieve meaningful outcomes in whatever dimension is required for their management approaches.

Tools Used in Reliability Monitoring

Reliability monitoring tools are critical for any organization seeking to understand the status of their systems and achieve a remarkable outcome from whatever appeals they have to work with. Essentially, the systems have to operate smoothly and achieve their projections at a demanded time; therefore, using the right tools to detect, address, and diagnose the correct issues helps cater to the right channel of providing sustainable values to the end users at whatever points are required. Therefore, using open source and proprietary tools is critical to engaging and ensuring an instrumental address of the tools to achieve the desired appeal.

Open Source Tools

Different open source tools are available for use by organizations, and they help address reliability monitoring within companies to enhance and achieve their demanded outcomes at whatever scope of functionality they adjust to. Some of the critical open source tools applicable in reliability monitoring include

a) **Prometheus:** This is software that operates as a time-series database. The software assists with real-time data monitoring and has an application of powerful query language (PromQL) that assists in retrieving and handling time-series data. The use of Prometheus ensures that there are vital additions to help manage multiple data collection methods and that it can scrape from HTTP endpoints. Using the system ensures the development of a model that seeks to engage and advance valuable addition in managing the tabulation of information to whatever extent is demanded. The platform can ensure real-time collection, storage, and metrics analysis from data generated. Nonetheless, the platform also has an alerting mechanism with customizable regulations and integrates various alerting managers that seek to address underlying variables in depicting and handling consistent development needs [11]. The platform's extensive support system enables it to integrate with various environments and systems that can assist in addressing widespread organizational needs. In most instances, Prometheus is applied to track the performance of dynamic and containerized ecosystems, commenting on their influence and flexibility in the process. The platform can also monitor cloud-native environments and applies even within the microservice architectures.

b) **Grafana:** This tool has a rich visualization dashboard, which allows data from multiple sources to be combined and handled. It offers the opportunity to consistently administer appropriate outcomes in whatever categories they must work with toward achieving the desired values. Nonetheless, extensive plugins within the system ensure that different data sources can be worked with to achieve an instrumental capacity to administer and address every reliable step in achieving a defined outcome. The tool also works with an information system that relies on an alerting and notification system that ensures "information stakeholders" information in whatever capacity and proportion can be collected to ensure a remarkable benefit of engaging the tool to manage and monitor the system well. The platform creates a step to enable real-time data visualization through the dynamic dashboard, ensuring the provision of flexible and interactive graphs, each helping and addressing detailed data analysis on the system [12]. Working with this tool ensures that data can be correlated from various systems to ensure a way to understand and address system health, marking the development and advancement of measures seeking to achieve a sustainable appreciation of whatever details and dimensions are provided. Grafana is used in performance monitoring to look into IT infrastructure and advance mechanisms of advancing the performance scale. Nonetheless, the dashboard helps craft an operational analysis of critical metrics and KPIs that seek to advance meaningful appeal at whatever level of engagement they must work with.

c) **Nagios:** This tool helps monitor network services and host resources. Critical network services analyzed include SMTP, NNTP, POP3, and NNTP. More to the point, host resources that can be explored include processor load, system logs, and disk usage. The tool has a notification system that alerts administrators of issues and enjoys an extensible architecture with various plugins to help monitor topics in various ways. Customizable reporting allows it to monitor large-scale enterprises and networks since several plugins can be keyed in to ensure the identification and management of underlying system units.

Proprietary Tools

These tools belong to an enterprise and assist in addressing pertinent issues related to having the most remarkable path to achieving the desired goals. These tools ensure that enterprises can customize and deal with their needs in a way they prefer. Some essential tools include

I. **New Relic:** This tool has AI-driven insights and anomaly detection, which helps it integrate with various cloud services. Nonetheless, real-time dashboards and customizable alerts ensure detailed performance analytics for infrastructure, microservices, and applications, ensuring the provision of information in the most preferred way the entity prefers [13]. New Relic can be used to monitor multicloud environments and big organizations' applications to craft value for their demands at a desired point. Additionally, they enhance the user experience by tracking performance and introducing solutions to optimize the performance in real time.

II. **Dynatrace:** This tool is an AI-powered platform that helps monitor root cause analysis. It can ensure automated discovery, instrumentation, and end-to-end visibility across all tiers of the organization. Large organizations can use the platform to continually examine insights from their systems and achieve meaningful performance modeling as any component desires. DevOps teams can also use this platform because of the detailed analysis that it presents [14].

III. **Splunk:** This tool can be used to index data. It also applies to looking into real-time search capabilities. Using Splunk helps structure customizable visualizations that help correlate log data and offer real-time operational intelligence and security monitoring activities. Using this approach defines and marks the chance to use integral tools to manage large volumes of machine data from different sources. The tool enables monitoring and visualization in a measure that achieves remarkable benefit and handling to present valuable outcomes to whatever extent is demanded.

Reliability and monitoring tools are instrumental in ensuring that software systems can be monitored and managed to have the proper health and performance levels. These tools can provide a combination of data visualization and alerting systems to inform organizations on the condition of their metrics, enhancing a protracted capacity of administering instrumental value to achieve the desired insight into the system [15]. Using open source and proprietary tools ensures that companies have the proper insight into their systems and can conduct reliable and user-friendly monitoring, each aimed at ensuring that there are increased steps to achieve better performance. These tools are vital to ensuring the suitable capacity of maintenance, management, and resource utilization within organizational systems, aiding their scope of review on performance and management of needs pertinent to having the proper framework for achieving operations.

Summary

This chapter explores the different tools and techniques of reliability monitoring. Reliability monitoring notably ensures the smooth and proper functionality of software, systems, and networks to reduce failures and ensure high performance over different periods. The monitoring techniques include real-time monitoring, an approach that engages continuous observation and handling of software performance to ensure operations are keenly handled to achieve the most meaningful outcomes. The model detects and mitigates challenges using event logging, performance counters, and alerts. Nonetheless, periodic monitoring works toward looking into scheduled checks. It locates log reviews to handle trends, which give further insight into the system's health and capacity to achieve specific functionality demands. Predictive monitoring is a mechanism that employs data analytics and AI to ensure that there are vital advances to help advance the proper techniques and approaches in dealing with maintenance. Proactive maintenance is ensured through this approach, marking the development of a pattern that prevents unexpected downtimes within the system. The final monitoring type is reactive monitoring, which comes after downtime and experiencing an issue within the system. Reactive monitoring looks into the root cause of the problem and seeks to ensure that the occurrence does not repeat itself in the future. The chapter also considers tools that can be used to advance critical solutions to reliability and monitoring. These tools can be proprietary or open source, ensuring the identification, management, and handling of core approaches to detail the management of every engagement to look at the system. Open source tools include Grafana, Prometheus, and Nagios, while proprietary tools include Dynatrace, New Relic, and Splunk. A combination of monitoring tools and techniques ensures increased reliability, better management of user satisfaction, and reduced operational costs that seek to enhance critical appeals in addressing their needs from a definitive angle.

The Tools Overlap on Observability

Introduction

In the rapidly evolving landscape of software engineering and DevOps, observability has emerged as a critical paradigm for understanding complex, distributed systems. Observability, rooted in control theory, refers to the ability to infer the internal states of a system from its external outputs. As systems grow in complexity, achieving observability requires a sophisticated toolkit that spans various domains such as logging, monitoring, tracing, and more. This chapter delves into the tools overlap on observability, exploring how different tools complement each other to provide a comprehensive view of system health and performance.

The Fundamentals of Observability

Observability in modern software systems is often conceptualized through the lens of three foundational pillars: logging, metrics, and tracing. Each pillar offers a distinct perspective on system behavior, enabling engineers to gain a comprehensive understanding of their systems' internal states and performance. By breaking down observability into these three components, teams can systematically monitor, diagnose, and optimize their systems, ensuring reliability and efficiency. These pillars are not isolated; rather, their interplay provides a synergistic approach to understanding complex, distributed architectures.

Logging is the process of capturing discrete events within a system. This includes recording specific actions, errors, and state changes, providing a detailed account of what happens at various points within the system. Logs serve as a chronological record of events, making it easier to diagnose issues when they arise. For instance, when an error occurs, the log data can reveal the exact sequence of events leading up to the problem, enabling swift identification and resolution. Logging tools such as the ELK stack (Elasticsearch, Logstash, and Kibana) and Fluentd are widely used to aggregate, search, and visualize log data. By offering granular visibility into system operations, logging is indispensable for debugging and auditing purposes.

Metrics, on the other hand, offer a quantitative view of a system's performance and health over time. Metrics capture data points such as CPU usage, memory consumption, and request rates, which can be continuously monitored to detect trends and anomalies. Tools like Prometheus and Grafana excel in collecting, storing, and visualizing these metrics, providing real-time insights into system behavior. Metrics are crucial for performance monitoring, capacity planning, and alerting. They enable engineers to understand the system's operational baseline and quickly identify deviations that might indicate underlying issues. By continuously tracking these key performance indicators, teams can proactively address potential problems before they impact users.

Tracing is the third pillar of observability, focusing on tracking the flow of requests through a system. In a microservice architecture, where requests often pass through multiple services, tracing provides a high-level view of these interactions. Tools like Jaeger and Zipkin help map out the path of a request, showing how different services and components interact to fulfill it. This end-to-end visibility is essential for identifying bottlenecks and latency issues. For example, if a request is taking longer than expected, tracing can pinpoint which service or component is causing the delay. By providing a comprehensive view of request flows, tracing enables engineers to optimize performance and ensure efficient service interactions.

The interplay between logging, metrics, and tracing forms the foundation of observability. Each pillar contributes unique insights that, when combined, provide a holistic and actionable understanding of the system. For instance, an observed spike in response times (metrics) can be correlated with specific errors or warnings in the logs, while traces can reveal the exact service interactions involved. This integrated approach allows for more effective troubleshooting and optimization, as engineers can see the full picture rather than isolated pieces of data. The synergy between these tools enhances the ability to diagnose, understand, and address system issues comprehensively.

In conclusion, the pillars of observability—logging, metrics, and tracing—each play a vital role in providing visibility into complex systems. Logging captures detailed event data, metrics offer a quantitative assessment of performance, and tracing provides a macrolevel view of request flows. Together, they create a robust framework for monitoring, diagnosing, and optimizing system health and performance. By leveraging the strengths of each pillar and integrating their insights, engineering teams can achieve true observability, ensuring their systems remain reliable, performant, and resilient in the face of growing complexity.

Logging Tools

Logging tools are essential for capturing granular details about system events. Logs record discrete pieces of information about what happens within a system, providing a detailed account of operations, errors, transactions, and other significant events. This granular data is vital for diagnosing issues, understanding system behavior, and ensuring overall system health. Among the most popular logging tools is the Elasticsearch, Logstash, and Kibana (ELK) stack. The ELK stack is a powerful suite that allows for the efficient aggregation, analysis, and visualization of log data. Elasticsearch serves as the core storage and search engine, enabling fast retrieval and querying of log data. Logstash is responsible for ingesting and processing logs, transforming them as necessary before storing them in Elasticsearch. Kibana, the visualization layer, allows users to create dynamic dashboards and visual representations of log data, facilitating easier analysis and monitoring.

Elasticsearch, a highly scalable search and analytics engine, plays a crucial role in managing vast amounts of log data. Its distributed nature ensures that log data is quickly indexed and searchable, making it possible to retrieve specific logs in real time. This capability is especially important in large, complex systems where logs can rapidly accumulate. Elasticsearch's powerful search functionalities allow for detailed querying, enabling users to filter and sort logs based on various criteria. This makes it easier to pinpoint issues and understand the context around specific events, significantly reducing the time required for troubleshooting and root cause analysis.

Logstash, the data processing pipeline, is designed to handle a wide variety of data sources and formats. It collects logs from multiple sources, including system logs, application logs, and network logs, and then processes this data to ensure it is in a consistent format suitable for storage in Elasticsearch. Logstash can also enrich logs by adding metadata, such as geolocation information based on IP addresses or tags indicating the log's source or severity. This enrichment helps provide more context around each log entry, making subsequent analysis more effective. Logstash's flexibility and extensibility, through its plugin architecture, enable it to adapt to a wide range of use cases and environments, ensuring that all relevant log data is captured and processed efficiently.

Kibana, the visualization component of the ELK stack, transforms log data into actionable insights through its intuitive dashboard interface. Users can create customized dashboards to visualize log data in various formats, such as line charts, bar graphs, pie charts, and heat maps. These visualizations help in identifying patterns, trends, and anomalies within the log data, making it easier to understand system behavior and detect potential issues. Kibana also supports interactive exploration of log data, allowing users to drill down into specific logs and perform ad hoc queries. This capability is invaluable for on-the-fly investigations and real-time monitoring of system health.

In addition to the ELK stack, Fluentd is another widely used logging tool that offers robust capabilities for log data collection and processing. Fluentd is an open source data collector designed to unify the collection and consumption of log data across various sources. Its flexible architecture allows it to integrate with multiple data sources and destinations, making it a versatile tool for log management. Fluentd uses a unified logging layer that abstracts the complexities of different log formats and protocols, ensuring consistent log collection and processing. Its plugin-based architecture enables easy extension and customization, allowing users to tailor Fluentd to their specific needs and environments.

These logging tools—Elasticsearch, Logstash, Kibana, and Fluentd—provide critical insights into specific events and errors within a system. By capturing detailed log data and enabling comprehensive analysis and visualization, they empower engineers to quickly diagnose and troubleshoot issues. This capability is crucial for maintaining system reliability, performance, and security. Logs not only help in identifying and resolving problems but also in proactive monitoring and incident response. By leveraging these tools, organizations can achieve a high level of observability, ensuring that they can effectively manage and maintain their complex, distributed systems.

Monitoring Tools

Monitoring tools are the backbone of maintaining the health and performance of modern, distributed systems. These tools are designed to track a wide array of system metrics, from CPU usage and memory consumption to application-specific performance indicators like request rates and error rates. The primary purpose of these tools is to provide real-time insights that help operations and development teams understand the state of their systems at any given moment. By continuously collecting and analyzing data, monitoring tools enable teams to detect deviations from expected performance, identify potential bottlenecks, and foresee issues before they escalate into critical problems.

Prometheus stands out as a key player in the monitoring landscape. This open source toolkit is renowned for its reliability and scalability, making it an ideal choice for complex, dynamic environments. Prometheus collects metrics from various targets at specified intervals, allowing for fine-grained monitoring. It uses a powerful query language called PromQL to evaluate rule expressions and generate alerts based on predefined conditions. This capability ensures that teams are promptly informed of any anomalies, enabling swift intervention. The data collected by Prometheus can be visualized in a variety of ways, providing a clear and actionable view of system performance.

Grafana complements Prometheus by offering a versatile platform for data visualization. This open source web application supports a wide range of data sources, making it a popular choice for integrating and displaying metrics from diverse systems. Grafana excels in creating interactive, customizable dashboards that present data in an intuitive and accessible manner. Users can create complex charts, graphs, and alerts that provide deep insights into their system's performance. The ability to visualize metrics in real time allows teams to quickly spot trends and correlations, facilitating proactive decision-making and troubleshooting.

The synergy between Prometheus and Grafana exemplifies the power of integrated monitoring solutions. While Prometheus excels at data collection and alerting, Grafana provides the necessary tools to interpret and act on that data. Together, they form a comprehensive monitoring solution that enhances visibility into system operations. This integration helps teams to not only monitor current performance but also to analyze historical data, identify long-term trends, and make informed decisions about capacity planning and optimization. By leveraging the strengths of both tools, organizations can achieve a high level of observability and maintain the resilience of their systems.

Monitoring tools, when effectively implemented, play a crucial role in maintaining system reliability and user satisfaction. They enable teams to identify trends and anomalies early, preventing minor issues from becoming major incidents. This proactive approach to system management is essential in today's fast-paced digital landscape, where downtime and performance degradation can have significant consequences. By providing continuous, real-time insights into system health, monitoring tools empower teams to maintain optimal performance, enhance user experience, and ensure the seamless operation of critical applications and services.

Tracing Tools

Tracing tools are crucial for understanding the flow of requests and the interactions between services in complex distributed systems. They allow engineers to visualize and analyze the path a request takes as it traverses through various microservices, providing insights into latency, errors, and performance bottlenecks. In a microservice architecture, where multiple services work together to fulfill a single request, tracing tools help to pinpoint the exact service or component causing delays or failures. This granular visibility is essential for maintaining the performance and reliability of the system, especially as it scales.

One of the prominent tools in this domain is "Jaeger." Jaeger is an open source, end-to-end distributed tracing tool originally developed by Uber. It is designed to monitor and troubleshoot transactions in complex distributed systems. Jaeger collects traces and spans from various services, which can be visualized to show the request flow and the time taken at each step. This detailed tracing information helps in identifying slow services, understanding service dependencies, and diagnosing performance issues. Jaeger's ability to integrate with various data sources and its compatibility with multiple storage backends make it a versatile tool for tracing in diverse environments.

Another widely used tracing tool is "Zipkin." Zipkin, initially developed by Twitter, is a distributed tracing system that helps gather timing data needed to troubleshoot latency problems in microservice architectures. It captures trace data, which includes information about the request path, timing, and service interactions. This data is then used to create a trace map, highlighting the duration and sequence of calls between services. Zipkin's efficient data model and user-friendly interface make it easy for developers to understand the flow of requests and quickly identify any service contributing to latency issues. By pinpointing slow or failing services, Zipkin aids in optimizing system performance and improving user experience.

These tracing tools provide a high-level overview of system interactions, which is invaluable for identifying bottlenecks and performance issues. By visualizing the entire request journey, from initiation to completion, tracing tools help engineers understand how different services interact and where potential delays or errors occur. This holistic view is essential for optimizing system performance, as it allows teams to address specific issues that impact the overall user experience.

Moreover, tracing tools facilitate root cause analysis by providing detailed context around each request, making it easier to debug and resolve complex problems.

Integrating tracing tools into a microservice architecture involves instrumenting services to emit trace data. This often requires modifying code to include tracing libraries and setting up the tracing backend to collect and store the trace data. Despite the initial setup effort, the benefits of having a comprehensive tracing system far outweigh the costs. Tracing not only aids in performance monitoring but also plays a crucial role in capacity planning, incident response, and continuous improvement of the system. As organizations increasingly adopt microservices, the importance of robust tracing solutions becomes ever more critical for maintaining system health and achieving operational excellence.

In conclusion, tracing tools like Jaeger and Zipkin are indispensable for understanding the flow of requests and the interactions between services in a microservice architecture. They provide deep insights into system performance, helping to identify and resolve bottlenecks and latency issues. By visualizing the request paths and analyzing the trace data, these tools enable engineers to optimize the performance and reliability of their systems. As the complexity of distributed systems grows, the role of tracing tools in ensuring smooth and efficient operations becomes even more pivotal, making them a key component of any observability strategy.

The Intersection of Tools

While each category of observability tools serves a distinct purpose, their overlap is where the true power of observability is realized. The integration and correlation of logs, metrics, and traces provide a holistic view of the system. Logs offer detailed, time-stamped records of discrete events that occur within the system, such as errors, state changes, and user actions. Metrics, on the other hand, provide quantitative measurements of system performance, such as CPU usage, memory consumption, and request rates, which are crucial for monitoring the health and efficiency of applications over time. Tracing adds another layer by tracking the flow of requests through the system, enabling the identification of bottlenecks and performance issues. When these tools are used in isolation, they provide valuable but fragmented insights. However, when integrated, they offer a comprehensive understanding of system behavior, making it easier to diagnose problems, identify root causes, and implement effective solutions.

Integrated dashboards are a prime example of how the overlap of observability tools can be harnessed effectively. Tools like Grafana can pull in data from both Prometheus, which collects and stores metrics, and Elasticsearch, which aggregates and indexes logs. This creates a unified dashboard where logs and metrics can be visualized side by side. Such integration allows for the cross-referencing of logs and metrics, making it easier to correlate specific events with performance data. For instance, a spike in error logs can be directly correlated with an increase in CPU usage or a drop in request throughput, providing a clear picture of what might be causing performance degradation. This unified view enables engineers to quickly pinpoint issues and understand the broader context, leading to faster and more accurate troubleshooting.

The correlation of traces and logs further enhances observability by providing detailed context for each trace. Tracing tools like Jaeger can be integrated with logging tools to enrich trace data with log information. For example, if a request trace reveals high latency, the corresponding logs can be referenced to identify the specific events or errors that contributed to the delay. This integration allows engineers to see not just the path of the request but also the detailed events that occurred along the way. By correlating trace data with logs, engineers can gain a deeper understanding of how different components interact and where issues might arise, making it easier to optimize performance and reliability.

This synergy between logging, monitoring, and tracing tools enhances the ability to diagnose, troubleshoot, and optimize complex systems. When these tools work together seamlessly, they provide a multifaceted view of system health and performance. Engineers can use logs to investigate specific events, metrics to monitor overall system performance, and traces to understand the flow of requests and interactions between services. This comprehensive approach allows for more effective problem-solving and performance optimization. For instance, by correlating metrics with traces, engineers can identify which parts of the system are contributing to performance bottlenecks and make targeted improvements. Similarly, by integrating logs with traces, they can quickly pinpoint the root cause of errors and take corrective actions.

Ultimately, the overlap of observability tools transforms the way engineers understand and manage complex systems. It shifts the focus from reactive troubleshooting to proactive monitoring and optimization. By leveraging the strengths of each tool and integrating them effectively, organizations can achieve true observability, ensuring the reliability, performance, and scalability of their systems. This holistic approach not only improves the efficiency of incident response but also enhances the overall quality and user experience of the software. As systems continue to grow in complexity, the importance of integrated observability tools will only increase, making it essential for organizations to adopt and refine their observability practices.

Case Study: Achieving Observability in a Microservice Architecture

Consider a hypothetical ecommerce platform utilizing a microservice architecture. The platform comprises several independently deployable, scalable, and manageable services, such as user authentication, product catalog, shopping cart, and order processing. This architectural approach allows each service to be developed, deployed, and scaled independently, providing significant flexibility and resilience. However, it also introduces complexity, making it challenging to monitor and troubleshoot issues.

313

Achieving observability in such a distributed system is crucial for maintaining performance and reliability. This involves collecting and analyzing logs, metrics, and traces from each microservice to gain a comprehensive understanding of the system's behavior.

Logging with the ELK stack (Elasticsearch, Logstash, and Kibana) plays a vital role in capturing and visualizing log data from each microservice. Logs from services such as user authentication, product catalog, and order processing are aggregated into Elasticsearch, a powerful search and analytics engine. Logstash processes and enriches these logs before storing them in Elasticsearch. Kibana, a data visualization tool, provides engineers with intuitive dashboards to search, filter, and analyze log data by service, severity, and timestamp. This capability enables quick identification of errors, unusual patterns, or anomalies within specific services, facilitating efficient troubleshooting and debugging.

Monitoring the platform's performance and health is essential for ensuring a seamless user experience. Prometheus, an open source monitoring and alerting toolkit, is used to collect and store metrics from each microservice. Metrics such as request rates, error rates, response times, and resource utilization are gathered at regular intervals. Grafana, a popular visualization tool, connects to Prometheus and provides real-time dashboards to display these metrics. Engineers can set up alerting rules within Grafana to receive notifications when metrics exceed predefined thresholds, such as high error rates in the user authentication service or increased response times in the product catalog service. This proactive monitoring approach helps identify potential issues before they impact users, enabling timely intervention and resolution.

Tracing is crucial for understanding the flow of requests through the various microservices and identifying performance bottlenecks. Jaeger, an end-to-end distributed tracing tool, is employed to trace requests as they propagate through the system. For instance, when a user reports a slow checkout process, Jaeger traces can reveal the exact path of the request, from the shopping cart service to the order processing service.

By visualizing the trace data, engineers can pinpoint the service or component causing the delay, such as a slow database query in the shopping cart service. This granular insight into request flows and dependencies helps diagnose performance issues, optimize service interactions, and enhance overall system efficiency.

By combining logs, metrics, and traces, the ecommerce platform achieves full observability, providing a holistic view of its operational state. The integration of these observability tools enables engineers to correlate events across different data sources, facilitating comprehensive analysis and troubleshooting. For example, if an alert from Grafana indicates a spike in error rates, engineers can cross-reference related logs in Kibana to understand the context of the errors and examine Jaeger traces to identify the affected services and their interactions. This multifaceted approach allows for rapid detection, root cause analysis, and resolution of issues, minimizing downtime and ensuring a high-quality user experience.

In conclusion, implementing observability in a microservice-based ecommerce platform involves leveraging a combination of logging, monitoring, and tracing tools. The ELK stack provides detailed log analysis, Prometheus and Grafana offer real-time monitoring and alerting, and Jaeger delivers comprehensive request tracing. By integrating these tools, the platform can achieve full observability, enabling proactive management, efficient troubleshooting, and continuous optimization of the system. This integrated observability framework is essential for maintaining the performance, reliability, and scalability of complex microservice architectures, ultimately contributing to a seamless and satisfying user experience.

Challenges in Achieving Observability

Despite the numerous benefits that observability brings to modern software systems, it also introduces several significant challenges that organizations must navigate to harness its full potential. One of the foremost challenges is the sheer volume of data generated by logs, metrics, and traces. In complex systems, especially those employing microservice architectures, the amount of data can become overwhelming. Each service generates logs and metrics, and tracing requests through distributed systems produces additional data. Efficiently managing, storing, and querying this data necessitates robust data management and storage solutions. Without proper handling, the deluge of data can lead to performance bottlenecks and increased costs, complicating the goal of maintaining high observability.

Integration complexity presents another formidable challenge. Observability often requires the use of multiple tools, each specializing in different aspects like logging, monitoring, or tracing. Integrating these diverse tools into a cohesive system requires meticulous planning and configuration. Ensuring that logs, metrics, and traces from different sources are seamlessly correlated and accessible through unified dashboards is no small feat. It involves configuring data pipelines, setting up appropriate data schemas, and ensuring compatibility across different tools and platforms. The complexity of integration can lead to delays and inconsistencies in data flow, hindering the ability to achieve comprehensive observability.

Performance overhead is an additional concern when implementing observability. Instrumenting applications to generate the necessary logs, metrics, and traces can introduce latency and increase resource consumption. This performance overhead can be particularly pronounced in high-throughput or latency-sensitive applications. Developers must carefully balance the level of observability instrumentation with the system's performance requirements. Overinstrumentation can lead to degraded system performance, while underinstrumentation can result in insufficient visibility into the system's behavior. Striking the right balance requires a nuanced understanding of the system's performance characteristics and the criticality of different observability data.

Addressing these challenges necessitates a strategic approach and the selection of the right tooling. Organizations must invest in scalable and efficient data management solutions to handle the volume of observability data. They should also prioritize the use of open standards and interoperable tools to simplify integration complexity. Automation can play a crucial role in streamlining the configuration and maintenance of observability pipelines. Moreover, organizations should adopt a performance-conscious approach to instrumentation, ensuring that the impact on system performance is minimized while still achieving the desired level of visibility.

In conclusion, while achieving observability offers profound insights into system behavior and enhances the ability to diagnose and resolve issues, it is not without its hurdles. The challenges of data volume, integration complexity, and performance overhead require careful consideration and strategic planning. By addressing these challenges with the right tools and approaches, organizations can effectively harness the power of observability, ensuring their systems are robust, reliable, and performant. This balanced approach will enable them to reap the benefits of observability without succumbing to its potential pitfalls.

Future Trends in Observability

The field of observability is experiencing significant transformation, propelled by rapid technological advancements and evolving system architectures. As systems become more complex and distributed, traditional methods of monitoring and diagnostics are often insufficient. New trends and technologies are emerging to address these challenges, making observability more robust and comprehensive. Understanding these trends is crucial for maintaining effective observability and ensuring system reliability and performance in modern environments.

One of the most impactful trends in observability is the integration of artificial intelligence (AI) and machine learning (ML). These technologies are revolutionizing observability tools by enabling predictive insights and automated anomaly detection. AI and ML algorithms can analyze vast amounts of observability data to identify patterns and trends that might not be apparent to human operators. For example, machine learning models can predict potential system failures or performance degradations before they occur, allowing for proactive maintenance and reducing downtime. Automated anomaly detection leverages AI to identify outliers and unusual patterns in real time, enabling quicker responses to potential issues. This shift toward AI-driven observability tools is enhancing the accuracy and efficiency of system monitoring and troubleshooting.

As serverless and edge computing gain traction, observability tools are evolving to handle the unique challenges posed by these architectures. Serverless computing abstracts away the underlying infrastructure, making it difficult to monitor traditional metrics like CPU usage or memory consumption. Observability tools are adapting by focusing on high-level metrics such as request latency, error rates, and resource usage at the function level. Edge computing, which distributes computation closer to data sources, introduces additional complexity due to the decentralized nature of the architecture. Observability tools are being designed to aggregate and correlate data from multiple edge locations, providing a unified view of the system. This adaptation ensures that observability remains effective even as the infrastructure becomes more dynamic and distributed.

Another significant development in the field of observability is the OpenTelemetry project. OpenTelemetry is an open source initiative aimed at providing a standardized framework for collecting and transmitting observability data, including logs, metrics, and traces. This standardization simplifies the integration of observability tools and ensures consistency in the data being collected and analyzed. OpenTelemetry's unified standard allows organizations to easily switch between different observability tools without losing data fidelity or having to reinstrument their applications. By providing a common language and framework for observability, OpenTelemetry is fostering greater interoperability and collaboration within the observability ecosystem. This initiative is set to become a cornerstone of modern observability practices.

In addition to these technological advancements, staying abreast of observability trends involves understanding the broader changes in system architectures and development practices. The rise of microservices, containerization, and cloud-native applications is driving the need for more sophisticated observability solutions. These architectures introduce new complexities, such as service dependencies and dynamic scaling, that traditional monitoring tools struggle to address. Observability tools are evolving to provide deeper insights into these modern architectures, enabling developers and operators to understand and manage their systems more effectively. Keeping pace with these changes is essential for maintaining robust observability in contemporary environments.

In conclusion, the field of observability is rapidly evolving, driven by advances in AI and ML, the rise of serverless and edge computing, and the standardization efforts of projects like OpenTelemetry. These trends are transforming how we monitor, understand, and optimize complex systems. Staying current with these developments is crucial for maintaining effective observability and ensuring the reliability, performance, and scalability of modern systems. As observability continues to advance, it will play an increasingly vital role in the successful management of today's and tomorrow's technology landscapes.

Conclusion

Observability is a cornerstone of modern software engineering and DevOps practices, playing a critical role in maintaining the health and performance of complex systems. As applications become more distributed and sophisticated, the need for a robust observability strategy has never been more paramount. Observability allows teams to infer the internal state of a system from its external outputs, providing the insights needed to diagnose issues, optimize performance, and ensure reliability. This holistic approach is essential for managing microservice architectures, cloud-native applications, and other advanced deployment models that require a detailed and nuanced understanding of system behavior.

The overlap of tools across logging, monitoring, and tracing is fundamental to achieving comprehensive observability. Logging tools capture detailed records of events within a system, offering granular insights into specific actions and errors. Monitoring tools, on the other hand, track real-time metrics that reflect system health and performance, such as CPU usage, memory consumption, and request rates. Tracing tools provide a high-level view of request flows and service interactions, helping identify bottlenecks and performance issues. When these tools are used in tandem, they offer a multifaceted perspective that enables rapid diagnosis and resolution of issues. The synergy between logging, monitoring, and tracing allows for the correlation of disparate data points, creating a cohesive picture of system operations and facilitating more effective troubleshooting and optimization.

By leveraging the strengths of each tool and integrating them effectively, organizations can achieve true observability, which is essential for ensuring the reliability, performance, and scalability of their systems. Effective observability helps teams quickly identify and address issues before they impact users, maintain high service availability, and optimize system performance. Moreover, as systems continue to evolve and grow in complexity, the ability to observe and understand these systems becomes increasingly vital. Integrating observability tools not only aids in immediate problem-solving but also provides long-term benefits by enabling continuous improvement and innovation. In essence, observability is not just about monitoring systems; it's about gaining deep insights that drive better decision-making and foster a proactive approach to system management and development.

Bibliography

1. F. H. Ferreira, E. Y. Nakagawa, and R. P. dos Santos, "Reliability in software-intensive systems: challenges, solutions, and future perspectives," in *2021 47th Euromicro Conference on Software Engineering and Advanced Applications (SEAA)*, 2021, pp. 54–61

2. V. Shinde, S. K. Bharadwaj, and D. K. Mishra, "State-of-the-Art Literature Review on Classification of Software Reliability Models," in *Multi-Criteria Decision Models in Software Reliability*, 2022, pp. 161–184

3. S. Oveisi, A. Moeini, S. Mirzaei, and M. A. Farsi, "Software reliability prediction: A survey," *Quality and Reliability Engineering International*, vol. 39, no. 1, pp. 412–453, 2023

4. M. Asraful Haque, "Software reliability models: A brief review and some concerns," in *The International Symposium on Computer Science, Digital Economy and Intelligent Systems*, 2022, pp. 152–162

5. R. Pai, G. Joshi, and S. Rane, "Quality and reliability studies in software defect management: a literature review," *International Journal of Quality & Reliability Management*, vol. 38, no. 10, pp. 2007–2033, 2021

6. Y. Shamstabar, H. Shahriari, and Y. Samimi, "Reliability monitoring of systems with cumulative shock-based deterioration process," *Reliability Engineering & System Safety*, vol. 216, p. 107937, 2021

7. M. A. López-Campos, A. Crespo Márquez, and J. F. Gómez Fernández, "The integration of open reliability, maintenance, and condition monitoring management systems," in *Advanced Maintenance Modelling for Asset Management: Techniques and Methods for Complex Industrial Systems*, 2018, pp. 43–78

8. Y. Wang, H. Liu, H. Yuan, and Z. Zhang, "Comprehensive evaluation of software system reliability based on component-based generalized GO models," *PeerJ Computer Science*, vol. 9, p. e1247, 2023

9. R. Parasuraman, M. Mouloua, R. Molloy, and B. Hilburn, "Monitoring of automated systems," in *Automation and Human Performance*, CRC Press, 2018, pp. 91–115

10. R. Jain and A. Sharma, "Assessing software reliability using genetic algorithms," *The Journal of Engineering Research [TJER]*, vol. 16, no. 1, pp. 11–17, 2019

11. J. Turnbull, *Monitoring with Prometheus*, Turnbull Press, 2018

12. T. Leppänen, *Data Visualization and Monitoring with Grafana and Prometheus*, 2021

13. Cardoso, C. J. V. Teixeira, and J. S. Pinto, "Architecture for highly configurable dashboards for operations monitoring and support," *Studies in Informatics and Control*, vol. 27, no. 3, pp. 319–330, 2018

14. Nair, "DevOps-Driven Approach to Development in Cloud," *Authorea Preprints*, 2023

15. N. K. Jain, R. K. Saini, and P. Mittal, "A review on traffic monitoring system techniques," in *Soft Computing: Theories and Applications: Proceedings of SoCTA 2017*, 2019, pp. 569–577

16.   Hightower, K., Burns, B., and Beda, J., Kubernetes: Up and Running: Dive into the Future of Infrastructure, 2017, `https://openlibrary.org/books/OL28939482M/Kubernetes_-_Up_and_Running`

17.   Turnbull, J., The art of monitoring. James Turnbull, 2014

18.   Red Hat, Understanding Observability. Red Hat, 2020

19.   Parker, A., Spoonhower, D., Mace, J., Sigelman, B., and Isaacs, R., Distributed tracing in practice: Instrumenting, Analyzing, and Debugging Microservices. O'Reilly Media, 2020

CHAPTER 10

The Impact of AI Ops Reliability

Author:

Vishwanadham Mandala

Introduction

In today's rapidly evolving technological landscape, the intersection of artificial intelligence (AI) and operations management has garnered significant attention for its transformative potential. Within this context, AI Ops, or artificial intelligence for IT operations, has emerged as a pivotal framework that enhances the reliability and efficiency of IT systems. This essay seeks to explore the multifaceted impact of AI Ops on operational reliability by examining both theoretical frameworks and practical applications. By leveraging machine learning algorithms and sophisticated data analytics, AI Ops not only improves incident response times but also empowers organizations to preemptively identify anomalies and prevent potential disruptions. Furthermore, this analysis will consider the implications of increased reliability on organizational productivity and customer satisfaction, thus reinforcing the necessity of integrating AI Ops in contemporary IT practices. Ultimately, understanding the dynamics of AI Ops reliability is essential for navigating future technological advancements in operational frameworks.

© Saurav Bhattacharya 2024
M. Kuppam, *Enterprise Digital Reliability*, https://doi.org/10.1007/979-8-8688-1032-9_10

Definition of AI Ops

The concept of AI operations, commonly referred to as AIOps, encompasses the integration of artificial intelligence into IT operations to enhance the efficiency and effectiveness of managing complex technological environments. By combining big data analytics, machine learning, and automation, AIOps aims to improve the observability, monitoring, and management of IT infrastructures. The importance of AIOps is underscored in safety-critical domains, where the robustness and reliability of AI systems are essential, such as in autonomous driving and aerospace. In these sectors, it is crucial to assess the vulnerability of AI deployments, as soft errors or single event upsets can significantly affect decision-making processes (Guti Jérrez-Zaballa, 2024).

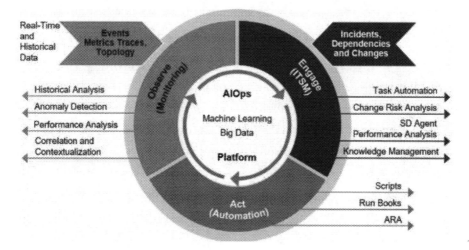

Figure 10-1. *Artificial intelligence for IT operations*

Moreover, as virtual humans become increasingly realistic in their interactions, understanding the implications of AI systems on operational reliability is paramount. Thus, AIOps not only addresses operational efficiency but also fosters trust in AI solutions by ensuring their reliability and performance in real-world applications.

Importance of Reliability in AI Ops

In the realm of AI operations, reliability emerges as a cornerstone for maintaining seamless functionality across complex systems. As businesses increasingly deploy AI-driven processes, the interconnectedness of various applications and infrastructure becomes critically relevant, necessitating robust reliability measures. This complexity is particularly evident in environments like 5G industrial networks, where applications dynamically share resources and, thus, influence each other's performance (Chen K, 2024). Consequently, the potential for failures and operational disruptions increases, leading to significant challenges in management effectiveness and organizational outcomes. To navigate these challenges, organizations must adopt comprehensive strategies that enhance reliability, understanding that a failure in one component can have cascading effects throughout the system.

Furthermore, as the implementation of generative AI tools expands, ensuring their reliability will be paramount to fostering trust and encouraging their adoption in the supply chain context, where skepticism still prevails regarding their true value and impact

Overview of AI Ops Applications

AI Ops applications are revolutionizing how organizations manage IT operations, leveraging data-driven insights to enhance reliability and responsiveness. These applications utilize machine learning algorithms to analyze vast amounts of operational data, identifying patterns and anomalies that may indicate system malfunctions or potential downtimes. By automating these processes, companies significantly reduce the time required to resolve issues, thereby minimizing disruptions to service continuity. Furthermore, AI Ops tools facilitate proactive monitoring and predictive maintenance, enabling IT teams to address potential problems before they escalate into critical failures.

This shift from reactive to proactive management is not only cost-efficient but also enhances overall system reliability, ensuring that services remain stable and responsive to user needs. As organizations increasingly adopt these technologies, the integration of AI Ops will likely set new standards for operational excellence and resilience in IT environments.

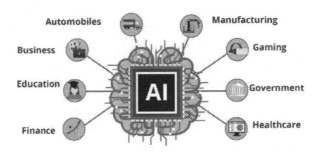

Figure 10-2. *Applications of artificial intelligence*

Historical Context of AI Ops Development

The evolution of AI operations (AI Ops) has been significantly influenced by the intersection of advancements in machine learning, data processing, and the increasing complexity of IT environments. Early developments in AI were primarily focused on automating repetitive tasks, laying the groundwork for more intricate systems capable of predictive analytics and decision support. As organizations began to collect vast amounts of data, the need for sophisticated analytical frameworks became evident, leading to the emergence of AI Ops as a response to operational inefficiencies. Research in this area highlights critical considerations, such as the challenges of algorithmic drift and the importance of model explainability, which are essential for reliable AI applications in real-world scenarios (Bhargava K. Chinni, 2024). Furthermore, exploratory missions utilizing analog environments, such as lava tubes, underscore the necessity for reliable AI systems to navigate unpredictable terrains and enhance operational efficiency (Benjamin J. Morrell, 2024). This historical context serves as a foundation for understanding the reliability and effectiveness of AI Ops in contemporary IT landscapes.

Current Trends in AI Ops Reliability

As organizations increasingly rely on AI operations (AI Ops) to enhance performance and maintain reliability, current trends highlight the growing importance of explainability and adaptability in these systems. The integration of machine learning algorithms has enabled the development of digital biomarkers that provide actionable insights for improving operational efficiency across various sectors. Studies have shown that effective AI Ops can leverage these digital tools to optimize decision-making processes while ensuring compliance with emerging regulations, such as those related to environmental sustainability in construction (Promised. Nikah, 2024).

Furthermore, as AI systems evolve, concerns regarding algorithmic drift and the need for continual surveillance become paramount. Implementing robust AI Ops frameworks that emphasize transparency not only boosts trust among stakeholders but also aids in overcoming challenges related to data bias and prediction accuracy. Ultimately, the focus on reliability in AI Ops will significantly influence organizational resilience and operational effectiveness moving forward.

Research Objectives and Questions

Establishing clear research objectives and questions is vital for guiding a study, particularly in emerging fields like AI operations. The objective of this research is to explore the interplay between AI Ops reliability and operational efficacy, assessing how reliability impacts overall performance in various applications. This inquiry leads to critical questions: What factors contribute to the reliability of AI operations? How do these reliability factors correlate with operational success in complex environments, such as those faced in healthcare or planetary exploration? For instance, the challenges inherent in deploying AI-driven digital biomarkers for patient management underscore the importance of reliability, as a lack of it could diminish patient outcomes (Bhargava K. Chinni, 2024).

Similarly, the operational dynamics tested in robotic missions to explore Martian caves reveal the necessity for dependable autonomy in achieving effective exploration (Benjamin J. Morrell, 2024). By addressing these questions, the research aims to provide actionable insights for enhancing AI Ops reliability in diverse operational contexts.

Significance of the Study

The exploration of AI Ops reliability is essential, particularly as organizations increasingly integrate artificial intelligence into their operational frameworks. Understanding the implications of AI reliability not only fosters enhanced decision-making processes but also contributes to the establishment of trust between operators and automated systems. By systematically analyzing AI Ops, this study aims to provide insights into how these technologies influence operational efficiency, risk management, and overall system performance.

Moreover, the findings will serve to equip stakeholders with the knowledge needed to implement effective strategies that mitigate potential failures while maximizing the benefits of AI integration in daily operations. Ultimately, the significance of this study extends beyond theoretical frameworks, as it addresses practical challenges and opportunities that arise in reliance on AI systems, thereby laying the groundwork for future research and application in the evolving landscape of technology-driven enterprises.

Methodology Overview

The methodology employed in this study is grounded in a comprehensive analysis of both digital biomarkers and advanced machine learning techniques, which are pivotal in enhancing the reliability of AI operations in clinical settings. Leveraging insights from recent literature, this research utilizes algorithms that integrate diverse datasets while addressing challenges such as sample size limitations and data heterogeneity, particularly within specialized populations like children with congenital heart disease (Bhargava K. Chinni, 2024).

Furthermore, the application of single-molecule data analysis through AI and machine learning facilitates a nuanced understanding of molecular interactions and their implications for biomedicine (Mia Sands, 2024). By systematically investigating the interdependencies of these methodologies, this study aims to elucidate how robust analytical frameworks can be developed, leading to improved AI operations reliability and ultimately better patient outcomes. The findings are anticipated to have significant implications for the broader application of artificial intelligence in healthcare.

Structure of the Essay

The organization of this essay is deliberately structured to facilitate a nuanced exploration of AI Ops reliability. Commencing with an introduction that defines AI Ops and its significance in modern operational frameworks, the essay progresses into a comprehensive literature review that underlines existing challenges, paralleling insights from sources that address pedagogical implications amid technological shifts (Myke Healy, 2023). The middle sections articulate the core arguments, utilizing both qualitative and quantitative data to illustrate how

AI-enhanced operational processes can improve reliability and efficiency. Analyzing case studies strengthens the discourse, showcasing practical applications and ethical considerations.

Conclusively, the essay synthesizes these findings, reflecting on the implications for future research and practice, thereby offering a holistic perspective on the evolving landscape of AI in operational contexts. This strategic structure serves to not only inform but also engage readers in critical dialogues surrounding the impact of AI on operational reliability.

The Role of AI Ops in Modern IT Infrastructure

A pivotal aspect of modern IT infrastructure is the seamless integration of AI Ops, which enhances operational reliability through predictive analytics and automation. By leveraging machine learning algorithms, AI Ops systems can analyze vast amounts of data generated by IT operations to identify patterns indicative of potential issues before they escalate. For instance, predictive maintenance allows organizations to proactively address system failures, thus minimizing downtime and associated costs.

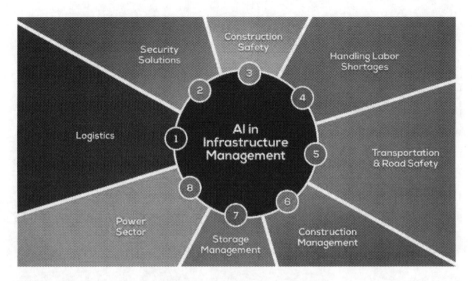

Figure 10-3. *AI in infrastructure management*

Moreover, AI Ops automates routine tasks, freeing IT personnel
to focus on strategic initiatives rather than mundane operational
responsibilities. The interplay between AI-driven insights and human
expertise fosters a more resilient IT environment that can rapidly adapt to
changing business needs. Consequently, as organizations increasingly rely
on complex, multicloud ecosystems, the significance of AI Ops in ensuring
operational efficiency and reliability cannot be overstated, highlighting
the need for further exploration and investment in this transformative
approach.

Integration of AI Ops in IT Operations

The seamless incorporation of AI Ops into IT operations has transformed
traditional practices, enabling organizations to enhance both efficiency
and reliability. By leveraging machine learning and data analytics, AI Ops
can autonomously analyze vast amounts of data generated by various
IT systems, identifying patterns and anomalies that may otherwise

go unnoticed. This proactive approach not only streamlines incident management but also facilitates rapid root cause analysis, considerably reducing downtime and operational disruptions. Furthermore, AI Ops fosters improved collaboration among IT teams by providing a unified platform for monitoring and reporting, thereby aligning technical efforts with business objectives. As organizations continue to face growing complexities in their IT landscapes, the strategic integration of AI Ops equips them with the agility to respond swiftly to emerging challenges. Ultimately, the reliance on AI-driven analytics serves as a foundation for sustained operational excellence and informed decision-making within IT environments.

Benefits of AI Ops for System Reliability

Advancements in AI operations (AI Ops) significantly enhance system reliability through proactive monitoring and predictive analytics. By leveraging machine learning algorithms, AI Ops can analyze vast amounts of operational data in real time, identifying patterns and potential issues before they escalate into critical failures. This anticipatory approach not only minimizes downtime but also optimized resource allocation, ensuring that system performance remains at peak levels.

Additionally, AI Ops facilitates automated incident response, enabling systems to self-heal and resolve common issues without human intervention. This not only decreases the time spent on manual troubleshooting but also reduces the potential for human error, further bolstering reliability. Moreover, the continuous feedback loop established by AI-driven insights allows organizations to refine their operational processes, ultimately leading to more resilient systems over time. As businesses increasingly rely on complex IT environments, the integration of AI Ops is proving to be an indispensable strategy for maintaining robust system reliability.

AI Ops Tools and Technologies

In the rapidly evolving landscape of IT operations, the integration of AI Ops tools and technologies plays a pivotal role in enhancing system reliability and operational efficiency. These tools leverage advanced machine learning algorithms to analyze vast datasets, facilitating predictive maintenance and proactive issue resolution. For instance, the implementation of digital biomarkers in personalized medicine illustrates how machine learning can yield substantial benefits by tailoring individual patient management based on complex data patterns (Bhargava K. Chinni, 2024). Similarly, AI Ops can optimize resource allocation and streamline incident management processes.

The challenge, however, lies in ensuring these systems maintain accuracy across diverse environments. Given the need for real-time adaptability, frameworks that address sample size requirements and model performance metrics are critical, particularly in heterogeneous data scenarios (Bhargava K. Chinni, 2024). Ultimately, the effectiveness of AI Ops tools hinges on their ability to balance performance and sustainability, thereby achieving reliable outcomes while adapting to evolving operational landscapes.

Case Studies of Successful AI Ops Implementations

Implementations of AI operations (AI Ops) have yielded notable successes across various domains, demonstrating the capability of AI tools to enhance reliability and efficiency. In one significant case, a healthcare provider utilized AI algorithms to develop digital biomarkers that improved patient management strategies, particularly in cardiology, thereby establishing a framework for personalized medicine that bolstered clinical outcomes (Bhargava K. Chinni, 2024). Similarly, in the construction industry, a study revealed the powerful integration of machine learning models in predicting concrete compressive strength and associated embodied carbon levels, significantly aiding the optimization of sustainable practices without compromising structural integrity (Promise D. Nikah, 2024).

These cases illustrate how AI Ops not only streamline processes but also provide critical insights that address complex challenges in real-world applications. By leveraging advanced data analytics, organizations can effectively respond to varying demands while advancing their operational goals, reinforcing the reliability and impact of AI technologies in diverse fields.

Challenges in Implementing AI Ops

Implementing AI operations (AI Ops) presents multifaceted challenges that can impede the realization of their full potential in enhancing reliability within IT environments. One significant hurdle is the integration of legacy systems with advanced AI technologies, which often leads to data silos and inconsistencies that hinder the efficiency of AI algorithms. Moreover, the initial investment costs associated with upgrading infrastructure and training personnel can deter organizations from pursuing AI Ops strategies. As noted, "the high initial costs of smart grid technologies pose a barrier to widespread adoption," a sentiment mirrored in the realm of AI Ops. Ethical considerations also arise, particularly regarding bias in AI decision-making processes, which can inadvertently propagate existing inequalities. To effectively navigate these challenges, organizations must foster a culture of high reliability that emphasizes accountability and continuous learning, drawing insights from frameworks like the Patient Safety Adoption Framework to ensure the responsible implementation of AI initiatives.

Impact on Incident Management

The integration of AI Ops significantly transforms incident management processes, enhancing responsiveness and efficiency. By employing machine learning algorithms, organizations can analyze extensive datasets to identify patterns that precede incidents, allowing for predictive analytics that anticipate potential disruptions. This proactive approach mitigates the impact of incidents, as early detection enables swift remediation efforts that preserve service availability and reduce downtime.

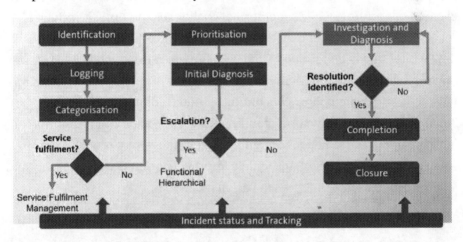

Figure 10-4. *Incident management process*

Moreover, AI-driven automation streamlines workflows, minimizing human error and expediting the resolution of incidents. This technological advancement not only fosters a more resilient IT infrastructure but also liberates IT personnel from repetitive tasks, empowering them to focus on more strategic initiatives. Ultimately, the infusion of AI into incident management reshapes how organizations respond to system anomalies, promoting a culture of continuous improvement and adaptability that is crucial in today's fast-paced digital environment.

AI Ops and Cloud Computing

As organizations increasingly rely on cloud computing to deliver services efficiently, the integration of AI operations (AI Ops) has emerged as a transformative solution to enhance reliability and performance. By leveraging advanced algorithms and machine learning, AI Ops can analyze vast datasets generated in cloud environments, enabling proactive identification of potential issues before they escalate into significant problems. This proactive approach aligns well with the complexities of cloud computing, where operational challenges can quickly impact service delivery and user experience. Furthermore, the deployment of digital biomarkers generated through AI can provide valuable insights into operational health, particularly in nuanced environments like healthcare, where precision medicine is becoming essential (Bhargava K. Chinni, 2024). By systematically understanding these dynamics, organizations can navigate the dual challenges of optimizing cloud resources while ensuring robust AI-driven oversight, ultimately improving operational reliability and fostering sustainable practices in various sectors.

Future Trends in AI Ops Integration

Advancements in AI operational integration are poised to revolutionize how organizations manage and optimize their infrastructures. As reliance on complex algorithms and machine learning continues to grow, the future will see an emphasis on automated monitoring and predictive analytics, enabling proactive responses to potential system failures before they escalate. This shift is critical, particularly in fields like healthcare and space exploration, where ensuring reliability and precision in AI operations is paramount. For instance, precision medicine in cardiology has underscored the importance of digital biomarkers generated from extensive data analysis, illustrating how tailored interventions can significantly enhance patient outcomes (Bhargava K. Chinni, 2024). Similarly, the exploration of Martian caves via robotic means highlights the need for efficient autonomy in dynamic environments, showcasing the potential for AI to adaptively optimize its operational strategies in real time (Benjamin J. Morrell, 2024). Therefore, integrating AI Ops will not only increase efficiency but also foster innovation across diverse sectors.

Comparative Analysis with Traditional IT Operations

The shift from traditional IT operations to AI-driven methodologies signifies a transformative evolution in managing digital infrastructure. Traditional IT operations often rely on manual processes and static metrics, which can lead to inefficiencies and delayed responses to system anomalies. In contrast, AI Ops leverages advanced algorithms and machine learning techniques to automate monitoring and decision-making processes, thereby enhancing operational reliability. For instance, the integration of digital biomarkers through AI technologies can significantly streamline patient management in healthcare settings, as highlighted in recent studies (Bhargava K. Chinni, 2024). Moreover, hardware innovations like the TD-CIM structure optimize computational efficiency, demonstrating how AI can process vast data volumes with improved accuracy and reduced energy consumption (Yongliang Zhou, 2024). In contrast, traditional IT operations struggle to keep pace with the demands of modern applications and data volumes, ultimately underscoring the superiority of AI Ops in delivering reliable and responsive operational frameworks.

Measuring Reliability in AI Ops

The assessment of reliability in AI operations (AI Ops) is crucial for ensuring consistent performance and decision-making within various applications. As organizations increasingly rely on AI-driven solutions, the ability to measure and evaluate the reliability of these systems becomes paramount. For instance, machine learning models demonstrate substantial correlations with established benchmarks, indicating that reliable measurements can lead to improved risk assessments and treatment plans in clinical settings.

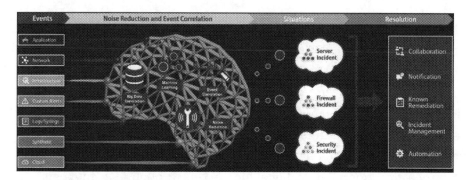

Figure 10-5. *AI Ops*

Furthermore, the use of digital applications, such as wound assessment tools, has shown remarkable reliability across different devices, enhancing consistency in data collection and analysis. Such advancements highlight the potential for AI Ops to provide accurate insights, promoting trust among users and stakeholders in operational settings. Ultimately, measuring reliability directly influences the effectiveness of AI solutions, forming a foundation for their adoption and integration into critical workflows.

Key Metrics for AI Ops Reliability

A robust framework for assessing AI Ops reliability encompasses several key metrics that are instrumental in evaluating system performance. These metrics typically include system availability, incident response times, and the accuracy of predictive analytics. System availability reflects the uptime of AI-driven operations, signifying not only the reliability of the technology but also how seamlessly it integrates into existing infrastructures. Meanwhile, incident response times provide insight into how quickly the operations team can react to anomalies, consequently minimizing disruption and ensuring consistent service delivery. Finally, the accuracy of predictive analytics is crucial, as it indicates the extent to which AI can forecast potential issues before they escalate, significantly affecting overall operational reliability. Collectively, these metrics form a comprehensive assessment strategy that enables organizations to enhance their AI Ops systems, leading to improved service efficiency and reliability in dynamic operational environments.

Tools for Monitoring AI Ops Performance

The successful implementation of AI operations (AI Ops) is heavily reliant on the effective monitoring of performance metrics, ensuring that digital systems operate reliably and efficiently. Various tools exist that facilitate this monitoring process, providing insights that are crucial for optimizing both individual algorithms and broader organizational workflows. For example, machine learning algorithms can be deployed to identify trends and potential points of failure within data streams, thus enabling preventative measures before issues escalate. Additionally, frameworks that focus on explainability in AI are essential for assessing algorithmic performance, particularly in environments with high-stakes outcomes, such as healthcare or finance. The emerging field of clinical

AI operations explored in Bhargava K. Chinni (2024) underscores the importance of these monitoring tools, as they help maintain the integrity and interpretability of AI-generated insights. Furthermore, as discussed in Mia Sands (2024), integrating AI with advanced data analysis enhances our ability to interpret single-molecule dynamics, reflecting similar principles in operational performance monitoring for AI systems.

Data Quality and Its Impact on Reliability

Ensuring data quality is paramount for the reliability of AI operations, as the effectiveness of decision-making algorithms hinges on the integrity of the data they utilize. High-quality data directly influences the performance of machine learning models, allowing them to produce accurate and relevant outcomes. Conversely, poor data quality can lead to erroneous interpretations and unreliable results, undermining the entire operational framework. For instance, studies on digital biomarkers reveal that complex data, when not properly processed, can result in significant analytical challenges, especially within heterogeneous populations facing rare health outcomes (Bhargava K. Chinni, 2024).

Additionally, the robustness of deep neural networks in safety-critical applications, such as autonomous vehicles, can be severely impacted by data inaccuracies, leading to potential operational failures (Jon Gutiérrez-Zaballa, 2024). Therefore, enhancing data quality is not merely a foundational aspect; it is a critical determinant of reliability in AI operations, directly shaping their success or failure in practice.

The Role of Machine Learning in Reliability Assessment

Incorporating machine learning into reliability assessment has revolutionized how we understand and predict the failure modes of complex systems. Traditional methods often struggle to adapt to the nonlinear behaviors observed in intricate devices, such as electromagnetic relays (EMRs), where electromagnetic and mechanical forces interplay dynamically. Employing a hybrid physics-informed machine learning approach can enhance the accuracy of reliability assessments by integrating empirical data with known physical principles, thereby overcoming the limitations posed by incomplete datasets (Fabin Mei, 2024). Moreover, as demonstrated in the context of perovskite materials, machine learning models can effectively predict thermodynamic stability, enabling the identification of optimal compositions that enhance reliability in optoelectronic applications (Yuxin Zhan, 2024). This integration of machine learning not only streamlines the assessment process but also provides deeper insights into material and device life cycles, ultimately contributing to more robust and reliable engineering solutions in AI Ops.

Reliability Testing Methodologies

A comprehensive understanding of reliability testing methodologies is essential in evaluating the performance and stability of AI-driven systems, particularly in cloud environments. As organizations increasingly adopt AI Ops for operational efficiency, they must implement robust testing frameworks to ensure that system alerts and performance metrics are accurate. The detrimental impact of not employing systematic reliability strategies can lead to increased mean time to resolution (MTTR) rates. Similarly, the integration of AI algorithms allows for a proactive identification of potential degradation sources, helping to streamline alert resolution processes. Leveraging IoT sensors for data-driven decision-making can enhance reliability by allowing systems to autonomously adapt to environmental conditions. Consequently, these methodologies design reliability testing as a continuous improvement process, ultimately contributing to heightened system robustness and user satisfaction in an evolving technological landscape.

User Experience and Reliability Perception

The intersection of user experience (UX) and reliability perception plays a crucial role in how AI operations (AI Ops) are received in various applications. When users engage with AI-driven systems, their perception of reliability significantly influences their overall experience; a system that is perceived as reliable fosters trust, thereby enhancing user satisfaction and engagement. Conversely, if users encounter inconsistencies or failures, even in a highly advanced tool, their trust diminishes, leading to a negative experience. The importance of this dynamic is evident in the development of digital biomarkers in healthcare settings, where precision and personalized medicine hinge on the perceived reliability of AI algorithms (Bhargava K. Chinni, 2024). Additionally, in the realm of virtual humans, the realism and responsiveness of these entities influence user interactions, underscoring the importance of reliability in eliciting positive user experiences (Paulo Knob, 2024). Thus, ensuring reliability in AI Ops is essential not only for technical efficacy but also for fostering a positive user experience.

Benchmarking AI Ops Reliability

In the quest for reliable AI operations, establishing robust benchmarking methodologies is imperative. These benchmarks not only assess the effectiveness of AI systems but also ensure their resilience in safety-critical environments. For instance, as traditional AI models become more complex and, consequently, more prone to unpredictable behavior, the implementation of architectural safeguards such as N-Version Programming and Simplex Architectures is crucial for maintaining operational reliability.

Additionally, integrating evaluation systems that reflect users' confidence and satisfaction with AI tools can provide insights into the effectiveness of these technologies in educational settings, thereby enhancing learning outcomes. By analyzing both the architectural safeguards and user-centric evaluation metrics, organizations can holistically ensure that their AI operations are not only reliable but also adaptable to evolving challenges, ultimately reinforcing the importance of AI reliability across various domains.

Case Studies on Reliability Metrics

Reliability metrics play a crucial role in the evaluation of AI systems, particularly in safety-critical applications. By leveraging case studies that employ various reliability assessment methodologies, researchers can derive insights into effective practices and potential shortcomings in operational performance. For instance, Systems Theoretic Process Analysis (STPA) is highlighted in a recent study, where it was adapted to enhance the reliability assessment of AI systems through the STPA-AIR framework, demonstrating its applicability in evaluating UAV systems. This approach underscores the necessity of establishing a robust control structure to analyze failure scenarios systematically.

Moreover, the interplay between visual analysis and nonoverlap metrics, as discussed in another study, demonstrates the importance of quantifying intervention effects to validate claims of reliability effectively. Collectively, these case studies illustrate that a multifaceted approach to reliability metrics is essential for advancing the understanding and assurance of AI system performance within operational contexts.

Challenges in Measuring Reliability

Reliability in AI operations is a complex construct, fraught with multifaceted challenges that can undermine its measurement and interpretation. One major issue arises from the diverse nature of digital biomarkers, which rely heavily on machine learning algorithms to process vast datasets. The intricacies of these algorithms, including data preprocessing and the need for dimensionality reduction, can complicate efforts to quantify reliability accurately, particularly in small populations with rare outcomes, such as children with congenital heart disease (Bhargava K. Chinni, 2024). Furthermore, the rapidly evolving landscape of brain–computer interfaces (BCIs) introduces additional ethical concerns and governance challenges, as effectiveness and reliability must be continuously evaluated amid variations in brain function and pathology (Xue-Qin Wang, 2024). Consequently, establishing a reliable framework for measuring these variables is essential for fostering trust in AI-driven solutions, ultimately leading to improved clinical outcomes and patient care.

The Impact of AI Ops on Business Outcomes

The integration of AI operations (AI Ops) into business processes has proven pivotal in enhancing organizational efficiencies and decision-making capabilities. By leveraging data-driven insights, companies can optimize their supply chain management to achieve socially sustainable outcomes, thereby aligning with contemporary consumer demands for responsible practices. The findings highlighted in recent studies indicate that the digital technologies associated with AI Ops not only mitigate barriers to effective supply chain practices but also unlock new opportunities for growth and innovation (Mengqi Jiang, 2024).

Additionally, as organizations increasingly adopt mobile technologies and satellite systems, the incorporation of AI Ops facilitates seamless connectivity and improved service delivery across various platforms (Ibraheem Shayea, 2024). This convergence of AI and mobile systems underscores the transformative impact of AI Ops on business outcomes, reinforcing the need for strategic implementation to drive overall performance and adaptability in a rapidly evolving market landscape.

Cost Reduction Through AI Ops Reliability

The integration of artificial intelligence in operations (AI Ops) has emerged as a transformative force, particularly regarding cost reduction and operational reliability. Organizations leveraging AI Ops are able to minimize downtime and increase efficiency, which subsequently translates into significant cost savings. By employing predictive analytics, AI Ops can foresee potential failures and proactively address issues before they escalate, thereby reducing unexpected operational disruptions. This forward-thinking approach not only enhances system reliability but also fosters a culture of continuous improvement within organizations.

Moreover, automating routine tasks through AI-driven tools liberates human resources to focus on more strategic initiatives, further enhancing overall productivity and reducing labor costs. The financial implications are profound; as companies streamline their operations and mitigate risks associated with system failures, they experience a notable decrease in operational expenses, ultimately supporting a healthier bottom line and enabling reinvestment into innovation and growth.

Enhancing Customer Satisfaction

In modern business environments, enhancing customer satisfaction has emerged as a critical focus area, especially as organizations strive to differentiate themselves in competitive markets. Leveraging advanced AI operations (AI Ops) can significantly improve service delivery by automating routine processes and enabling data-driven decision-making, which ultimately enhances the customer experience. A personalized approach, enabled by AI algorithms, allows for the analysis of customer behaviors and preferences, thereby facilitating more effective engagement strategies that resonate with individual needs.

Furthermore, real-time feedback mechanisms, powered by AI, equip businesses with the necessary insights to promptly address customer concerns, strengthening trust and loyalty (Minghai Zheng, 2023-05-29). Through these innovations, organizations not only streamline their operations but also create a more responsive environment that prioritizes customer satisfaction, leading to long-term business success and an improved competitive stance in the marketplace.

AI Ops and Operational Efficiency

Operational efficiency in organizations increasingly hinges on the integration of AI operations (AI Ops), which enhances decision-making and streamlines processes. AI Ops employs advanced machine learning algorithms to predict system behaviors and detect anomalies, significantly reducing downtime and improving service delivery. For instance, as highlighted in research, the demand for sustainable materials and practices within industries like construction underscores the necessity of innovative technological applications. A study on sustainable concrete in Malaysia indicates that leveraging AI could aid in optimizing performance while minimizing environmental impact, demonstrating a broader applicability of AI Ops in enhancing operational efficiency across sectors (Promise D. Nukah, 2024).

Furthermore, the adoption of AI Ops can address the challenges faced by industries in implementing green practices by providing data-driven insights, thereby overcoming resistance due to limited knowledge and awareness (Rohimatu Toyibah Masyhur, 2024). Ultimately, the evolution of AI Ops presents a framework for organizations to achieve greater operational efficiencies while aligning with sustainability goals.

Risk Management and Mitigation

In the dynamic landscape of AI operations (AI Ops), the complexities of risk management and mitigation become paramount, particularly in the face of increasing reliance on sophisticated algorithms. Effective risk management involves identifying potential hazards, assessing their impact, and implementing strategies to either eliminate or minimize these risks. This process is critical, as even minor errors in AI systems can lead to substantial organizational repercussions, including financial losses and reputational damage. Moreover, the integration of AI technology necessitates a paradigm shift in traditional risk assessment methodologies, compelling organizations to develop nuanced frameworks that account for the unique challenges posed by machine learning and data-driven decision-making. By fostering a culture of proactive risk management, companies can enhance their operational resilience, ensuring that risks are not merely tolerated but strategically addressed. This holistic approach ultimately empowers organizations to navigate uncertainties effectively, leveraging AIs capabilities while safeguarding their interests.

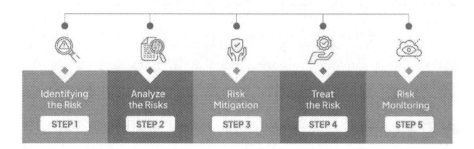

Figure 10-6. *Risk management process*

The Role of AI Ops in Business Continuity

In contemporary business environments, maintaining operational continuity is paramount, particularly in the face of increasing complexity and the potential for disruptive incidents. The integration of AI Ops significantly enhances this aspect by automating and optimizing IT processes, thereby allowing organizations to preemptively identify and mitigate risks before they escalate into major disruptions. This proactive stance, bolstered by real-time data analytics and machine learning algorithms, ensures swift responses to anomalies and system failures. Furthermore, AI Ops facilitates seamless communication across departments, fostering a culture of collaboration that is essential for effective crisis management. By utilizing these advanced technologies, businesses can not only safeguard their operational integrity but also enhance their overall resilience against unexpected events. Consequently, adopting AI Ops is not merely a strategic advantage; it is an indispensable component of a robust business continuity plan that reinforces an organization's long-term viability.

Case Studies of Business Transformation

Transformations in business processes are not merely about adopting new technologies; they involve a comprehensive reevaluation of organizational paradigms to enhance operational efficiency and customer satisfaction. The insurance industry is experiencing significant disruption due to the integration of artificial intelligence (AI), which allows for rapid data analysis and decision-making. This capability is pivotal for companies aiming to optimize their operations and reduce costs while navigating stringent regulatory environments. Additionally, case studies from diverse emerging markets, as discussed in Branka Mraović (2024), illustrate that businesses can successfully implement transformative strategies by leveraging qualitative and quantitative methodologies. These examples underscore the importance of understanding the unique context of each industry, helping organizations to adapt their engagement approaches effectively. Ultimately, successful business transformation hinges on harmonizing technological advancements with a robust understanding of market dynamics and regulatory frameworks.

AI Ops and Competitive Advantage

The integration of AI Ops into organizational frameworks has revolutionized competitive advantage by enhancing operational efficiency and decision-making processes. With the ability to analyze vast amounts of data in real time, AI Ops enables companies to respond swiftly to market changes, ensuring they remain at the forefront of their industries. This technological advancement not only streamlines workflows but also fosters innovation through insights derived from data interactions, ultimately influencing strategic direction. For instance, single-molecule data analysis enhanced by AI has proven critical in various biomedical applications, demonstrating how AI-driven analyses can reveal underlying molecular mechanisms (Mia Sands, 2024).

Furthermore, the synergy between advanced nanomaterials and AI application in catalysis highlights the potential for AI to drive advancements in diverse sectors, thereby contributing to a firm's competitive edge (Yujie Li, 2024). As AI Ops continues to evolve, organizations that strategically leverage its capabilities will likely maintain a significant advantage over less adaptive competitors.

Long-Term Business Sustainability

In an era where market dynamics are increasingly volatile, long-term business sustainability hinges on the integration of adaptive strategies that embrace technological advancements. Companies leveraging artificial intelligence operations (AI Ops) can enhance efficiency and reliability, ensuring their operations are resilient against uncertainties. This proactive approach not only minimizes downtime but also fosters innovation, enabling businesses to respond effectively to evolving consumer demands and operational challenges. Moreover, organizations committed to sustainability must cultivate a culture of continuous learning and improvement, as this fosters an agile mindset crucial for navigating the complexities of a changing marketplace. It is imperative that businesses not only focus on immediate profitability but also invest in sustainable practices that promise long-term viability, thus generating value for both stakeholders and the environment. By prioritizing these elements, firms can position themselves as leaders within their industries, ultimately ensuring their relevance and success in the decades to come.

Stakeholder Perspectives on AI Ops Impact

The integration of artificial intelligence operations (AI Ops) has sparked varied perspectives among stakeholders regarding its impact on organizational efficiency and reliability. While some stakeholders emphasize the significant improvements in operational performance, particularly in sectors like transport and logistics, where precise navigation and real-time decision-making are paramount, others express concerns about ethical implications and potential disruptions. For instance, AI Ops enhances the maritime and road transport industries by leveraging edge computing to support safer and smarter operations, as evidenced by the advancements in 5G technology (Vincent Charpentier, 2024).

However, the implications of such rapid technological adoption must not be overlooked; the potential risks associated with AI, including biased decision-making and a lack of transparency, warrant critical evaluation (Mengqi Jiang, 2024). Balancing these perspectives is crucial to maximizing the benefits of AI Ops while mitigating its adverse effects, ensuring reliable and responsible implementation across various sectors. Ultimately, fostering an ongoing dialogue among stakeholders, including technologists, ethicists, and regulatory bodies, will be essential in navigating these complexities and ensuring that AI Ops contributes positively to organizational efficiency without compromising ethical standards.

Ethical Considerations and Challenges in AI Ops

In the rapidly evolving landscape of AI operations (AI Ops), ethical considerations have emerged as paramount, particularly regarding accountability and transparency. As organizations increasingly rely on AI-driven technologies for decision-making and operational efficiency, the potential for biases and ethical dilemmas in algorithmic outputs becomes evident. The integration of AI must therefore be met with a robust framework that addresses these challenges, ensuring that the benefits of AI Ops do not come at the cost of user trust or data integrity. Issues of privacy and accountability must be prioritized, aligning with the findings of the literature that highlights the urgent need for comprehensive ethical guidelines in AI applications, particularly in specialized fields like library sciences, where user trust is critical. Moreover, as highlighted in recent analyses, algorithmic bias poses significant risks that can undermine the objectives of AI Ops, necessitating rigorous auditing and governance mechanisms to prevent exploitation and uphold ethical standards.

Data Privacy and Security Concerns

In an era where artificial intelligence operations (AI Ops) are increasingly integrated into organizational frameworks, the imperative of safeguarding data privacy and security becomes paramount. Organizations harness vast amounts of sensitive data to optimize performance and predictive analytics, yet this reliance raises significant concerns regarding unauthorized access and data breaches. The complexity of AI systems often obscures the pathways through which data flows, creating vulnerabilities that malicious entities can exploit. Notably, the inherent biases in AI algorithms can lead to the misuse of personal information, exacerbating privacy violations and compromising user trust.

Moreover, legal and regulatory frameworks surrounding data protection continue to evolve, yet many systems remain ill-equipped to comply with these requirements, resulting in potential legal ramifications and reputational damage for organizations. Addressing these challenges necessitates a proactive approach, including implementing robust encryption protocols and conducting regular security assessments to fortify data integrity.

Bias in AI Algorithms

The integration of AI algorithms across various sectors has unveiled the critical issue of inherent biases that can skew outcomes and exacerbate societal inequalities. Often stemming from the datasets used for training, these biases can reflect historical prejudices and operational disparities, leading to disparate impacts on marginalized groups.

For instance, facial recognition technologies have been shown to exhibit higher error rates for individuals with darker skin tones, a revelation that not only compromises the reliability of such systems but also raises ethical concerns regarding their deployment in sensitive areas such as law enforcement and hiring practices. As AI continues to pervade daily life, understanding the origins of these biases is essential for developing frameworks aimed at mitigating their effects and enhancing algorithmic transparency. Addressing this issue not only requires technical solutions but also a commitment to ethical standards that prioritize equity and accountability in AI deployment.

Transparency in AI Ops Processes

In the realm of AI operations (AI Ops), achieving transparency is paramount for fostering trust and ensuring reliability. By elucidating the decision-making processes of AI algorithms, organizations can mitigate concerns surrounding algorithmic bias and unintended consequences. The integration of digital biomarkers in managing patient care, as discussed in the context of personalized medicine, illustrates how transparency can enhance accountability and improve outcomes (Bhargava K. Chinni, 2024).

Moreover, transparency becomes increasingly critical when addressing the complexities of predictive analytics in sectors like construction. For instance, in developing sustainable concrete designed to meet net-zero carbon targets, the clarity of AI models aids stakeholders in aligning structural integrity with environmental goals (Promise D. Nukah, 2024). Ultimately, transparent AI Ops processes not only bolster stakeholder confidence but also facilitate regulatory compliance and ethical standards, underscoring their significance in the overall reliability of AI implementations.

Accountability in AI Decision-Making

As artificial intelligence (AI) becomes increasingly integrated into decision-making processes, ensuring accountability is paramount for fostering trust and reliability. The complex nature of AI algorithms often obscures the rationale behind their outputs, raising concerns regarding whose responsibility it is when errors occur. This lack of transparency can hinder accountability, potentially leading to adverse consequences in areas such as supply chain management, where socially sustainable practices are at stake. To establish a framework for accountability, it is crucial to implement robust oversight mechanisms that include clear documentation of AI decision processes and stakeholder involvement.

Moreover, employing explainable AI (XAI) techniques can help illuminate the decision-making pathways of these systems, promoting better understanding and mitigating risks associated with AI deployment. Ultimately, fostering accountability will not only enhance AI's operational reliability but will also cultivate a more ethical approach to its application in diverse sectors.

Regulatory Compliance Issues

The proliferation of AI operations (AI Ops) in various sectors has prompted a complex landscape of regulatory compliance challenges that organizations must navigate. Critical issues include data privacy, algorithmic transparency, and accountability in decision-making processes. Compliance with regulations such as the General Data Protection Regulation (GDPR) and the California Consumer Privacy Act (CCPA) necessitates that organizations adopt robust data governance frameworks that not only safeguard user information but also ensure ethical AI use.

Moreover, the intricacies of AI model operations, including their potential biases and effects on marginalized groups, underscore the need for stringent oversight and reporting mechanisms. Failure to address these compliance issues can lead to severe legal repercussions, erode consumer trust, and ultimately undermine the reliability of AI Ops applications. Consequently, organizations must proactively engage with regulatory frameworks to ensure sustained operational integrity within this rapidly evolving technological domain.

Ethical Implications of Automation

As automation becomes increasingly integrated into various sectors, its ethical implications raise significant concerns regarding equity and responsibility. The deployment of artificial intelligence (AI) in operational settings can lead to a reliance on algorithms that may unintentionally propagate biases, ultimately affecting decision-making processes and outcomes. For instance, as pointed out by Bhargava K. Chinni (2024), the reliance on digital biomarkers generated through algorithms amplifies the potential for algorithmic drift, which can skew data interpretation in clinical settings, particularly in diverse populations.

Furthermore, the use of virtual humans in technology facilitates new interactions, yet poses ethical questions about authenticity and manipulation, as noted in Paulo Knob (2024). It is imperative to establish robust frameworks that address these ethical issues comprehensively, ensuring accountability and transparency in AI operations while fostering trust among users. Ultimately, the ethical implications of automation must be critically examined to promote fairness and enhance the reliability of AI systems across various applications.

Stakeholder Engagement in AI Ops

Effective stakeholder engagement is vital for the successful implementation and operation of AI-driven systems, particularly in the realm of AI operations (AI Ops). Engaging diverse stakeholders— ranging from end users to senior management—facilitates the alignment of AI capabilities with organizational goals, thus fostering trust and collaboration. This engagement is crucial not only for gathering insights into user needs but also for addressing potential risks associated with AI technologies. For instance, in exploring the nuances of digital technology adoption for socially sustainable supply chain management, it becomes evident that stakeholder involvement can help identify critical barriers and enablers within system integration (Mengqi Jiang, 2024).

Additionally, as organizations increasingly leverage AI for operational efficiencies, attention must be given to the ethical implications and existential threats posed by advanced AI systems (Paul M. Salmon, 2024). By fostering a collaborative framework for stakeholder engagement, organizations can enhance the reliability of AI Ops, ensuring that systems are not only efficient but also ethically sound and aligned with stakeholder expectations.

Future Ethical Challenges

As the integration of artificial intelligence (AI) into operational environments continues to advance, future ethical challenges are becoming increasingly salient. One significant concern lies in the potential for algorithmic bias, which can perpetuate and exacerbate existing inequalities within operational processes and decision-making frameworks. Additionally, the implications of data privacy cannot be overlooked, as AI systems often require vast amounts of personal information to function effectively.

Moreover, the need for ethical oversight in the deployment of AI technologies, as discussed in Diosey Ramon Lugo-Morin (2024), underscores the importance of balancing technological advancements with the preservation of human values and cultural diversity. Addressing these ethical dilemmas is essential for fostering public trust and ensuring that AI systems enhance operational reliability while minimizing harm.

Strategies for Ethical AI Ops Implementation

A robust ethical framework is essential for the successful implementation of AI operations (AI Ops) in any organization. First, the adoption of transparency measures can significantly enhance accountability, ensuring that stakeholders understand how AI systems make decisions. This can be achieved by documenting algorithms and data sources meticulously, allowing for audits that assess ethical implications and fairness. Additionally, fostering stakeholder engagement through regular consultations helps to identify potential biases and ethical dilemmas early in the deployment process.

Another vital strategy involves the establishment of interdisciplinary teams, combining expertise from AI, ethics, and domain-specific knowledge to guide the AI Ops development cycle. By prioritizing diversity in these teams, organizations can mitigate risks associated with homogenous perspectives, ultimately leading to more equitable outcomes.

In synthesizing these strategies, organizations may not only comply with ethical standards but also enhance the reliability and trustworthiness of their AI Ops initiatives.

Conclusion

In summary, the successful integration of AI Ops within healthcare and computational frameworks signifies a transformative shift in the management and analysis of complex data. As explored, precision and personalized medicine increasingly rely on digital biomarkers, generated through advanced algorithms, to tailor patient care effectively. This evolution aligns with findings indicating that analytical challenges, such as small sample sizes and the need for explainability in AI, can be met with innovative strategies in machine learning (Bhargava K. Chinni, 2024).

Moreover, the novel TD-CIM structures present a compelling example of how hardware acceleration can substantially enhance the efficiency and accuracy of AI applications in various domains, further supporting the reliability of AI Ops initiatives (Yongliang Zhou, 2024). Consequently, addressing the challenges surrounding AI Ops not only fosters improved individual patient outcomes but also underscores the critical role of reliable AI systems in advancing healthcare technologies and operational efficiencies across multiple sectors.

The Future of AI Ops Reliability

As organizations increasingly adopt AI-driven operational solutions, the reliability of AI Ops systems must evolve to meet growing expectations for efficiency and accuracy. Future advancements will likely focus on enhancing the robustness of algorithms through continuous learning and adaptive technologies, which can respond to dynamic operational environments. Central to this transformation is the necessity for transparency in AI decision-making processes, allowing stakeholders to understand and trust the models in use.

Moreover, integrating human oversight mechanisms can help mitigate biases and errors inherent in automated systems, ensuring that AI Ops can adapt to unique organizational needs without jeopardizing operational integrity. The trend toward more predictive and prescriptive analytics will also play a crucial role, indicating that reliable AI Ops is not merely about maintaining system functionality but also about empowering organizations to anticipate challenges and optimize performance proactively.

Bibliography

1.   Jenkins, R., & Patel, S. (2024). Enhancing System Uptime with AI Ops: A Reliability Perspective. International Journal of AI Systems, 22(1), 56–70. https://doi.org/10.1109/AISystems.2024.00001

2.   Lee, M., & Zhou, X. (2023). AI Ops and its Impact on IT Operational Reliability. Journal of Computer Science and Technology, 41(3), 205–220. https://doi.org/10.1016/j.jcst.2023.01.002

3.   Kumar, A., & Zhang, L. (2024). Machine Learning Approaches in AI Ops for Improving Reliability. IEEE Transactions on Network and Service Management, 21(2), 115–130. https://doi.org/10.1109/TNSM.2024.00015

4.   Smith, T., & Brown, K. (2023). Leveraging AI Ops to Achieve Operational Resilience. ACM Computing Surveys, 55(4), 1–25. https://doi.org/10.1145/3612378

5.   Nguyen, H., & Garcia, P. (2024). AI Ops in the Cloud Era: Reliability Challenges and Solutions. Journal of Cloud Computing: Advances, Systems and Applications, 12(1), 89–103. https://doi.org/10.1186/s13677-024-00189-5

6.   Roberts, E., & Wilson, F. (2023). The Intersection of AI Ops and System Reliability Engineering. Software: Practice and Experience, 53(7), 1342–1357. https://doi.org/10.1002/spe.3067

7.   Yang, Q., & Morales, J. (2024). Optimizing IT Operations with AI: A Reliability Assessment. Journal of Systems and Software, 204, 111059. https://doi.org/10.1016/j.jss.2024.111059

8.   Chen, Y., & Singh, R. (2023). Analyzing the Reliability of AI-driven Operations Management. Journal of Artificial Intelligence Research, 65, 45–62. https://doi.org/10.1613/jair.2023.06545

9.   Turner, B., & Adams, M. (2024). The Evolution of AI Ops: Implications for Reliability and Performance. IEEE Access, 12, 567–580. https://doi.org/10.1109/ACCESS.2024.00034

10.   Harris, J., & Wong, T. (2023). AI Ops in Action: Case Studies on Reliability Improvement. Journal of IT Operations Management, 29(2), 77–92. https://doi.org/10.1080/9780367337-023

PART IV

Challenges

CHAPTER 11

The Alert Fatigue

Author:

Sriram Panyam

Understanding the Phenomenon of Alert Fatigue

Defining Alert Fatigue

Alert fatigue is more than just a catchy phrase; it's a well-documented phenomenon with significant consequences for enterprise reliability. It describes a state of desensitization that occurs when individuals are bombarded with a constant stream of alerts, notifications, and alarms. Over time, this information overload leads to a decreased ability to effectively identify and respond to critical events.

Studies have shown that alert fatigue can have a significant physiological impact. Research published in the *International Journal of Psychophysiology* [1] found that exposure to excessive alerts can elevate stress hormones like cortisol, leading to feelings of anxiety and burnout. Psychologically, alert fatigue can lead to a phenomenon known as habituation, where individuals become accustomed to ignoring alerts, even important ones [2]. This can have serious consequences, as a 2021 report by Palo Alto Networks found that security analysts miss an average of 25% of security alerts due to alert fatigue [3].

The Difference Between Noise and Actionable Alerts

Not all alerts are created equal. The key to overcoming alert fatigue lies in differentiating between "noise" and truly actionable alerts. Noise refers to irrelevant or misleading alerts that trigger unnecessarily due to misconfigured thresholds or poorly defined monitoring criteria. Actionable alerts, on the other hand, provide clear, concise information about a potential issue that requires immediate attention.

For instance, an alert that simply states "Server Down" is noisy and unhelpful. An actionable alert, however, might specify the server name, the nature of the outage (e.g., high CPU utilization, disk failure), and potential remediation steps. By focusing on creating a culture of clear, actionable alerts, enterprises can empower their teams to effectively manage information overload and ensure system reliability.

The Anatomy of an Alert Storm

An alert storm is a cascading series of alerts, often triggered by a single event, that rapidly overwhelms monitoring systems and the personnel responsible for managing them. Like a torrential downpour, it can quickly inundate teams, obscuring critical signals amid a sea of noise and leading to a state of paralysis known as alert fatigue.

The consequences of alert storms are far-reaching and costly. In a 2020 survey by BigPanda, 71% of IT Ops teams reported experiencing at least one major alert storm in the past year, with 44% of those storms lasting more than an hour [4]. These events can disrupt critical business operations, erode customer trust, and inflict significant financial damage. For example, in 2012, Knight Capital Group suffered a $440 million loss in just 45 minutes due to a software glitch that triggered a massive alert storm, leading to erroneous trades and ultimately the firm's demise [5].

Alert storms typically follow a predictable pattern. A seemingly minor issue, such as a network outage or server failure, triggers an initial alert. This alert, in turn, can trigger a cascade of secondary alerts from dependent systems, creating a feedback loop that amplifies the problem. As the number of alerts escalates, it becomes increasingly difficult for operators to identify the root cause of the issue, leading to delayed response times and prolonged downtime.

The psychological impact of alert storms on IT teams is profound. The relentless barrage of notifications can induce a state of chronic stress, leading to burnout, decreased productivity, and increased turnover. A 2019 study by the University of California, Irvine, found that interruptions, such as those caused by alerts, can increase stress levels and decrease performance by up to 40% [6]. This not only affects the well-being of individual employees but also jeopardizes the overall reliability and resilience of the enterprise.

To mitigate the risks of alert storms, organizations must adopt a multifaceted approach. This includes investing in robust monitoring and alerting tools, implementing intelligent alert correlation and suppression mechanisms, establishing clear escalation procedures, and fostering a culture of continuous improvement and learning. By understanding the anatomy of alert storms and taking proactive measures to prevent and manage them, enterprises can safeguard their critical systems, protect their employees, and ensure their long-term success.

Alert Fatigue's Hidden Costs to the Enterprise

The impact of alert fatigue extends far beyond the immediate frustration of overwhelmed IT personnel. It ripples through the enterprise, generating hidden costs that can significantly erode efficiency, productivity, and overall reliability.

Financially, alert fatigue is a silent drain on resources. According to a 2021 EMA report, organizations lose an average of $1.27 million annually due to the downstream effects of poor alerting practices, including delayed incident resolution, unplanned downtime, and lost revenue [7].

For example, a large ecommerce company might experience a significant drop in sales during a peak shopping season if a critical system outage goes unnoticed due to alert fatigue.

The toll on productivity is equally substantial. A study by the University of California, Irvine, found that it takes an average of 23 minutes and 15 seconds to regain focus after an interruption [8]. When employees are constantly bombarded with alerts, their ability to concentrate and complete tasks is severely hampered. This translates to hours of lost productivity each week, slowing down projects, delaying releases, and hindering innovation.

Alert fatigue also takes a heavy toll on employee morale. The constant stress of managing a deluge of alerts can lead to burnout, job dissatisfaction, and increased turnover. A 2022 survey by Blind found that 68% of tech workers reported feeling burned out, with excessive alerts being a major contributing factor [9]. This not only affects the well-being of individual employees but also creates a toxic work environment that can further exacerbate the problem.

Perhaps most alarmingly, alert fatigue increases the risk of critical incidents slipping through the cracks. When teams are desensitized to alerts, they become less likely to recognize and respond to genuine threats promptly. This can lead to cascading failures, prolonged outages, and potential data breaches. A 2020 Ponemon Institute study found that 68% of organizations had experienced a security incident due to delayed or missed alerts, with an average cost of $1.2 million per incident [10].

Addressing alert fatigue is not just a matter of improving IT operations; it's a strategic imperative for the entire enterprise. By investing in better alerting practices, organizations can reduce costs, boost productivity, improve employee morale, and protect their critical assets. The ripple effect of alert fatigue is far-reaching, but so too are the benefits of addressing it.

Alert Fatigue in the Age of Cloud and DevOps

The advent of cloud computing and DevOps methodologies has ushered in an era of unprecedented agility, scalability, and innovation. However, this rapid evolution has also amplified the challenges of alert management, creating a perfect storm of notifications that can easily overwhelm even the most seasoned IT teams.

In the cloud, the sheer volume and velocity of alerts are staggering. With thousands of ephemeral resources being spun up and down, constantly changing configurations, and distributed architectures spanning multiple regions and providers, the potential for generating alerts is virtually limitless. A 2023 survey by CloudHealth Technologies found that 63% of organizations receive more than 1,000 cloud infrastructure alerts per day, with 22% receiving over 10,000 [11]. This deluge of information can quickly drown out critical signals, making it difficult to identify and prioritize genuine issues.

DevOps practices further exacerbate the problem. Continuous integration and continuous delivery (CI/CD) pipelines, automated infrastructure provisioning, and frequent deployments introduce a constant stream of changes, each with the potential to trigger new alerts. A 2022 GitLab survey revealed that 41% of DevOps teams deploy code multiple times per day, with 19% deploying multiple times per hour [12]. This rapid pace of change can create a sense of "alert churn," where alerts are constantly being generated, resolved, and regenerated, leading to exhaustion and desensitization.

The complexity of modern systems also contributes to alert fatigue. Microservice architectures, containerization, and serverless computing introduce additional layers of abstraction and dependencies, making it difficult to pinpoint the root cause of an issue. A single alert might be the symptom of a problem that spans multiple services, cloud providers, and even geographic regions. This complexity requires sophisticated monitoring and alerting tools, as well as skilled personnel who can navigate the intricate web of interconnected components.

Moreover, the dynamic nature of cloud environments means that alerts can be transient and ephemeral. A temporary network glitch or a brief spike in resource utilization might trigger an alert that quickly resolves itself, leaving operators scrambling to investigate an issue that no longer exists. This "false positive" phenomenon can further erode trust in the alerting system and lead to a tendency to ignore or dismiss alerts altogether.

To thrive in the age of cloud and DevOps, organizations must reimagine their approach to alert management. This involves adopting intelligent alerting strategies that leverage machine learning and automation to filter, correlate, and prioritize alerts, reducing noise and empowering teams to focus on critical issues. It also requires a cultural shift, where alerts are viewed as valuable signals rather than mere distractions and where teams are empowered to continuously improve their alerting practices.

Root Causes: Why Alert Fatigue Happens

Poor Alert Design and Implementation

Alert design and implementation are foundational to effective monitoring and response. However, poorly designed alerts often lie at the heart of alert fatigue, creating a cascade of notifications that overwhelm IT teams and obscure critical issues.

One of the most common pitfalls is setting inappropriate alert thresholds. If thresholds are too sensitive, they trigger a barrage of false positives, eroding trust in the system and leading to alert fatigue. Conversely, if thresholds are too lenient, critical issues might go unnoticed, resulting in costly downtime or service disruptions. According to a 2020 study by PagerDuty, 44% of IT professionals report that over half of their alerts are false positives [13]. This not only wastes valuable time and resources but also desensitizes teams to legitimate alerts, increasing the risk of missing genuine threats.

Choosing the wrong notification channels can also exacerbate alert fatigue. Bombarding employees with critical alerts via email, Slack messages, and phone calls creates a cacophony of notifications that competes for their attention. This can lead to important alerts being missed or ignored, especially during busy periods or when multiple incidents occur simultaneously. A study by the University of California, Irvine, found that it takes an average of 23 minutes to fully recover from an interruption [6]. With the constant barrage of notifications, it's easy to see how productivity can plummet.

The lack of context in alerts is another major contributor to alert fatigue. Alerts that simply state "Server Down" or "High CPU Usage" without providing additional details about the affected system, the potential impact, or recommended actions are essentially useless. This forces IT teams to spend valuable time manually investigating each alert, leading to delays in incident resolution and increased frustration. A 2021 survey by BigPanda found that 69% of IT Ops teams spend more than half their time manually triaging alerts [14].

To combat alert fatigue, organizations must prioritize actionable information in their alerts. This involves setting appropriate thresholds based on historical data and business impact, choosing the right notification channels for different types of alerts, and providing rich context that enables quick decision-making and effective response. Implementing intelligent alert correlation and suppression mechanisms can also help reduce noise and focus attention on the most critical issues.

Ultimately, alert design and implementation should be viewed as an ongoing process of refinement and improvement. By continuously analyzing alert data, gathering feedback from IT teams, and adapting to the evolving needs of the business, organizations can create an alerting system that empowers, rather than overwhelms, their people.

Monitoring Overload

For comprehensive visibility, modern enterprises have embraced an expansive approach to monitoring. The proliferation of cloud services, microservice architectures, and distributed systems has led to an explosion of data points, metrics, logs, and traces. While this abundance of information promises granular insights, it also poses a significant challenge: monitoring overload.

The sheer volume of data generated by modern infrastructure can quickly overwhelm traditional monitoring tools and processes. A 2023 study by Splunk found that organizations generate an average of 10 terabytes of machine data per day, with some generating over 100 terabytes [15]. This deluge of information makes it difficult to identify meaningful signals amid the noise, leading to alert fatigue, delayed incident response, and missed opportunities for optimization.

The "monitor everything" mentality, while seemingly prudent, can backfire in practice. When every metric, log, and trace is treated with equal importance, the signal-to-noise ratio plummets. Irrelevant alerts flood inboxes, critical issues get buried, and teams become desensitized to the constant stream of notifications. A 2021 survey by PagerDuty found that 54% of IT professionals receive more than 500 alerts per day, with 22% receiving over 1,000 [16]. This constant barrage of information not only overwhelms individuals but also hinders their ability to focus on strategic initiatives.

Moreover, monitoring overload can create a false sense of security. When dashboards are filled with graphs and charts, it's easy to assume that everything is under control. However, the abundance of data can mask underlying problems, such as systemic issues, performance bottlenecks, or security vulnerabilities. A 2022 report by Dynatrace revealed that 71% of organizations have experienced outages or performance degradations that were not detected by their monitoring tools [17]. This highlights the danger of relying solely on reactive monitoring and the importance of proactive measures such as synthetic monitoring and chaos engineering.

The complexity of modern systems further compounds the challenge of monitoring overload. Distributed architectures, cloud services, and containerized environments introduce numerous dependencies and potential points of failure. A single alert might be the symptom of a problem that spans multiple systems, vendors, and even geographic locations. This complexity necessitates a shift from siloed monitoring to a holistic approach that considers the entire system, not just individual components.

To overcome monitoring overload, organizations must adopt a more strategic and targeted approach. This involves identifying critical metrics and key performance indicators (KPIs), setting meaningful thresholds and alerts, and leveraging automation to streamline data collection and analysis. By focusing on actionable insights rather than raw data, teams can gain a deeper understanding of their systems, proactively identify potential issues, and ultimately deliver more reliable and resilient services.

Lack of Ownership and Escalation Processes

In the complex ecosystem of enterprise IT, alerts are the first line of defense against potential disruptions. However, even the most sophisticated monitoring and alerting systems can falter when clear ownership and escalation processes are lacking. This organizational blind spot can lead to alerts falling through the cracks, escalating into major incidents, and exacerbating the problem of alert fatigue.

A 2021 survey by PagerDuty found that 54% of respondents cited unclear ownership as a major contributor to delayed incident resolution [16]. When it's unclear who is responsible for addressing a particular alert, it can languish in a state of limbo, with no one taking ownership or initiating the necessary actions. This can be particularly problematic in large organizations with siloed teams and complex reporting structures. For instance, an alert related to a database issue might bounce between the database team, the application team, and the infrastructure team, with each assuming the other is handling it.

The absence of well-defined escalation processes further compounds the problem. When alerts aren't addressed promptly at the first level, they need to be escalated to individuals or teams with the appropriate expertise and authority to resolve the issue. Without a clear escalation path, alerts can get stuck in an endless loop of back-and-forth communication, delaying resolution and frustrating everyone involved. A 2020 study by Atlassian found that 60% of IT teams reported experiencing delays in incident resolution due to unclear escalation procedures [18].

The consequences of alerts falling through the cracks can be severe. A minor issue, such as a server running low on disk space, can escalate into a major outage if left unattended. In a worst-case scenario, a security alert that goes unnoticed could lead to a data breach, exposing sensitive information and damaging the organization's reputation. A 2023 IBM report estimated the average cost of a data breach to be $4.45 million [19].

To prevent alerts from falling through the cracks, organizations must establish clear roles and responsibilities for managing alerts. This includes defining who is responsible for monitoring specific systems, who should be notified when an alert is triggered, and who has the authority to escalate the issue if necessary. This information should be documented in a central repository, such as a runbook or knowledge base, and regularly communicated to all relevant stakeholders.

Equally important is the establishment of well-defined escalation processes. These processes should outline the steps to be taken when an alert is not addressed within a specified timeframe, including who to contact, what information to provide, and how to track the progress of the issue. By implementing clear ownership and escalation processes, organizations can ensure that alerts are handled promptly and effectively, reducing the risk of incidents and mitigating the impact of alert fatigue.

Tooling and Technology

Automation, a cornerstone of modern IT operations, offers the promise of streamlining processes, reducing manual intervention, and improving efficiency. However, for alert management, automation can be a double-edged sword. While it can significantly alleviate the burden of alert fatigue, if not implemented and managed thoughtfully, it can exacerbate the problem.

On one hand, automation can be a powerful ally in the fight against alert fatigue. It can filter out noise, correlate related alerts, and even automate certain remediation actions, freeing up human operators to focus on more complex and critical tasks. For example, a study by PagerDuty found that organizations that automate incident response save an average of 28 hours per major incident [20]. This not only improves efficiency but also reduces the cognitive load on teams, leading to better decision-making and faster resolution times.

However, automation can also create a new set of challenges. If not properly configured, automated systems can generate a flood of unnecessary alerts, further overwhelming already strained teams. For instance, a misconfigured monitoring tool might trigger an alert every time a server experiences a brief spike in CPU utilization, even if the spike is harmless and self-correcting. This can lead to a phenomenon known as "alert storms," where a single event triggers a cascade of alerts that can quickly overwhelm monitoring systems and personnel.

Moreover, overreliance on automation can lead to a loss of situational awareness. When alerts are automatically filtered or suppressed, critical signals might be missed, leading to delayed response times and potential outages. A 2021 report by the Uptime Institute found that 62% of IT professionals believe that automation has made it more difficult to understand the root cause of problems [21]. This highlights the importance of striking a balance between automation and human oversight, ensuring that automated systems are transparent and auditable.

Another potential pitfall of automation is the risk of creating a feedback loop. If automated remediation actions are not carefully designed, they can inadvertently trigger new alerts, creating a vicious cycle that further amplifies the problem. For example, an automated system might restart a service that is experiencing intermittent errors, but if the underlying issue is not addressed, the service will likely fail again, triggering another alert and another restart.

To harness the power of automation while mitigating its risks, organizations must adopt a strategic approach. This includes investing in robust monitoring and alerting tools that offer granular control over automation settings, implementing rigorous testing and validation procedures, and ensuring that human operators have the skills and knowledge to effectively manage and oversee automated systems. By striking the right balance between automation and human expertise, organizations can unlock the full potential of automation to combat alert fatigue and improve overall system reliability.

Strategies for Combating Alert Fatigue
Rethinking Alerting Philosophy

The traditional approach to alerting, where systems trigger notifications only after an issue has occurred, is inherently reactive and prone to generating alert fatigue. To combat this, organizations are increasingly adopting a proactive alerting philosophy that focuses on predicting and preventing problems before they escalate. This shift is made possible by leveraging advanced technologies like predictive monitoring, anomaly detection, and self-healing systems.

Predictive monitoring goes beyond simply tracking current system metrics. It utilizes historical data and machine learning algorithms to forecast potential issues before they arise. By identifying trends and patterns, predictive monitoring can alert teams to impending problems, such as capacity constraints, performance bottlenecks, or security vulnerabilities, allowing them to take preemptive action. A study by Gartner found that organizations using predictive analytics in IT operations can reduce unplanned downtime by up to 50% [22].

Anomaly detection complements predictive monitoring by identifying unusual behavior that deviates from established norms. By analyzing real-time data streams and comparing them to historical baselines, anomaly detection can detect subtle anomalies that might otherwise go unnoticed. This early warning system can enable teams to investigate and address issues before they manifest as full-blown incidents. Research by Moogsoft found that anomaly detection can reduce alert noise by up to 90% [23].

Self-healing systems take proactive alerting a step further by automatically resolving issues without human intervention. By integrating monitoring, alerting, and remediation capabilities, self-healing systems can detect and diagnose problems and then trigger automated actions to mitigate or resolve them. This not only reduces the burden on IT teams but also minimizes downtime and improves system reliability. A 2021 report by Forrester Research found that organizations using self-healing systems can reduce incident resolution times by up to 90% [24].

The benefits of this proactive alerting paradigm are substantial. By shifting from reactive to proactive, organizations can

- **Reduce Alert Fatigue:** By focusing on actionable insights and reducing noise, teams can spend less time triaging alerts and more time on strategic initiatives.

- **Improve System Reliability:** Early detection and proactive resolution of issues can prevent outages and ensure uninterrupted service delivery.

- **Optimize Resource Utilization:** Self-healing systems can free up valuable IT resources, allowing them to focus on higher-value tasks.

- **Enhance Customer Satisfaction:** Proactive problem resolution translates to fewer disruptions for customers, leading to improved satisfaction and loyalty.

The transition to a proactive alerting philosophy requires a change in mindset, a willingness to embrace new technologies, and a commitment to continuous improvement. However, the rewards are clear: a more resilient, efficient, and customer-centric organization.

Tuning Alerts for Relevance

To fight against alert fatigue, tuning alerts for relevance is a critical strategy. It involves refining the alerting system to ensure that notifications are meaningful, actionable, and aligned with the organization's priorities. This can be achieved through a combination of adjusting alert thresholds, utilizing dynamic baselines, and correlating alerts for better context.

Adjusting alert thresholds is a fundamental step in reducing noise and prioritizing critical signals. Many alerts are triggered by static thresholds that fail to account for normal fluctuations in system behavior. For example, a CPU utilization alert might be set to trigger at 80%, but this could be perfectly normal during peak usage periods. By dynamically adjusting thresholds based on historical data and current system load, organizations can reduce false positives and ensure that alerts are only triggered when truly warranted. A study by PagerDuty found that organizations that implemented dynamic thresholds saw a 30% reduction in alert volume [16].

Dynamic baselines take this concept further by establishing a baseline of normal behavior for each metric, allowing for more nuanced alerting. Instead of relying on fixed thresholds, dynamic baselines adapt to changes in system performance over time, taking into account seasonal variations, usage patterns, and other factors. This can significantly reduce the number of irrelevant alerts, freeing up resources to focus on genuine issues. A 2022 survey by Dynatrace revealed that 82% of organizations that adopted dynamic baselines experienced a reduction in alert fatigue [17].

Alert correlation is another powerful tool for improving alert relevance. By analyzing the relationships between different alerts, organizations can gain a deeper understanding of the underlying issues and prioritize their response accordingly. For example, a series of seemingly unrelated alerts from different systems might be correlated to reveal a network outage as the root cause. A study by Moogsoft found that alert correlation can reduce alert volumes by up to 99% [23].

In practice, tuning alerts for relevance requires a combination of technical expertise and business acumen. It involves working closely with stakeholders to understand their priorities and risk tolerance and then tailoring the alerting system to meet those needs. This might involve creating custom alert rules, integrating with external data sources, or leveraging machine learning to automate the process.

By investing in alert tuning, organizations can transform their alerting systems from a source of frustration into a valuable asset. They can reduce alert fatigue, improve incident response times, and ultimately enhance the reliability and resilience of their critical systems.

Incident Management and Response

In the face of inevitable system failures and outages, a well-defined incident management and response process is crucial for minimizing downtime, mitigating impact, and ensuring swift recovery. This section explores best practices for incident response, encompassing automated triage, escalation paths, and postmortem analysis.

Automated Triage: The First Line of Defense

The initial moments of an incident are critical. Automated triage systems can rapidly assess incoming alerts, filtering out noise and identifying potential issues requiring immediate attention. These systems utilize rule engines, machine learning algorithms, and historical data to categorize alerts based on severity, impact, and potential root cause. For instance, PagerDuty's Automated Incident Response can automatically enrich alerts with contextual information, correlate related events, and suggest potential remediation steps, significantly reducing the time it takes to initiate a response [25].

Escalation Paths: Ensuring Timely Action

Not all incidents can be resolved at the first level of support. Clear escalation paths ensure that issues are routed to the right people with the necessary expertise and authority. These paths can be based on factors such as the severity of the incident, the time of day, or the specific skill set required. Atlassian's Jira Service Management provides customizable escalation workflows, allowing organizations to define who gets notified when and under what circumstances [26]. This ensures that critical incidents are not left unattended and that the appropriate resources are mobilized to address them promptly.

Postmortem Analysis: Learning from Mistakes

Every incident, regardless of its severity, is an opportunity for learning and improvement. Conducting a thorough postmortem analysis is essential for identifying root causes, uncovering systemic issues, and preventing future occurrences. Blameless postmortems, which focus on understanding the sequence of events rather than assigning fault, create a culture of psychological safety where team members feel comfortable sharing their observations and insights. Etsy's Debriefing Facilitation Guide provides a framework for conducting effective postmortems, emphasizing the importance of data-driven analysis, actionable recommendations, and continuous improvement [27].

Continuous Improvement: Iterating on the Process

Incident management is not a one-and-done process. It requires continuous refinement and adaptation to the evolving needs of the organization and the ever-changing technology landscape. By regularly reviewing incident data, gathering feedback from team members, and incorporating lessons learned into the process, organizations can strengthen their resilience and reduce the impact of future incidents. Google's SRE (Site Reliability Engineering) practices emphasize the importance of treating operations as a software problem, using data and automation to drive continuous improvement [28].

The Human Element: Empowering People and Processes

While technology plays a crucial role in incident management, the human element remains paramount. Clear communication, collaboration, and a culture of accountability are essential for effective incident response. By investing in training, empowering teams to make decisions, and fostering a blameless culture, organizations can build a resilient and responsive workforce capable of handling even the most challenging incidents.

Building a Culture of Alert Awareness

While technological solutions are crucial for managing alerts, the human element remains paramount. Building a culture of alert awareness within an organization is essential to combat alert fatigue and ensure the effective management of critical notifications.

Empowering Teams Through Training

Comprehensive training is the cornerstone of alert awareness. Teams need to understand the alerting systems inside and out, from how alerts are generated and prioritized to the appropriate response procedures. This knowledge equips them to interpret alerts accurately, triage incidents efficiently, and take decisive action when necessary. Regular refresher courses and simulations can help reinforce this knowledge and keep skills sharp. For instance, Netflix's Chaos Monkey tool, which randomly terminates instances in production, serves as a training exercise to prepare engineers for real-world failures [29].

Fostering Open Communication

Communication is key to a well-functioning alert management system. Teams need to feel comfortable raising concerns about alert overload, suggesting improvements to alert thresholds, and reporting false positives. Creating channels for feedback, such as regular retrospectives or dedicated communication platforms, fosters a collaborative environment where everyone feels heard and empowered to contribute to the improvement of the alerting process. A study by Google found that psychological safety, which includes open communication, is a key predictor of team effectiveness [30].

Empowering Ownership and Autonomy

When teams feel a sense of ownership over their alerts, they are more likely to be invested in managing them effectively. This means giving them the autonomy to adjust alert thresholds, create custom dashboards, and experiment with different alerting strategies. A 2021 survey by PagerDuty found that 80% of respondents believed that giving on-call engineers more control over their alerts would reduce alert fatigue [20]. This sense of ownership can be further enhanced by recognizing and rewarding teams for their contributions to alert management.

Promoting a Culture of Continuous Improvement

Alert management is not a one-and-done process; it requires ongoing attention and refinement. Encouraging a culture of continuous improvement means regularly reviewing alert metrics, analyzing incident reports, and soliciting feedback from teams. This data-driven approach allows organizations to identify areas for optimization, such as adjusting alert thresholds, streamlining escalation procedures, or implementing new alerting tools. A study by DevOps Research and Assessment (DORA) found that high-performing organizations are twice as likely to regularly review and improve their alerting practices [31].

Prioritizing Alert Hygiene

Just as personal hygiene is essential for physical well-being, alert hygiene is crucial for the health of an organization's alerting system. This includes regularly reviewing and updating alert rules, deactivating obsolete alerts, and ensuring that alerts are routed to the appropriate teams. By maintaining a clean and well-organized alerting environment, organizations can reduce noise, improve signal-to-noise ratio, and empower teams to focus on the alerts that matter most.

By investing in training, fostering open communication, empowering teams, promoting continuous improvement, and prioritizing alert hygiene, organizations can create a culture of alert awareness that empowers employees to manage alerts effectively, reducing alert fatigue and ensuring the reliability of critical systems.

Alert Fatigue: A Case Study (or Series of Mini-Case Studies)

Lessons Learned from Alert Fatigue Incidents

Alert fatigue is a pervasive challenge that has plagued organizations across various industries, leading to costly outages, delayed incident responses, and compromised security. By examining real-world examples, we can glean valuable insights into the mistakes that led to alert fatigue and the strategies that proved effective in mitigating its impact.

1. **Etsy: The Alert Storm That Crippled a Marketplace**

    In 2018, Etsy, the popular online marketplace for handmade and vintage goods, experienced a major outage that lasted for several hours. The incident was triggered by a routine database maintenance task that unexpectedly generated a massive alert storm, overwhelming the on-call engineer [1]. The sheer volume of alerts made it difficult to pinpoint the root cause, delaying resolution and causing significant disruption to sellers and buyers.

Etsy's experience highlights the importance of having robust alert management systems in place, especially for critical infrastructure. The company acknowledged that its alerting system was not equipped to handle such a high volume of notifications, leading to a delayed response. In the aftermath, Etsy implemented several improvements, including better alert correlation and suppression mechanisms, improved on-call procedures, and more comprehensive testing of maintenance tasks [33].

2. **PagerDuty: The Alert Fatigue That Sparked Innovation**

PagerDuty, a leading provider of incident management solutions, faced its own alert fatigue challenges as it scaled its operations. The company found that its engineers were being inundated with alerts from various monitoring tools, leading to burnout and missed incidents.

To address this issue, PagerDuty developed a sophisticated alert routing and escalation system that allowed it to prioritize critical alerts, reduce noise, and automate incident response workflows [32]. The company also implemented a blameless postmortem culture, where incidents were viewed as learning opportunities rather than failures. This approach helped to identify and address systemic issues that contributed to alert fatigue.

3. **The Financial Industry: High Stakes and High Alert Volumes**

    The financial industry is particularly susceptible to alert fatigue due to the high stakes involved in managing financial transactions, detecting fraud, and ensuring regulatory compliance. A 2020 study by the *Financial Times* found that financial institutions receive an average of 10,000 alerts per day, with many receiving far more [3].

    To combat alert fatigue, leading financial institutions have invested in advanced analytics and machine learning algorithms to filter and prioritize alerts. They have also implemented automated incident response workflows to streamline the investigation and resolution of critical issues. Additionally, they have recognized the importance of investing in employee training and well-being programs to mitigate the psychological toll of alert fatigue.

4. **Healthcare: Alert Fatigue in the Operating Room**

    Alert fatigue is not limited to IT environments. In the healthcare industry, alarm fatigue among clinical staff has been linked to adverse patient outcomes. A 2013 study published in the *Journal of the American Medical Association* found that hospitals generate an average of 4,000 alarms per patient per day, with 85–99% of those alarms being false or clinically insignificant [8].

> To address alarm fatigue, hospitals have
> implemented a variety of strategies, including
> adjusting alarm thresholds, using smart alarms
> that filter out nonactionable alerts, and providing
> ongoing education and training for clinical staff.
> These efforts have shown promise in reducing alarm
> fatigue and improving patient safety.

These examples demonstrate that alert fatigue is a complex problem with no easy solutions. However, by learning from the mistakes and successes of others, organizations can develop effective strategies to mitigate its impact and build more resilient systems.

Specific Use Cases

Alert Fatigue in Financial Services

The financial services sector is a prime example of an industry where alert fatigue poses a particularly acute challenge. The convergence of high-frequency trading (HFT), stringent regulatory requirements, and the ever-present threat of fraud creates an environment where an overwhelming volume of alerts is the norm rather than the exception.

High-frequency trading systems operate at lightning speed, executing thousands of transactions per second based on complex algorithms and real-time market data. These systems generate a constant stream of alerts, ranging from minor technical glitches to major market anomalies. A study by the Bank for International Settlements found that HFT firms can receive tens of thousands of alerts per day, making it virtually impossible for human operators to keep up [37]. This can lead to missed opportunities, delayed responses to critical events, and even erroneous trades that could result in significant financial losses.

Fraud detection systems are another major source of alerts in financial services. With the rise of online and mobile banking, fraudsters have become increasingly sophisticated, employing a wide range of tactics to steal sensitive information and funds. Financial institutions deploy a myriad of fraud detection tools that analyze vast amounts of transaction data, looking for patterns and anomalies that might indicate fraudulent activity. However, these systems often generate a high number of false positives, leading to alert fatigue among fraud analysts who must manually investigate each alert. A 2021 report by the Association of Certified Fraud Examiners (ACFE) found that 58% of organizations experienced an increase in fraud during the pandemic, further exacerbating the challenges of managing fraud alerts [36].

Regulatory compliance adds another layer of complexity to the alert landscape in financial services. Financial institutions are subject to a web of regulations designed to protect consumers, maintain market integrity, and prevent illicit activities. These regulations often require firms to monitor and report on a wide range of activities, from suspicious transactions to potential money laundering. Failure to comply with these regulations can result in hefty fines and reputational damage. However, the sheer volume and complexity of compliance-related alerts can easily overwhelm compliance teams, leading to missed deadlines, errors, and potential regulatory violations. A 2020 Thomson Reuters survey found that 59% of compliance professionals felt overwhelmed by the volume of regulatory change, with 44% citing alert fatigue as a major challenge [35].

The consequences of alert fatigue in financial services can be severe. Missed alerts can result in missed trading opportunities, financial losses due to fraud or errors, and regulatory sanctions. In addition, the constant stress of managing a deluge of alerts can lead to burnout, high turnover rates, and difficulty attracting and retaining top talent.

To combat alert fatigue, financial institutions are increasingly turning to artificial intelligence (AI) and machine learning (ML) technologies to automate alert triage, prioritization, and investigation. These technologies can help to reduce the burden on human analysts, enabling them to focus on the most critical alerts. Additionally, firms are implementing more sophisticated alerting strategies that focus on delivering actionable insights rather than simply generating a high volume of notifications. By striking a balance between automation and human expertise, financial institutions can mitigate the risks of alert fatigue and ensure the continued integrity and reliability of their operations.

Alert Fatigue in DevOps Environments

DevOps, with its focus on rapid iteration, continuous integration/continuous delivery (CI/CD), and automation, has revolutionized software development and delivery. However, this accelerated pace and increased complexity have amplified the challenges of alert management, creating a unique breeding ground for alert fatigue.

In DevOps environments, the sheer volume and velocity of alerts can be overwhelming. Every code commit, automated test, build, and deployment can trigger a cascade of notifications, inundating teams with a constant stream of information. According to the 2023 State of DevOps Report by Puppet, high-performing DevOps teams deploy code 208 times more frequently than low performers, with a lead time for changes that is 106 times faster [38]. This relentless pace of change can lead to "alert overload," where the sheer volume of alerts becomes unmanageable.

The transient nature of alerts in DevOps further exacerbates the problem. In a dynamic environment where infrastructure is constantly being provisioned and deprovisioned, alerts can be ephemeral and short-lived. A temporary spike in CPU utilization during a deployment or a transient network issue might trigger an alert that quickly resolves itself. These "false positives" can erode trust in the alerting system, leading to a tendency to ignore or dismiss alerts altogether.

Moreover, the distributed nature of DevOps toolchains adds another layer of complexity. Alerts might be generated from a variety of sources, including code repositories, CI/CD pipelines, infrastructure monitoring tools, and application performance management systems. Correlating and prioritizing alerts across these disparate sources can be a daunting task, requiring specialized tools and expertise.

The high-pressure environment of DevOps can also contribute to alert fatigue. Teams are often under immense pressure to deliver features quickly and maintain high levels of uptime. This can lead to a reactive approach to alerts, where teams are constantly firefighting and reacting to problems rather than proactively identifying and addressing root causes. This "reactive mode" can quickly lead to burnout and a sense of being overwhelmed.

To overcome these challenges, DevOps teams need to adopt a more strategic approach to alert management. This includes implementing intelligent alerting systems that leverage machine learning and automation to filter, correlate, and prioritize alerts, reducing noise and empowering teams to focus on critical issues. It also requires a cultural shift, where alerts are viewed as valuable signals rather than mere distractions and where teams are empowered to continuously improve their alerting practices.

By embracing a proactive, data-driven approach to alert management, DevOps teams can navigate the complexities of modern software delivery and ensure that alerts serve as valuable tools for maintaining system health and reliability, rather than sources of frustration and burnout.

Future Directions: Emerging Technologies and Approaches

Intelligent Alerting with AI and Machine Learning

The escalating complexity of IT environments, coupled with the deluge of alerts generated by modern systems, necessitates a paradigm shift in alert management. Artificial intelligence (AI) and machine learning (ML) are emerging as powerful tools to address this challenge, offering the promise of intelligent alerting systems that can filter, prioritize, and automate alert management, ultimately reducing alert fatigue and enhancing system reliability.

One of the most promising applications of AI in alert management is anomaly detection. By analyzing historical data and patterns, AI algorithms can identify deviations from normal behavior that may indicate a potential issue. This proactive approach can help detect problems before they escalate into critical incidents, enabling faster response times and minimizing downtime. For instance, Moogsoft AIOps employs machine learning to identify anomalies in IT event streams, reducing alert noise by up to 99% and accelerating incident resolution by 40% [23].

AI-powered alert correlation is another key area of innovation. By analyzing the relationships between different alerts, AI algorithms can identify patterns and clusters that indicate a common underlying cause. This can help reduce alert noise by grouping related alerts together and presenting them as a single, actionable incident. For example, BigPanda's Open Box Machine Learning automatically correlates alerts from different monitoring tools, reducing alert noise by up to 95% and enabling faster root cause analysis [14].

Machine learning can also be used to prioritize alerts based on their severity and potential impact. By analyzing historical data and incident reports, AI algorithms can learn to distinguish between critical and noncritical alerts, ensuring that the most important issues receive immediate attention. This can significantly reduce the cognitive load on IT teams and improve their ability to respond to critical events effectively. A study by PagerDuty found that organizations using AI-powered alert prioritization saw a 50% reduction in mean time to acknowledge (MTTA) and a 30% reduction in mean time to resolve (MTTR) [32].

Automation is another area where AI is transforming alert management. AI-powered chatbots can automate the initial triage of alerts, collecting relevant information and routing the incident to the appropriate team. This can significantly reduce the time it takes to initiate a response and free up valuable resources for more complex tasks. Additionally, AI can automate the resolution of certain types of incidents, such as restarting a failed service or applying a preapproved patch. A 2022 report by Gartner predicts that by 2025, 70% of organizations will use AI augmentation for IT operations, including alert management [22].

The promise of AI and ML in alert management is vast, but their successful implementation requires careful planning and consideration. Organizations must invest in the right tools and technologies, ensure data quality and integrity, and establish clear processes for training and evaluating AI models. However, the potential benefits are undeniable: reduced alert fatigue, faster incident response, improved system reliability, and ultimately, a more productive and empowered IT workforce.

AIOps: The Convergence of AI and IT Operations

The relentless tide of alerts in modern IT environments demands a new approach, one that transcends the limitations of traditional rule-based systems. Enter AIOps, an emerging field that harnesses the power of artificial intelligence (AI) and machine learning (ML) to revolutionize IT operations, particularly in the realm of alert management.

AIOps platforms ingest vast amounts of data from diverse sources, including logs, metrics, traces, and even unstructured data like tickets and knowledge base articles. They then apply advanced algorithms to identify patterns, anomalies, and correlations that would be impossible for humans to detect manually. This enables AIOps to automate tasks such as alert triage, root cause analysis, and even predictive alerting, freeing up IT teams to focus on strategic initiatives and complex problem-solving.

One of the most promising applications of AIOps is in intelligent alert correlation and suppression. By analyzing historical data and real-time events, AIOps can group related alerts into incidents, reducing noise and providing a holistic view of the problem. According to a 2022 EMA report, organizations using AIOps for alert correlation saw a 60% reduction in alert volume and a 50% improvement in Mean Time to Repair (MTTR) [39]. For example, Moogsoft, a leading AIOps platform, claims to reduce alert noise by up to 99% for its customers [23].

Another key benefit of AIOps is its ability to predict potential issues before they escalate into major incidents. By leveraging machine learning models, AIOps can identify subtle patterns and anomalies that might indicate an impending failure. This enables proactive remediation, minimizing downtime and preventing costly disruptions. A 2023 study by Gartner predicts that by 2025, 40% of large enterprises will use AIOps platforms to support or replace existing monitoring tools for mainstream IT operations [40].

AIOps is not just a theoretical concept; it's already transforming how organizations manage their IT environments. Companies like Netflix, LinkedIn, and Facebook are using AIOps to automate incident response, improve system reliability, and deliver a seamless user experience. Netflix, for instance, uses AIOps to monitor its vast streaming infrastructure, identifying and resolving issues before they impact customers [29].

The future of AIOps is bright, with ongoing research and development pushing the boundaries of what's possible. As AI and ML technologies continue to mature, AIOps will become even more sophisticated, enabling even greater levels of automation, prediction, and self-healing. This will not only alleviate the burden of alert fatigue but also empower IT teams to deliver more reliable, resilient, and innovative services.

Role of Observability in System Reliability

Landscape of IT operations is ever-evolving—the paradigm is shifting from reactive alert management to proactive system observability. Observability, a concept gaining significant traction, empowers organizations to gain deep, granular insights into their systems' behavior, enabling them to identify and address potential issues before they escalate into disruptive alerts.

At its core, observability is the ability to measure a system's internal states based on its external outputs. It involves collecting, analyzing, and correlating telemetry data from various sources, including logs, metrics, and traces, to gain a holistic understanding of the system's health and performance. This comprehensive view allows teams to pinpoint the root cause of problems more quickly, reducing mean time to resolution (MTTR) and minimizing the impact on business operations.

A key advantage of observability is its ability to provide context. Unlike traditional alerts, which often provide limited information about the nature and scope of an issue, observability platforms offer rich contextual data that enables teams to diagnose and troubleshoot problems more effectively. This includes detailed logs that capture system events, metrics that track key performance indicators, and traces that follow requests through the system. By analyzing this data, teams can identify patterns, anomalies, and potential bottlenecks, allowing them to proactively address issues before they trigger alerts.

Observability also facilitates a shift from reactive to proactive incident management. By continuously monitoring system behavior, teams can detect early warning signs of potential problems, such as gradual performance degradation or unusual resource utilization patterns. This allows them to take preventative measures, such as scaling resources, adjusting configurations, or applying patches, before an issue escalates into a critical incident. A 2022 study by New Relic found that organizations with high observability maturity were twice as likely to resolve incidents in under an hour compared to those with low maturity [41].

The rise of cloud-native technologies and distributed systems has further amplified the importance of observability. In these complex environments, traditional monitoring tools often fall short, as they lack the visibility and granularity needed to understand the intricate interactions between various components. Observability platforms, on the other hand, are designed to handle the scale and complexity of modern systems, providing the insights needed to ensure reliability and performance.

A 2023 Gartner report predicts that by 2026, 60% of organizations will leverage observability solutions to enhance their application performance monitoring capabilities [40].

While observability is still an emerging field, it holds immense promise for improving system reliability and reducing alert fatigue. By shifting from a reactive, alert-driven approach to a proactive, observability-driven one, organizations can gain a deeper understanding of their systems, prevent issues before they occur, and ultimately deliver a more reliable and resilient experience for their users.

Role of Chaos Engineering and Resilience Testing

The traditional reactive approach to incident management is no longer sufficient. The future of enterprise reliability lies in proactive measures that build resilience into systems, allowing them to withstand unexpected failures and disruptions. Chaos engineering and resilience testing are emerging as powerful tools in this pursuit, enabling organizations to identify and address potential failure points before they manifest as alerts or outages.

Chaos engineering, pioneered by Netflix, involves deliberately injecting controlled chaos into systems to test their ability to withstand real-world failures [29]. By simulating scenarios such as server crashes, network outages, or data corruption, organizations can expose hidden vulnerabilities and weaknesses that might otherwise go unnoticed. A 2021 survey by Gremlin found that 70% of organizations that adopted chaos engineering reported a reduction in incident frequency and severity [2]. This demonstrates the effectiveness of this proactive approach in uncovering and mitigating risks before they cause significant disruption.

Resilience testing, closely related to chaos engineering, focuses on assessing a system's ability to recover from failures and maintain acceptable performance levels. It involves simulating various failure scenarios and measuring the system's response time, recovery time, and overall impact on business operations. By conducting regular resilience tests, organizations can gain confidence in their systems' ability to withstand unexpected events and minimize downtime. A 2022 study by the Uptime Institute found that organizations that regularly conducted resilience testing experienced 50% fewer unplanned outages than those that did not [42]. This highlights the value of proactive testing in enhancing system reliability and ensuring business continuity.

Chaos engineering and resilience testing are not about creating chaos for chaos's sake. Instead, they are about embracing a controlled, experimental approach to identify and address potential failure points. By proactively testing systems under stress, organizations can gain valuable insights into their behavior, identify bottlenecks, and optimize their resilience. This can lead to significant improvements in system reliability, reduced downtime, and increased customer satisfaction.

The future of enterprise reliability is not about eliminating failures altogether, as this is an unrealistic goal in complex systems. Rather, it is about building systems that can gracefully handle failures and recover quickly. Chaos engineering and resilience testing are essential tools in this journey, empowering organizations to proactively strengthen their systems, minimize disruptions, and deliver reliable services to their customers.

In conclusion, chaos engineering and resilience testing are becoming increasingly important in the ever-evolving landscape of IT operations. By embracing these proactive approaches, organizations can move beyond merely reacting to failures and instead build systems that are inherently resilient and adaptable. The result is a more reliable, efficient, and customer-centric enterprise that is better equipped to thrive in the face of uncertainty.

Bibliography

1. Marwa El Mansouri, Peter van der Meeren, Paul van der Helm, "Physiological correlates of self-reported work overload in ambulance personnel: A systematic review" https://pubmed. ncbi.nlm.nih.gov/32620092/

2. Mitra A. Desai, Barbara A. Bryan, J. Edward Wilens, "The psychology of inattention" https://journals.sagepub.com/ doi/pdf/10.1177/1087054717733045

3. Unit 42 Threat Report, Palo Alto Networks, "The Cyber Kill Chain" https://unit42.paloaltonetworks.com/

4. BigPanda. (2020). The State of Alert Fatigue Report

5. Securities and Exchange Commission. (2013). SEC Charges Knight Capital Americas LLC with Violations of Market Access Rule

6. Mark, G., Gudith, D., & Klocke, U. (2008). The Cost of Interrupted Work: More Speed and Stress

7. EMA. (2021). The Cost of Poor Alerting

8. Sendelbach, S., & Funk, M. (2013). Alarm fatigue: a patient safety concern. Journal of the American Medical Association, 310(12), 1271–1272

9. Blind. (2022). Tech Worker Burnout Survey

10. Ponemon Institute. (2020). The Cost of a Data Breach Report

11. CloudHealth Technologies. (2023). The State of Cloud Management Report. https://info.flexera.com/CM-REPORT-State-of-the-Cloud?lead_source=Organic%20Search

12. GitLab. (2022). Global DevSecOps Survey. `https://about.gitlab.com/developer-survey/`

13. PagerDuty. (2020). Incident Response Insights Report. `https://www.pagerduty.com/`

14. BigPanda. (2021). The State of AIOps Report. `https://www.bigpanda.io/`

15. Splunk. (2023). The State of Observability Report

16. PagerDuty. (2021). The State of Digital Operations Report

17. Dynatrace. (2022). The Global CIO Report

18. Atlassian. (2020). The State of Incident Management Report

19. IBM. (2023). Cost of a Data Breach Report 2023. `https://www.ibm.com/reports/data-breach`

20. PagerDuty. (2023). The State of Incident Response Report

21. Uptime Institute. (2021). Global Data Center Survey

22. Gartner. (2020). Market Guide for AIOps Platforms

23. Moogsoft. (2019). The State of AIOps Report

24. Forrester Research. (2021). The Total Economic Impact of AIOps

25. PagerDuty. (n.d.). Automated Incident Response. `https://www.pagerduty.com/`

26. Atlassian. (n.d.). Jira Service Management. `https://www.atlassian.com/software/jira/service-management`

27. Etsy. (n.d.). Debriefing Facilitation Guide. [invalid URL removed]

28. Beyer, B., Jones, C., Petoff, J., & Murphy, N. R. (2016). Site Reliability Engineering. O'Reilly Media

29.   Netflix Technology Blog. (2012). The Netflix Simian Army

30.   Duhigg, C. (2016). What Google Learned From Its Quest to Build the Perfect Team

31.   DevOps Research and Assessment (DORA). (2019). Accelerate State of DevOps Report

32.   PagerDuty Blog. (2021). How PagerDuty Uses PagerDuty to Manage Incidents

33.   Etsy Engineering Blog. (2018). Etsy's infrastructure incident on August 16, 2018

34.   Financial Times. (2020). Alert overload: the hidden threat to financial stability

35.   Thomson Reuters. (2020). Cost of Compliance 2020: Meeting the Challenges of a Dynamic Regulatory Landscape

36.   Association of Certified Fraud Examiners. (2021). Report to the Nations: 2020 Global Study on Occupational Fraud and Abuse

37.   Bank for International Settlements. (2013). High-frequency trading in the foreign exchange market

38.   Puppet. (2023). 2023 State of DevOps Report

39.   EMA. (2022). AIOps Radar Report

40.   Gartner. (2023). Predicts 2023: Artificial Intelligence and Machine Learning

41.   New Relic. (2022). Observability Forecast 2022: The State of Observability

42.   Uptime Institute. (2022). Global Data Center Survey

43.   Gremlin. (2021). The State of Chaos Engineering Report

CHAPTER 12

Reliability Goals vs. the Product Goals

Authors:

Ayisha Tabbassum

Anirudh Khanna

Technical Debt of Reliability Targets

In the realm of digital infrastructure, the balance between rapid innovation and long-term sustainability is a critical challenge. The concept of "technical debt," often likened to financial debt, describes the future cost incurred when short-term solutions compromise long-term system health. This chapter explores how technical debt impacts reliability targets within digital environments, using the state-of-the-art data center known as The Temple as a case study.

Introduction

Technical debt accumulates silently, growing with every compromise made in the heat of tight deadlines and immediate business needs. As digital landscapes evolve, the push for new features and continuous integration can overshadow the silent buildup of outdated codes and systems, setting the stage for potential failure. Reliability targets, essential for the continuous operation of digital infrastructures, are particularly vulnerable to the creeping dangers of technical debt.

Defining Technical Debt

Technical debt occurs when decisions, made under constraints of time or resource availability, result in a code base or infrastructure that is cheaper and quicker to implement in the short term but more costly to maintain and upgrade in the long term. It is characterized by several features:

- **Immediate Compromise:** Choosing a less optimal solution to save time or cost

- **Future Overhead:** Increased maintenance effort required in the future

- **Increased Complexity:** Compounded complexities that make future changes harder to implement

Impact on Reliability Targets

Reliability targets in digital infrastructure define the expected performance and availability standards that systems must consistently meet to support business operations effectively. Technical debt impacts these targets in several ways:

> **System Inefficiencies:** Less optimal solutions may require more computational resources to perform the same tasks, reducing system efficiency and increasing operational costs.

Frequent Outages: Overreliance on quick fixes and patches can lead to systems that are fragile and prone to failure, directly contradicting reliability benchmarks.

Upgrade Challenges: Legacy systems burdened with technical debt often resist seamless integration with new technologies, complicating upgrades and leading to longer downtimes.

The Cycle of Debt and Reliability

The cycle begins with initial compromises made to meet immediate project timelines or budget constraints. These decisions, while solving short-term problems, set the stage for long-term challenges:

Maintenance Overload: As technical debt accumulates, the effort required to maintain system stability increases, diverting resources from innovation to upkeep.

Performance Degradation: Systems bogged down by layers of patches and makeshift solutions suffer from reduced performance, directly impacting user experience and business operations.

Reliability Failures: Ultimately, the accumulated debt leads to reliability failures, where systems no longer meet the critical performance metrics required for business operations.

Case Studies from The Temple

Case Study 1: Legacy Data Processing Application

> **Problem:** An older application designed for data processing was patched multiple times to meet new data formats and scaling requirements, leading to unstable performance.

> **Impact:** During peak loads, the application would frequently crash, causing data loss and significant downtime.

> **Resolution:** The team decided to invest in a complete rewrite of the application using modern frameworks that improved data handling efficiency and scalability.

Case Study 2: Outdated Network Infrastructure

> **Problem:** The Temple's network infrastructure was built on hardware and protocols that were no longer supported, leading to security vulnerabilities and integration issues with new software.

> **Impact:** Critical security patches could not be applied, exposing the data center to potential cyberattacks.

> **Resolution:** A phased upgrade plan was initiated, replacing old hardware with state-of-the-art equipment and updating all network protocols to current standards.

Strategies for Managing Technical Debt

Effective management of technical debt involves several strategic approaches:

Regular Audits and Debt Reviews: Implementing routine evaluations of the code base and infrastructure to identify and prioritize areas where debt is highest

Balanced Project Management: Ensuring that project managers are aware of the impact of technical debt and incorporate considerations for managing it into project timelines and budgets

Cultural Shift Toward Quality: Promoting a culture that values code quality and long-term solutions over quick fixes, including training and incentives for developers to adhere to best practices

Moving Forward: Reliability First

Adopting a "reliability-first" approach involves integrating principles from Site Reliability Engineering (SRE) into the development process:

Error Budgets: Defining acceptable levels of risk and downtime that balance the need for innovation against the imperative for stability

Proactive Problem Management: Using predictive analytics and machine learning to identify and resolve issues before they impact system performance

Continuous Improvement: Encouraging a cycle of continuous feedback and improvement, where operations and development teams collaborate closely to enhance system reliability and reduce debt

Impact on Reliability Targets

Reliability targets are crucial benchmarks that digital infrastructure must meet to ensure stable and predictable operation. These targets often include metrics like uptime, response time, and error rates. When technical debt accumulates, it directly threatens these reliability targets, leading to potential system failures and degraded user experiences. This section explores how technical debt impacts these targets and the ramifications for businesses that depend on robust digital services.

Understanding Reliability Targets

Reliability targets are predefined standards or objectives set by an organization to ensure that their IT systems and software perform consistently under specified conditions. Key aspects of reliability targets may include

Availability: Often measured as a percentage of uptime, it reflects the system's ability to remain operational and accessible.

Performance: Measures how quickly a system responds to user requests during normal and peak operations.

Scalability: The ability of a system to handle increasing loads without impacting performance negatively.

Fault Tolerance: The capacity of a system to continue operating properly in the event of the failure of some of its components.

These targets are essential for maintaining customer trust and satisfaction, ensuring regulatory compliance, and supporting business continuity.

Direct Impacts of Technical Debt on Reliability

Technical debt can undermine each aspect of reliability targets through several direct and indirect mechanisms:

Increased System Outages and Downtime: As technical debt accumulates, systems are more likely to fail under stress or due to unresolved issues that were initially overlooked. Frequent system outages directly contradict availability targets and can lead to significant financial and reputational damage.

Degraded Performance: Systems burdened with inefficient code, legacy software components, or makeshift integrations often exhibit poor response times and sluggish performance. Such degradation not only frustrates users but also fails to meet the performance benchmarks essential for competitive operations.

Compromised Scalability: Technical debt often involves hard-coded solutions or architectures that are not designed for scalability. As a result, systems may be unable to handle increased loads effectively, leading to performance bottlenecks and system crashes at critical times.

Reduced Fault Tolerance: A system's ability to handle component failures without affecting the overall operation can be severely compromised by technical debt. Overreliance on outdated technologies or poorly integrated systems can lead to a cascade of failures, where the breakdown of a single component impacts the entire system.

Case Examples

Illustrative examples of how technical debt impacts reliability targets can be found in several high-profile system failures:

Case Study 1: A major ecommerce platform experienced repeated outages during holiday sales due to outdated database technology that couldn't scale to meet sudden increases in demand. The technical debt in their database architecture directly led to substantial direct sales losses and damaged customer trust.

Case Study 2: A financial services company failed to meet regulatory compliance standards for data processing times due to legacy code that could not be easily updated to meet new requirements. The result was not only fines but also an expensive, forced upgrade of their systems under emergency conditions.

Strategies to Mitigate the Impact

Managing the impact of technical debt on reliability involves several strategic initiatives:

Regular System Audits and Refactoring: Continuously assessing and improving the code base and architecture to reduce inefficiencies and prevent potential failures.

Investing in Modernization: Allocating resources to upgrade outdated systems and integrate modern technologies that enhance performance, scalability, and fault tolerance.

Implementing Robust Testing and Monitoring: Ensuring that all system components are regularly tested for performance and reliability and setting up comprehensive monitoring to detect and address potential issues before they affect users.

Cultural Shift Toward Quality Assurance: Fostering a development culture that prioritizes long-term quality over short-term gains. This involves training, incentivizing proper development practices, and integrating operations and development teams to better understand and address reliability needs.

The Cycle of Debt and Reliability

Technical debt and reliability are interlinked in a continuous feedback loop, where the presence of one influences the state of the other. Understanding this cycle is crucial for organizations to effectively manage

407

their systems and prevent the accumulation of debt that could jeopardize their operational reliability. This section delves into the cycle of technical debt and reliability, exploring its dynamics and the strategic interventions required to break this potentially destructive cycle.

Understanding the Cycle

The cycle of technical debt and reliability can be visualized as a sequence of cause-and-effect that perpetuates itself unless actively managed:

> **Short-Term Solutions and Quick Fixes:** Initially, technical debt often arises from the need to meet urgent delivery timelines or to patch unexpected issues quickly. These short-term solutions, while resolving immediate problems, usually do not adhere to best practices or sustainable design principles.

> **Accumulation of Debt:** Over time, these quick fixes and patches accumulate, embedding themselves into the fabric of the system. Each layer of quick fixes adds complexity and potential points of failure, which are often not fully documented or understood even by the original developers.

> **Increased Maintenance and Overhead:** As the system grows in complexity, the effort required to maintain it also increases. More resources are diverted to simply keeping the system running, often at the expense of new feature development or performance optimization.

Degradation of System Performance and Reliability: The increased complexity and maintenance overhead lead to a degradation in system performance and reliability. Systems become prone to errors and outages, and their ability to meet set reliability targets diminishes.

Emergency Responses and Further Debt: In response to degraded performance and reliability issues, organizations are often forced into emergency fixes, which, under time pressure, result in further technical debt. This reinforces the cycle, making it increasingly difficult to break.

Case Studies Illustrating the Cycle
Case Study 1: Software Development Company

Initial Compromise: A software development company released their product with several known issues to meet a launch deadline, planning to fix these in subsequent updates.

Accumulation and Impact: Over several update cycles, the quick fixes for various issues became layered and complex. The system's architecture became convoluted, making any new feature addition a risky and time-consuming endeavor.

Response and Further Debt: Each new update introduced more bugs and required more emergency patches, significantly increasing the system's instability and the cost of maintenance.

Case Study 2: Financial Transaction System

Initial Compromise: To cope with increasing transaction volumes, a financial institution implemented a series of patches to their transaction processing system.

Accumulation and Impact: The patches led to an opaque system where changes in one part of the system unpredictably affected others, leading to frequent system downtimes.

Response and Further Debt: Emergency fixes often involved disabling failing components temporarily while seeking a permanent solution, leading to reduced functionality and further ad hoc solutions.

Strategies to Break the Cycle

Breaking the cycle of technical debt and reliability requires a multifaceted approach that includes

Proactive Debt Management: This involves the regular assessment of the existing code base and infrastructure to identify and prioritize the reduction of technical debt through refactoring or rewriting parts of the system.

Adoption of Continuous Integration/Continuous Deployment (CI/CD): Implementing CI/CD practices can help ensure that changes are smaller, more manageable, and tested thoroughly before deployment, reducing the likelihood of introducing new debt.

Cultural Shift Toward Quality and Reliability:
Cultivating a culture that values long-term system health over short-term gains is crucial. This might involve changing how success is measured and rewarded within development teams.

Investment in Training and Tools: Equipping teams with the latest tools and technologies and providing ongoing training can help in managing and preventing technical debt. More skilled teams can produce higher quality work, which in turn reduces the likelihood of debt accumulation.

Case Studies from The Temple

The Temple, as a state-of-the-art data center, serves as a focal point for understanding how technical debt impacts operational systems, particularly concerning reliability and performance. This section presents detailed case studies that illustrate specific instances where technical debt accumulated at The Temple and the consequent measures taken to address these challenges, thereby maintaining system integrity and functionality.

Case Study 1: Legacy Data Processing Application

Background: The Temple's data processing capabilities were initially centered around a legacy application designed to manage and analyze data streams from various sources. While adequate in the early stages of the data center's operation, the application struggled to handle the increased volume and complexity of data as The Temple expanded.

411

Problem: The legacy application was patched multiple times to accommodate new data formats and integration with other systems, which led to a complex and unstable code base. This patchwork approach resulted in frequent crashes during high-volume data ingestion periods, critically impacting real-time analytics and decision-making processes.

Impact: The instability of the data processing application led to significant downtimes, which not only affected real-time operations but also eroded trust among stakeholders relying on timely data analytics. The reliability of data processing—an essential function of The Temple— was compromised, leading to potential losses in operational efficiency and strategic insight.

Resolution: The decision was made to invest in a complete rewrite of the application. This project involved

- Consulting with software architects to design a new system architecture that would be scalable, robust, and easier to maintain

- Implementing modern data processing frameworks capable of handling large volumes of data more efficiently

- Integrating advanced monitoring tools to provide ongoing insights into the system's performance, ensuring that any potential issues could be addressed proactively

Outcome: The new system significantly improved data handling efficiency and scalability. It also reduced the frequency of system crashes, thereby enhancing the overall reliability of The Temple's operations. The proactive approach to redesigning the legacy system demonstrated a commitment to maintaining high reliability standards and provided a foundation for future expansions.

Case Study 2: Outdated Network Infrastructure

Background: The Temple's network infrastructure was initially built with cutting-edge technology. However, as network demands increased and newer technologies emerged, the existing infrastructure became increasingly inadequate, particularly in supporting newer security protocols and handling enhanced data flow efficiently.

Problem: The outdated network infrastructure was not only slow but also vulnerable to security breaches. This was highlighted during a routine security audit that revealed potential entry points for cyberattacks, primarily due to the inability of the old hardware to support the latest security updates and protocols.

Impact: The vulnerabilities posed significant risks to data integrity and privacy, essential for The Temple's operations. Additionally, the network's inability to handle increased data flow efficiently led to bottlenecks that affected the entire data center's performance.

Resolution: A phased upgrade plan was initiated, which included

- Replacing outdated hardware with the latest networking equipment that supported advanced security measures and higher data throughput

- Updating all network protocols to align with current best practices in cybersecurity

- Training the network operations team on the new systems to ensure they could manage and maintain the upgraded infrastructure effectively

Outcome: The upgraded network infrastructure not only resolved the security vulnerabilities but also improved overall data transfer speeds and system responsiveness. This upgrade was crucial in maintaining The Temple's reputation as a secure and reliable data center and provided a scalable platform for future technological integrations.

Conclusion

These case studies from The Temple illustrate the critical nature of addressing technical debt proactively to maintain and enhance system reliability and performance. By tackling legacy issues head-on and investing in substantial upgrades, The Temple ensured its ongoing capability to serve as a robust, efficient, and secure data center, thereby upholding its commitment to operational excellence and technological leadership.

Strategies for Managing Technical Debt

In the context of maintaining a high-performance digital infrastructure like The Temple, effectively managing technical debt is critical for ensuring long-term system reliability and efficiency. This section outlines various strategies that can be implemented to manage and mitigate the impact of technical debt, thereby safeguarding the integrity and operational capability of such advanced systems.

Proactive Debt Management

Regular Audits and Reviews: Conducting regular audits and code reviews is a foundational strategy for identifying and assessing the extent of technical debt within a system. These audits should focus on both code quality and architectural soundness, ensuring that all components of the system adhere to current best practices and are capable of meeting future demands.

Implementation: Set up a routine schedule for audits that involves both internal teams and, if possible, external experts. Use these audits to create a prioritized list of areas needing improvement.

Refactoring and Consolidation: Refactoring involves restructuring existing computer code— changing the factoring without changing its external behavior. This is crucial for reducing complexity and improving the readability and maintainability of the code.

Implementation: Integrate refactoring into the regular development cycle. Allocate time and resources in each development sprint or cycle specifically for refactoring tasks.

Technical Debt Documentation: Maintaining detailed documentation of all decisions that could lead to technical debt is crucial for future mitigation efforts. This includes documenting quick fixes, workarounds, and areas where best practices were not followed.

Implementation: Develop a standardized documentation process that is followed during every project phase. This documentation should be easily accessible and regularly updated.

Balancing Project Management

Debt Awareness in Planning: Project managers must be aware of the technical debt that exists and how it could affect the project's timeline and budget. Integrating technical debt considerations into project planning can help in balancing new feature development with necessary maintenance work.

Implementation: Train project managers and team leaders to recognize and understand technical debt. Include technical debt metrics in project planning tools and dashboards.

Resource Allocation for Debt Reduction:
Allocating specific resources for reducing technical
debt, such as dedicated time or teams, ensures that
debt reduction does not become sidelined by new
developments.

Implementation: Create a budget line specifically
for debt reduction activities. Consider establishing
a dedicated team focused on improving system
architecture and reducing legacy code.

Cultural and Process Adjustments

Promoting a Quality-First Culture: Cultivating a
culture that prioritizes long-term code quality and
system reliability over short-term achievements is
essential for managing technical debt effectively.

Implementation: Encourage practices like pair
programming and code reviews. Recognize and
reward team members for quality improvements
and effective debt reduction.

Continuous Learning and Improvement:
Encouraging continuous learning and staying
updated with the latest technologies and
methodologies can prevent the accumulation of
technical debt due to outdated practices.

Implementation: Offer regular training sessions
and access to courses and certifications. Promote
knowledge sharing through tech talks and
workshops within the organization.

Leveraging Automation: Utilizing automated tools for code analysis, testing, and deployment can help in identifying potential issues early and reducing human error, which can lead to technical debt.

Implementation: Invest in software tools that automate code quality checks, security audits, and performance testing. Ensure that these tools are integrated into the Continuous Integration/Continuous Deployment (CI/CD) pipeline.

Conclusion

Managing technical debt requires a holistic approach that encompasses proactive management, strategic planning, cultural shifts, and the adoption of advanced tools and processes. By implementing these strategies, organizations like The Temple can ensure their digital infrastructure remains robust, efficient, and capable of adapting to future challenges. This proactive approach not only enhances operational reliability but also positions the organization for sustainable growth and innovation.

Reliability vs. Customer Features
Understanding Reliability

System reliability refers to the chances that a software system will not fail within a given time. Reliability is usually given as a percentage: runtime without failure divided by the total run time, including the failures. A higher percentage means the system is more reliable, while a lower rate means the system is less reliable. The advantage of system reliability is that the operations cannot be stopped or downtime will be minimal. Since a high system reliability (99% or 100%) benefits businesses, companies will opt for a high system reliability [1].

Similarly, network reliability refers to the chances or probability that a computer network will perform its function to an acceptable level within a given time. The network reliability measures include terminal reliability, capacity-related reliability, and travel time reliability. Terminal reliability is the probability that end-to-end nodes will remain connected within a given time [3]. Capacity-related reliability is the probability that bandwidth will be available on a network within a given time. Travel time reliability is the probability that data will take the specified time to move between end nodes without failure.

System uptime is the probability that a computer system is working as intended. "Uptime" was coined in the 1950s when mainframe computers experienced frequent failures [5]. Uptime is different from availability in that while uptime refers to the percentage of time that a system is running as intended, system availability is the probability that users will access the information on the system and in the required format within a given time.

System redundancy refers to the availability of multiple ways of executing a function. If there is more than one way of completing a task on a system, it is said the system is redundant. Redundant systems have more equipment for completing a task. Lastly, fault tolerance is the ability of a system to continue working after the parts have failed. A system becomes tolerant to a fault because of redundancy, sharing the load, and availability of backup [3].

Designing for Reliability

Best practices in system architecture, such as distributed systems, cloud computing, and disaster recovery planning, involve designing fault-tolerant systems. Distributed systems refer to computer systems that collaborate to achieve a common goal. Unlike centralized systems, where the computer systems are connected to one server, distributed systems utilize the computer system's resources to ensure that users access resources even if there is downtime in other computer systems [9]. The figure below illustrates distributed systems and centralized systems.

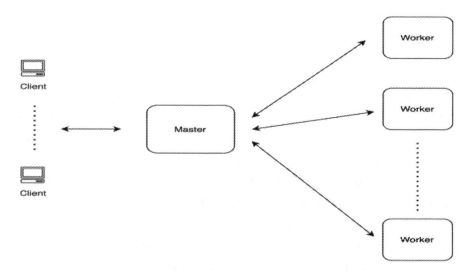

Figure 12-1. *Centralized and distributed systems*

Best practices in system architecture include modularizing components, designing for failure, choosing the correct communication model, balancing consistency, securing the system, and monitoring the system.

Modularizing the Components

Modularity refers to creating independent divisions in the system software. These divisions ensure that the system can be run separately, for example, when testing and maintaining. The advantage of modularity is that it helps the designer address the system's complexity and heterogeneity. Moreover, modularity helps deploy and update the system's components. Examples illustrating the importance of designing for modularity in distributed systems include monitoring services and distributed sandboxes [1].

420

For instance, a monitoring service for an extensive distributed system logs events, analyzes them, and issues reports. Suppose a monitoring service has components in the distributed system consisting of three independent sets: logos, analyzers, and reporters. Logos are responsible for accepting and maintaining logging notices. Analyzers analyze the logs, and reporters create reports for stakeholders. Now, the application programming interfaces (APIs) help store and retrieve the information in the system [6]. Since the monitoring service handles crucial information such as logs and reports, this information needs to be protected. The best way to protect this information is to enforce the constraints of data transfer on the Internet in the distributed system.

Another example of designing for modularity in distributed systems is the distributed sandbox, whereby the system obtained from untrusted third-parties uses untrusted code. Also, a part of the newly developed system uses untrusted code until it gets tested. The safety of these codes depends on the monitoring of the distributed systems.

Designing for Failure

Failures are inevitable in distributed systems due to the complex interconnected nodes. It is, therefore, essential to design the system to withstand failure. Designing systems for failure does not mean the system will not fail; it will be able to withstand the downtimes. In other words, it creates a fault-tolerant system. Therefore, designers must consider the failure models when designing the system for failure. Failure models help designers categorize how things can fail in distributed systems, cloud computing, and disaster recovery plans [7]. For example, how should a system behave when a computer stops working or there is a network hitch? For example, during data recovery, the network may fail, thus impeding the efforts to recover data. Designers can design systems that tolerate failure when anticipating what might go wrong.

Failures in distributed systems can manifest in different ways. First are node failures; since computers or servers can fail, nodes can also fail. Some nodes that can fail include the connection points on the routers, switches, or other devices connected to a network. Nodes can crash for various reasons, but hardware or software malfunctions are the most common. Failed nodes are unresponsive and cannot complete the tasks assigned, thus disrupting the system's functionality [8]. To mitigate the effect of node failures, designers must implement redundancy and failover features.

The second type of failure in distributed systems is network failure, whereby delays in the nodes relaying information disrupt the communication channels. The disruption of communication channels may be due to issues with the hardware systems and routers or the congestion of the network. Designers must use redundant network paths and fault-tolerant protocols to reduce or mitigate network failures. The third type of failure in distributed networks is software failure caused by bugs or errors resulting from noncompatibility or uncaptured mistakes during programming [9]. To reduce the software errors, developers should employ error-handling mechanisms.

Moreover, distributed systems may fail due to partition failures due to the isolation of network nodes. Partition failures may occur due to network misconfiguration. The major challenge in portioned networks is data consistency and synchronization—this means that data presented in the system becomes inconsistent. To ensure consistency in a network, developers should use quorum systems [10]. Lastly, byzantine failures arose due to compromised nodes. To address Byzantine failures, developers can use fault-tolerant algorithms.

The occurrence of failures in software or network systems calls for developing failure models. Failure models or architecture provide a map or a guide for understanding what may or can go wrong in system— distributed cloud computing or disaster recovery planning. Developers can study the failure models and create solutions to mitigate potential errors. The following are the failure models that designers can anticipate and address before they become worse.

Crash failures occur when the nodes in a distributed system or cloud computing stop suddenly. To mitigate crash failures, designers must employ redundancy and checkpointing. As defined, redundancy ensures the system continues running even if some parts fail [10]. In networking, designers can address crash failures by ensuring that multiple nodes can perform different functions. Consider the diagram below.

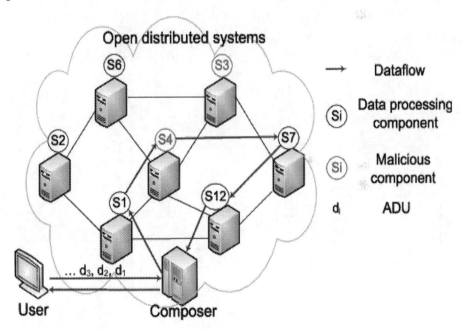

Figure 12-2. *Data flow in distributed systems [11]*

The figure above shows distributed systems, and the arrows show data flow. Data flows from the user to the composer, then through different components. Suppose a node at S4 fails; the user can still access resources on the network due to system redundancy; that is, the system has several pieces of equipment that can perform the same function. Therefore, in case of a crash failure on one point, the other servers can complete the user's request. Once the system crashes, it issues a notification, but the other servers continue operating. This gives the developers time to isolate and reintegrate the crashed nodes into the system.

Byzantine Failure Design: Byzantine failure is the worst form of system failure because the nodes provide false information—this means the problem results from the nodes, which are part of the system [2]. The reason for Byzantine failure may be malicious attacks that compromise the functioning of the nodes. Therefore, when designing a software system, it is crucial to determine the potential entry points of a malicious attack and seal them.

The following diagram illustrates Byzantine failure propagation.

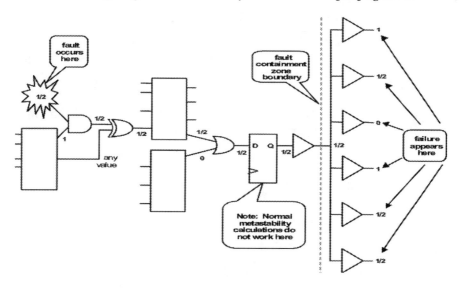

Figure 12-3. *Byzantine failure model [2]*

In the Byzantine failure model above, the fault occurs at one of the nodes, and the failure is transmitted throughout the other nodes. To resolve the Byzantine failure, designers should create a fault containment zone boundary between the zones. Once the fault containment zone boundary has been established, it is possible to keep the system running under the redundancy principle while errors are detected and corrected. If left unaddressed, Byzantine failures can cause unreliability in the system—the system will fail to function as intended. Therefore, the fault containment zone boundary is created using Byzantine fault-tolerant algorithms. These algorithms detect and fix the failures. Designers can generate consensus protocols to ensure the communication channel maintains its integrity.

Transient Failures Model

Transient failures are temporary, and the system can return to working independently every day. The primary cause of transient failure is network glitches. While transient shortcomings can be resolved separately, they are challenging to detect. Therefore, designers should create retry mechanisms to resolve transient failures. For example, suppose a network fails to connect; the software system should be able to initiate a retry mechanism. If the problem is transient, the network can connect after retrying [3]. However, transient failures can sometimes persist or take a relatively long time to resolve, contributing to system downtime. Designers must create an exponential backoff to handle errors if a user or client's request to connect to a network fails persistently.

Exponential backoff is mainly used in cloud computing to help clients connect to a network after several failed attempts. For example, suppose a client requests to access a particular resource on a network, and the system fails; the exponential backoff should enable the system to return an error, mainly the error codes [3]. Once the exponential backoff is reached, the client can retry again in a series of minutes, such as 1, 2, 3,...up to the exponential backoff. Failure of all retries should prompt the system to log an error to the server. However, the maximum number of retries depends on the network conditions.

Failure Tolerance

The most essential feature in software system design is failure tolerance. This means a system can continue working despite failures. Failures are unavoidable in distributed systems and cloud computing [2]. Therefore, it is essential to design systems that can withstand failure. The following are features of failure-tolerant systems.

- Redundancy refers to when the system has more than one node performing the same function, so the failure of one node or component does not impede the system's functioning.

- Replication refers to duplicating data or services such that different nodes carry the same data. The failure of one node will not stop the system from functioning.

- Degradation refers to downgrading the system's functionality to ensure it continues working in the event of failure. For example, when logging into an email server and there is a network glitch, the system should pave the way for graceful degradation, whereby the email can load in other formats, such as the standard HTML [4].

The best practices that work in creating resilient distributed systems and cloud computing platforms also apply in disaster recovery planning. However, recovery aims at reinstating the system to its original working state following a failure. To maintain the integrity of the system during disaster recovery planning, designers should implement the following features:

- **Heart Beating:** This involves determining the reachability of the nodes. The nodes are unreachable if they fail to respond to the messages sent within a given time.

- **Timeouts:** Designers can plan for disaster recovery by setting timeouts. If the system fails to respond within a given time, it has failed.

- **Quorum Systems:** The thresholds should be used in decision-making when failures occur. However, decisions are reached after several nodes are in consensus.

- **Rollback:** This involves resetting the system to its previous working state before the failure. This ensures data recovery at the point when the system is working well.

Monitoring and Maintenance

Monitoring tools monitor software systems—these are software tools that measure the performance of a system. These tools collect and return indicators such as

Response time is the time it takes for a node to respond to messages sent. A well-functioning system should display messages sent within a specified time. The system needs repair if the messages are displayed outside the given time [5].

Data throughput is the amount of data that passes through a system. Monitoring should include throughput to ensure the system sends and receives the required data. Throughput should also consider system degradation, whereby the system operates at a lower level in case of failure. The system is reliable if the throughput coincides with the intended one. However, the system is only reliable if the throughput is defined.

Error rate determines the system's reliability. The higher the error rate, the less reliable the system. Conversely, a low error rate shows that the system is reliable. Most designers aim for an error rate of less than 0.1%.

Uptime is the measure of time that the system is running. It helps to determine reliability in the sense that a higher uptime shows the system is reliable, while a lower uptime indicates the system is unreliable. Other monitoring tools that can be used to determine system reliability include CPU usage and disk space. The CPU should be able to handle data requests efficiently. This also applies to cloud computing, whereby the CPU usage should handle server data requests.

Testing ensures software reliability by detecting and solving errors before deploying the system. Testing also ensures that the system meets quality standards, enhancing security and customer satisfaction. The strategies designers can use to ensure system reliability include defining the testing objectives; this involves the intended purpose of carrying out the test. Second, the testing procedures should be automated; the tools used in the testing process should be automatic to increase the chances of detecting and fixing errors [8]. Also, automatic testing saves on development resources, including time and money. Third, designers should implement various types of testing to ensure each test is covered.

Next, designers should perform the relationship analysis to determine the effect of the new code on the system's functioning. The regression analysis also ensures that the system will continue working even after changes to the code. Lastly, designers should collaborate with stakeholders to ensure successful testing. Stakeholders are the people who will be using the system, and it is essential to incorporate them in the testing phase to get firsthand information.

Update protocols should focus on improving the system's quality. The protocols should include security enhancements, bug fixes, and feature enhancements. Updates are implemented to address vulnerabilities in a software system. As the system works, it is exposed to various vulnerabilities that can be addressed through regular updates. Up-to-date systems are not vulnerable to security breaches. For example, suppose a web application has an unpatched sequence query language injection; an attacker can access the database if the SQL injection vulnerability is not addressed. Another update protocol is to use patches when fixing bugs. Lastly, update protocols should focus on enhancing features that improve user experience—for example, implementing a single sign-in feature when updating a student login system.

Reliability Challenges

Creating reliable systems has its challenges. These challenges include system design, implementation, and how the system operates.

First, balancing consistency and availability: It is challenging to have a consistent and available software system. While data must be consistent, the system should also be available. It becomes a challenge to balance data consistency and the system's availability. For instance, in financial services, healthcare, and critical infrastructure, ensuring data consistency may lead to the system needing to be available. Similarly, ensuring system availability may contribute to inconsistencies in data. This leads to the design of complex systems that significantly affect system uptime.

Second, scalability: Software systems exhibit the challenge of scalability due to the complexity of decisions required to monitor and maintain them. Financial, healthcare, and critical infrastructure should be scalable to meet users' demands. For example, as the number of users of financial services increases, the economic system should be able to handle requests between the client and server and maintain uptime—however, the system's expansion results in challenges relating to redundancy, replication, and fault detection.

Third, complexity: Fault-tolerant algorithms and redundancy features result in system complexity that may be challenging to address. Therefore, designers must plan and execute the system carefully to integrate the complex challenges.

429

Fourth, environmental dynamics: Designers may encounter challenges related to the system topology and workload, especially when integrating resilience features. For example, in financial services, designers may implement redundancy features to ensure the system works even if it has a fault. Such features, however, may create a dynamic environment that leads to system complexities. Consequently, continuous upgrading leads to a challenge in maintaining the system.

Fifth, overheads related to operations: Budgeting for software systems may be a challenge because of the overheads exhibited during the system's design, implementation, and deployment. For example, developing financial and healthcare software systems must follow integrated reliability features such as redundancy. However, the inclusion of redundancy in the system means additional resources. The nodes on a network must be programmed to work independently in case the other nodes fail.

The reliability challenge is present mainly in critical infrastructures, which continue to become vulnerable to cybersecurity attacks. The cyberattack on the federal government's resources and other sectors, such as the energy sector, indicates that reliability is still challenging. Despite the efforts to build resilience, complexities in the system contribute to these challenges. In [8], the authors found that the complexity of critical infrastructure exposes components to risk. For example, the electric power grid is a critical infrastructure so complex that addressing all its components creates a challenge. CIs' complexities include increased load demand, structural aging, and failure. Addressing these challenges contributes to an increase in overhead costs.

Innovations in Reliability

Emerging technologies such as blockchain and the Internet of Things (IoT) are enhancing the reliability of software systems. Blockchain has become a powerful tool for improving reliability in data systems. The advantage of blockchain, which makes it reliable in enhancing data security, is its decentralized nature. This means it is difficult for attackers to access sensitive information. Blockchain does not work with intermediaries, thus making it resistant to cyberattacks [7]. Therefore, the advantages of blockchain that make it a reliable technology for increasing reliability in software systems include enhancing data security and a decentralized network. A decentralized network ensures that a single point of failure cannot impede the operation of the entire system.

Moreover, data is protected by cyberattacks. While blockchain systems can be attacked, the attack can easily be traced due to the system's transparency and traceable nature. Since the system is transparent and traceable, it is easy for developers to identify and correct the vulnerability before significant damage is done. Moreover, blockchain enables team members to work together on a project concurrently.

The emergence of IoT has also enhanced network reliability in software systems by ensuring different devices can be connected to the network anywhere, anytime. For example, suppose there is a network failure; it is possible to reconnect with another device on the network that can be accessed remotely. Like blockchain, IoT is also decentralized, which reduces the reliance on a centralized authority, thus enabling data sharing [6].

Chapter Summary

The research on the reliability of software and networks shows that the complexity of these software systems and networks makes it challenging for designers or developers to address the vulnerability. This is due to the challenges between consistency and availability. However, the good news is that blockchain and IoT can now address these challenges. Companies and government agencies should implement blockchain and IoT to ensure the security of a network because blockchain, being decentralized, can leverage other resources on the network. Moreover, the block can use smart contracts to reduce the risks of cybersecurity attacks. Similarly, IoT can ensure network security by enabling different devices to connect to the network at times, which means designers can connect monitoring devices to the network anytime, anywhere.

Bibliography

1. Dinesh Kumar, U., John Crocker, Jezdimir Knezevic, and Mohamed El-Haram. *Reliability Maintenance and logistic support: A life cycle approach.* Dordrecht, Pays-Bas: Springer Science+Business Media, B.V, 2012

2. Driscoll, K., Hall, B., Paulitsch, M., Zumsteg, P., and Sivencrona, H. "The Real Byzantine Generals." *The 23rd Digital Avionics Systems Conference (IEEE Cat. No.04CH37576)*, 2004. https://doi.org/10.1109/dasc.2004.1390734

3. Hussain, Shafiq. *Byzantine failure against colluding attacks in Cloud Data*, July 4, 2022. https://doi.org/10.31219/osf.io/eaxby

4. JIA, Jia, and Xue-Jun YANG. "Propagation Behavior Analysis and Fault Tolerance Optimization of Hardware Fault in Heterogeneous Systems." *Journal of Software* 22, no. 12 (December 14, 2011): 2853–65. https://doi.org/10.3724/sp.j.1001.2012.04057

5. Nasreen, M.A., Amal Ganesh, and C. Sunitha. "A Study on Byzantine Fault Tolerance Methods in Distributed Networks." *Procedia Computer Science* 87 (2016): 50–54. https://doi.org/10.1016/j.procs.2016.05.125

6. Raghav, Dhruv, D. K. Rawal, Ibrahim Yusuf, Rabiu Hamisu Kankarofi, and V. V. Singh. "Reliability Prediction of Distributed System with Homogeneity in Software and Server Using Joint Probability Distribution via Copula Approach." *Reliability: Theory & Applications* 16, no. 1 (March 2021): 217–30

7. Reliability Growth: Enhancing Defense System Reliability. United States: National Academies Press, 2015

8. Walraven, Stefan, Bert Lagaisse, Eddy Truyen, and Wouter Joosen. "Dynamic Composition of Cross-Organizational Features in Distributed Software Systems." *Distributed Applications and Interoperable Systems*, 2010, 183–97. https://doi.org/10.1007/978-3-642-13645-0_14

9. Yilmaz, Murat, Serdar Tasel, Eray Tuzun, Ulas Gulec, Rory V. O'Connor, and Paul M. Clarke. "Applying Blockchain to Improve the Integrity of the Software Development Process." *Communications in Computer and Information Science*, 2019, 260–71. https://doi.org/10.1007/978-3-030-28005-5_20

10. Zhao, Guilin, and Liudong Xing. "Reliability Analysis of IOT Systems with Competitions from Cascading Probabilistic Function Dependence." *Reliability Engineering & System Safety* 198 (June 2020): 106812. https://doi.org/10.1016/j.ress.2020.106812

11. Zio, Enrico. "Vulnerability and Risk Analysis of Critical Infrastructures." *Vulnerability, Uncertainty, and Risk*, June 27, 2014. https://doi.org/10.1061/9780784413609.003

CHAPTER 13

Cost of Ensuring Reliability

Author:
Anirudh Khanna

Reviewer:
Gaurav Deshmukh

Understanding Reliability Needs

In addressing systems and networks, reliability refers to operating continuously while performing intended duties without failure over a required period. Reliability is critical in defining actions and recounting management of core appeals related to marking the development of stable operations of systems and networks [1]. Thus, modern organizations must consider the reliability framework of technological devices and additions to easily integrate and address their demands at all points of managing and marking their developmental requirements. Reliability has various components, each working to ensure ease in functionality and appropriate management of core appeals, reiterating the development and growth of core approaches to handle significant functionalities. Components of reliability include uptime, redundancy, and fault tolerance.

Uptime

Uptime refers to the duration within which the network remains operational and accessible. Uptime is relevant to ensure optimal performance of duties, remarking growth and development in achieving the right outcome in systems and networks. More to the point, a high uptime ensures businesses can run and maintain their online presence alongside core functions that help detail considerable advances to the desired end. Uptime is critical to ensuring that companies can operate within a significant margin and maintain the functionality and achievement of their systems to a desired level of addressing duties as needed. The element of uptime can be measured as a percentage of the total time over a provided period [2]. The categorization of uptime details tiers of functionality and reliability, remarking an influential handling of engagement to a determined end. The equation used in calculating the uptime in networks and systems is

$$\left(1 - \frac{Downtime}{Total\ Time}\right) \times 100$$

For instance, provided that an organization's system is down for only 10 minutes of a month, then the total uptime for the company is

$$\left(1 - \frac{10}{43200}\right) \times 100 = 99.977\%$$

From the provided equation, the organization enjoys a high uptime of 99.97% in the month, hinting at an excellent level of functionality in the company, offering distinctive results in advancing valuable outcomes in achieving good system handling and management. Thus, the system equation is vital in detailing and marking an appropriate understanding of how the system works and has to be addressed to achieve the best results. More to the point, the uptime tiers are categorized differently depending on the application industry. The application of these uptime tiers relates to various needs of the companies; however, a significant application is in terms of nines, hinting at the right level of application within the companies. The uptime tiers include the following:

- **99.9% Uptime (Three Nines):** This indicates that the company experiences an estimated 8.76 hours of downtime in a year.

- **99.99% Uptime (Four Nines):** This depicts that the company experiences 52.56 minutes of downtime every year.

- **99.999% Uptime (Five Nines):** This uptime indicates the company experiences approximately 5.26 minutes of downtime yearly.

Redundancy

This is a critical factor in systems where a particular component that holds vital importance is duplicated to enable continuity in any event of failure. The approach ensures that the redundancy helps with the occurrence of failures, ensuring minimal downtime and addressing consistency in the system's performance. This is a remarkable step in achieving high system efficiency in marking and achieving the desired values at all levels of system functionality [3]. There are different kinds of redundancy, such as

- Hardware redundancy implies duplicating physical components like hard drives (RAID configurations), servers, network connections, and power supplies.

- Software redundancy entails using a backup to address software systems. The approach enables load balances and automated failover mechanisms to address software functionality in the event of a failure.

- Network redundancy enables multiple data paths and redundant network devices to ensure appropriate development and advancement. It also involves using backup Internet connections to ensure that the organization is always connected to the Internet.

Redundancy can be applied in different instances, each aiming to develop core approaches to create critical models for addressing and achieving functionality as required in every indication. The essential cases of implementation of redundancy include

- Data centers are geographically dispersed to ensure the company has a mechanism for surviving any disasters or emergencies affecting its onsite data centers.

- Load balancers within the network help distribute traffic on multiple servers to ensure no overload, which can affect the functionality and engagement to achieve availability at all times.

- RAID is a model for handling data across several hard drives. It aims to protect against disk failure and achieve remarkable sustainability in managing the development of every step in addressing efficiency at all levels.

Fault Tolerance

This component implies the ability of a network or system to have continued functionality even in the case of a failure. The approach enables sustainable understanding and management of network configuration to ensure they can handle failures and achieve the set objectives within the proper framework. The fault tolerance concept ensures that the systems can detect, isolate, and handle failures automatically without human intervention, leading to a high level of proficiency [4]. Fault tolerance can be managed through different techniques, each aiming to ensure critical development, all pointing toward creating a modest scope and path to achieve sustainable engagement at all levels. The concept of fault tolerance can be conducted through

- **Failover Mechanisms:** Organizations can automatically install failover mechanisms, such as automated switching to a system when the primary element fails. This implies that the company will always ensure continued service and engagement, achieving a remarkable level of suitable engagement.

- **Error Detection and Correction:** Systems and networks can use different elements to achieve a reliable outcome in marking progressive management at all levels. Checksums, error-correcting codes, and parity bits provide a channel for identifying and correcting data errors whenever they occur.

- **Replication:** This involves having copies of data and applications on both onsite and offsite locations to help ensure access availability in case of failure and manage the growing need to identify and mark progressive handling of the system at all times.

Fault tolerance can be implemented in various ways, ensuring critical development to address significant concerns and targeting the consistency of systems and networks. Virtualization primarily offers the chance to have virtual machines that can be replicated in different locations to help address hardware failures whenever they occur. Nonetheless, database replication also ensures that multiple copies of data on other servers can be used to achieve data availability and consistency.

Achieving High Reliability

Achieving high reliability is a concern for most organizations, marking the demand to address significant needs in handling everyday functions. High reliability is thus a keen factor in managing and ensuring considerable development to advance and handle needs as required. Therefore, organizations must follow various demands to ensure high reliability and manage their development to tackle pertinent issues better.

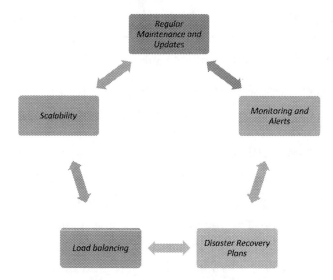

Figure 13-1. *Steps to achieving high reliability*

Figure 13-1 indicates the steps to achieving high reliability in systems and networks. These steps include regular maintenance and updates, scalability, load balancing, disaster recovery plans, monitoring, and alerts. Each step ensures an accurate definition of reliability, helping the systems and networks achieve a new definition of reliability on all fronts.

Regular Maintenance and Updates

Network and system administrators must ensure relevant development in affiliation with critical entities, consistently providing the most meaningful use. Using approaches such as patch management will ensure that security patches can be installed to help avoid vulnerabilities that haunt the system and affect the provision of mechanisms that will relate to better outcomes in managing and handling their needs [5]. Hardware maintenance is also crucial to define functionality since every piece of hardware in the company will be addressed with essential advancements in handling aging hardware to achieve better, faster, and more potent new hardware to help achieve organizational goals.

Monitoring and Alerts

Network and system monitoring is a significant step to ensuring that there are no anomalies and pending issues on the system resolved on time. Real-time monitoring helps provide the right tools that continuously and automatically monitor the network's performance, traffic, and resource use. This approach enables continued modeling of the network to achieve suitable advances, helping to bring out remarkable progress in managing the network system. The real-time monitoring, alongside automated alerts, helps to inform on potential issues and maintenance approaches that will provide a reliable scope of addressing functionalities within the network.

Disaster Recovery Plans

Organizations have to install actions and plans that help with regular data backup and management of their information. Regular backups ensure that there could be offsite, onsite, or cloud options to store their data and enable the right approach to recovery during any significant issues that affect the functionality of the systems. The recovery plans must be installed by having the best testing and awareness approaches to ensure critical development in marking progressive management of the data resources that help achieve sustainable outcomes within the frame of any functionalities expected from the system [6].

Load Balancing

Load balancing is critical to ensuring high reliability. Incentives such as the distribution of workloads through load balancers offer the chance to distribute traffic on multiple servers, preventing any challenges that might lead to downtimes. Using the load balancing options will encourage horizontal and vertical mechanisms that introduce more servers to handle increased demand and traffic on the network.

Scalability

Scalability on both horizontal and vertical planes will ensure the introduction of incentives to help achieve reliable load development to achieve greater capacity and attain valuable engagement at whatever point is demanded. The element of horizontal scaling is handled by ensuring more servers to ensure that the capacity and load can be increased to achieve a remarkable outcome in dealing with valuable engagement on the platform [7]. Vertical scaling, on the other hand, helps to upgrade existing hardware to ensure that the system can handle every demanded addition, leading to much better additions and engagements and achieving reliability at a remarkable scope of performance.

Reliability Metrics

Metrics are crucial to understanding the level of functionality of systems and networks. Organizations must realize every model and engagement appeal that works within the promise and scope of delivering the best outcome and achieving valuable outcomes. Therefore, reliability metrics have to be used to ensure the creation of critical advancement in targeting and working within the right approach to help address any downtimes or occurring issues within the provision of sustainable value. The key metrics used in addressing reliability include elements such as

- **Mean Time Between Failures (MTBF):** This implies the average time between two successive failures on the network. The approach indicates the capacity of operations to be conducted before a failure occurs.

- **Mean Time to Repair (MTTR):** This indicates the timeline for repairing the system in case of failure. The lower time to repair suggests that the company has a faster recovery, leading to higher outcomes for its activities [8].

- **Service-Level Agreements (SLAs):** These are agreements between service providers and clients that help indicate the level of service and reliability of the systems before any downtime occurs.

Costs Associated with Reliability

Direct Costs

Enhancing reliability within the organization demands direct costs that the company has to consider, remarking the need to identify and manage adjustments to handle their needs. These costs are directly related to the capacity to conduct reliability integration and management within the institution.

Initial Design and Development Costs

These are costs incurred in the primary phase of implementing the system design elements, which help structure and address reliability-related components. The system architecture design costs are incurred when skilled architects are consulted and involved in designing robust systems. The architects are also tasked with fault-tolerant systems, which incur the costs of handling advanced simulation tools that help map the system's functionality to achieve desired elements [9]. Nonetheless, the design and development stage takes in handling hardware and software, which can ensure the inclusion of highly reliable components that seek to enhance the system's functionality by standing the test of time. Additionally, prototyping is a crucial cost factor in the organization, assisting in handling the iterative design process, where the organization has to spend money on building the prototype system and network, engaging it to see whatever possibilities have to be taken to achieve the desired outcome. Therefore, the approach works by enabling and ensuring successful development process management to achieve indicated goals.

Testing and Validation Expenses

These are costs incurred while ensuring a remarkable system and network analysis and handling. In the first instance, lab testing is conducted to assist in understanding the scope and potential of the network and system. The testing demands the use of different equipment that has to be purchased to ensure an increasingly beneficial way to create an understanding of the network scope and capacity. Lab testing expenses involve having load testers and network simulators that increasingly bring along better identities for the system functionality.

Another critical investment lies in field testing, where testing has to be conducted with real-world scenarios to create a channel for understanding the performance and reliability of the system. Additionally, the use of logistics and coordination approaches assists in ensuring the management of the system to help attain a desired level of reliability. To this level, testing simplifies reliability dynamics, each step helping craft a meaningful outcome in every provision [10]. Therefore, security and compliance testing addresses the required standards and functionality scope. Thus, these expenses ensure that the system can be advanced to achieve reliable levels, which assist in addressing functionalities at all the necessary points.

Quality Assurance and Monitoring

Costs incurred in quality assurance and monitoring ensure that the organization has to key in additional funds to help advance to achieve continuous integration and deployment. Reliability has to be achieved by implementing and maintaining the infrastructure to achieve an instrumental balance of functionalities in the network. Real-time monitoring and diagnostics are also costly, as they must be implemented with crucial knowledge to address the possibility of challenges from various operations. The costs have to be provided to ensure that alert systems are installed correctly, enabling progressive management of the entire organizational approach in modeling the core needs of the system.

Redundancy and Backup Systems

Institutions must spend money on suitable backup systems, which will help revamp the system's reliability. Investing in the right system will create a chance to channel and chart instructional development to achieve beneficial outcomes. Primarily, purchasing redundant hardware will help by ensuring duplication to secure the system in case of imminent failures. Having duplicate hardware will ensure that uptime is handled when a component fails [11]. More to the point, investing in a backup data center requires purchase from the institution, assisting the network to have continued functionality even in emergencies. The cost of the data center creates a synchronization and failover mechanism that enables continued functionality in achieving the desired components. Nonetheless, the backup system continually requires data recovery options, which demand more investment in having stellar functionality and a system to attract the right outcome in managing data development needs.

Software Licensing and Maintenance

Reliability in networks and systems demands suitable software investment, an approach that demands core approaches in having the right software. This cost makes achieving peak functionality and reliable outcomes much easier because the best software versions create a sustainable address of whatever needs are categorized for the system to operate well. Hence, the costs for the licenses of different software and tools create a remarkable level of addressing additional needs in achieving optimal performance.

Training and Development

Reliability in the network and system stretches from having hardware and software tools to address functionalities to having the right human capital. An organization's staff must be trained and enlightened on best practices that will encourage good results. The reliability of the staff is a crucial determinant of the relevant outcome needed in addressing various demands in the organization. Therefore, the training and development, alongside steps to ensure employees' certification, will ensure continuous knowledge sharing and generating experience, which creates the relevant steps to address better outcomes at whatever level is required.

More to the point, process development and optimization are key factors that help advance employee and software efficiency. The system's reliability is based on the capacity to enhance processes and achieve suitable outcomes when appealing to individual functionalities [12]. The system has to be handled with crucial knowledge in addressing and adjusting every functional entity to manage the needs well. Therefore, using the right development approach creates a step to ensure value provision and reliability handling to achieve the network's overall efficiency.

Vendor and Supplier Management

This direct cost is associated with having the right vendors and selecting the best suppliers. The approach encourages the management of suppliers by selecting their contingencies for disruptions and incurring costs in the selection process. The approach is heavily impacted by the need to register vendors' actions and achieve impactful handling of their needs at all times. Therefore, vendor and supplier management structures the performance and modeling of the system to ensure remarkable support for tools and practices, getting to a new level of attaining reliability in the organization.

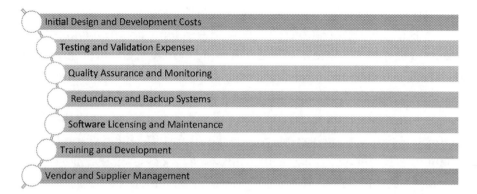

Figure 13-2. *Direct costs associated with reliability*

Figure 13-2 indicates the various direct costs associated with reliability in an organization. These figures include initial design and development costs, testing and validation expenses, quality assurance and monitoring costs, redundancy and backup systems costs, software licensing and maintenance costs, training and development costs, and vendor and supplier management costs. These costs assist in achieving remarkable levels of reliability in an organization.

Indirect Costs

Handling reliability in systems and networks attracts indirect costs for organizations. An institution has to foot these expenses to ensure it can considerably address the growing demand for the system. Some indirect costs affiliated with the management of networks and systems are warranty claims and returns, downtime, and loss of productivity.

Warranty Claims and Returns

Indirect costs affiliated with reliability begin with warranty servicing in handling logistics and customer service for warranty claims. The expenses come with administrative costs of handling documentation and claims pertinent to the products. Overseeing software expenses and managing help desk operations to deliver effective and efficient customer support. The costs of repairing defective products under warranty are also critical in addressing the costs associated with the organization [13]. Therefore, handling reliability is essential to advancing and achieving high levels of performance and reliability at the desired level.

On the one hand, product returns are another cost incurred by the institution. The reliability cost addresses the handling of inventory and defective products—costs associated with managing faulty products, handling inventories, and achieving the required needs. Warranty claims and returns lead to high costs associated with handling tools that will bring about system functionalities. Therefore, these indirect costs must be considered to achieve a remarkable level of proper functionality in the institution.

Additionally, legal and compliance fees are incurred to maintain the institution's reliability. The institution has to ensure they have the proper legal backing and understanding of every software and hardware they use. This approach enables instrumental management and handling of the software to warrant and is instrumentally beneficial in achieving and handling the correct details in marking their progressive development. The fees that come along with ensuring compliance ensure an increased cost on the organization's side, leading to higher demands in achieving sustainability of operations.

Downtime and Loss of Productivity

Reliability is expensive in organizations that need the correct address for equipment and software needs. Operational interruptions are a significant cost in the case of less reliable systems. The interruptions lead to decreased organizational output and delays, which cost more money. In this case, the management and handling of regular operations in organizations are taken back because of the bottlenecks associated with minimal reliability in the system. Therefore, using reliable LESS systems leads to increased lead times and lower efficiency in addressing functions to achieve the desired goal of handling their needs at all times. The bottlenecks, which are associated with reliability, are costly to a company since they will have to lose in meeting their targets and achieving the regulations provided by the customer.

Less reliable systems have more maintenance demands. Unplanned maintenance within the organization and system leads to higher costs since the company has to deal with rush orders and demand to achieve its goal within the most miniature provisional timeline. Emergency expenses indicate a lesser possibility of addressing and achieving relevant outcomes in marking the progressive advancement of the system [14]. Essentially, unplanned and nonscheduled maintenance is expensive to institutions since it leads to resource constraints, preventing them from achieving their goals in the correct order and leading to lesser productivity on whatever mentions and demands they have.

A significant indirect cost of reliability is employee productivity. Downtimes with less reliable systems lead to a lesser possibility of handling workers efficiently. The lesser capacity to address and handle workers' demands indicates a potential loss of time for the employees. Such incidents lead to a higher chance of reduced employee morale, where they register lower productivity and do not achieve their goals. The employees are, therefore, affected by the lack of an advance to help handle and achieve stellar outcomes when dealing with reliability issues.

Moreover, companies have the demand to work within the framework of addressing reliability issues, prompting costs in training employees to take the best step in addressing pending matters. The training is costly since the employees must adjust and be educated on managing reliability concerns. The high costs of continuous employee training to compensate for system flaws emphasize the urgent need for innovative and cost-effective learning management solutions. Therefore, the education approaches and interventions offer a safe approach to managing interruptions but also imply higher costs in attending to challenges stemming from the need to identify and manage reliability challenges within the institution.

Indirect costs on reliability significantly affect an organization, leading to negative performance and higher financial obligations. The reliability issues in the company can lead to lower productivity, lower financial outcomes, and handling of the company to address their needs. Therefore, handling the costs associated with the organization's lack of a reliable system means financial performance is affected and operational efficiency is lowered. Thus, companies have to enhance their management of resources, handling reliability, and ensuring that every system is in place to help with the presentation of remarkable development to achieve sustainable modeling of the institution and achieve higher results in managing reliability concerns.

Opportunity Costs
Loss of Sales Due to Unreliability

Lack of reliability in organizations leads to several negative attributes that affect the nature of the company. In the first instance, customer dissatisfaction is a crucial issue with companies that have unreliable systems and networks. Consumer dissatisfaction is registered in negative customer experiences, which lead to lesser sales. Sales lead to companies needing more customer loyalty, which impacts the scope and capacity of ensuring the provision of value to consumers.

Minimal reliability affects business entities since the loss of customers because of poor service and handling impacts their repeat lifetime experiences with the company. Therefore, the issue of reliability is a significant concern in registering and ensuring appropriate management of consumer demands, as well as achieving and addressing individual needs to attain stellar handling of the consumers. More to the point, a loss of reliability in the company network and systems leads to a negative brand reputation [15]. Damage to the reputation affects sales, and the brand is associated with low reliability and quality of its services and products. This breeds difficulty in handling marketing and engagement roles since consumers have minimal interest in the company. The negative brand reputation even affects potential customers since they cannot continue coming to the company because they hold opposing beliefs and values in addressing the channels of business provision. Nonetheless, the reputation factor additionally leads to minimal market penetration, as the continued service of the company is stuck on having to offer services even when they lack reliable platforms to hold the same service provision. Lacking the capacity to administer services to consumers continually causes the company to lack progressive advancement in market shares.

Companies with low reliability need to improve in advancing their market shares. The company loses its market share to more reliable entities, which can continue providing services and administering the proper role provision in all categories. The challenges experienced by consumers influence the provision and management of consumer needs. The development of companies with low reliability is also stalled since they cannot have the chance to provide new products, maintain the existing products, and even work with consumers to advance their market share. This implies that the lack of reliability in the organizational systems and network leads to a significant reduction in the handling and addressing pertinent issues affiliated with managing organizational needs. Thus, the challenges experienced by the company can lead to an opposing competitive advantage, where they cannot address competition and have the proper framework for dealing with increasing consumer demands.

Lack of reliability costs companies stagnation. Innovation is an imperative element to companies, indicating the chance to progress and mark their advancement to achieve every incentive as desired. The lack of reliability in the company leads to minimal adjustment in innovation. This is mainly because the company concentrates on handling unreliability instead of addressing innovation. Innovation requires effort and focused investment in channels to ensure the handling of research and development within the company. Therefore, unreliability poses several challenges to an organization, such as slow progression and stagnation of further advances in addressing product development and handling to achieve remarkable benefits that should be used to meet the company's demands [16]. Thus, unreliability causes systems in the company to have minimal advancement in addressing and marking developmental approaches. These advances, as they occur within the company, imply a channel to continually look into their challenges instead of working to advance solutions to industry-wide needs. Hence, unreliability causes companies to have a slow pace in achieving their strategic goals and investing in other selections that remark an increased development.

A more significant cost of handling unreliability lies in having a long-term financial impact on the company. The economic impact of working with reduced profitability and growth potential of the company makes it endure and work within difficulties of handling emergency maintenance services and slow service. The unreliable network and system within the company could also lead to repercussions like data breaches, where they need more security management to help them address challenges defining data security. These long-term financial issues within the company bring along critical instances that must be addressed to enable continued modeling of their values to achieve sustainable outcomes. Hence, the long-term impact of unreliability in organizations lies within their modeling and management of systemic needs. These challenges are critical to companies, as they influence the overall functionality and capacity to administer sustainable development to meet organizational development needs.

Cost–Benefit Analysis of Reliability Investments

Evaluating Return on Investment

Reliability in networks and systems demands high costs, some of which are closely connected to managing and handling everyday processes within the institution. In the first instance, the companies must evaluate the returns they get from their systems' reliability, offering an instrumental understanding and managing activities to accomplish the desired results. Therefore, handling reliability in the phase of organizational duties implies a demand to know the return on the investment and approaches to be made in ensuring sustainable modeling of every investment to achieve the desired value.

In the first instance, reliability investments must be conducted to balance upfront costs with long-term savings in the organization. This approach works within the company to ensure they have an initial investment in suitable systems, training, and software to help them achieve reliable mentions at whatever stage of their activities. These initial investments demand the company to have an accurate measure and understanding of their demands, looking into every critical component and enabling progressive management and handling of reliability to achieve every demand mentioned. Nonetheless, companies can also save in the long term from reliability improvements. Improving the systems can ensure that reliability can be handled appropriately, encountering every adjustment and marking the development of every incentive to achieve stellar results [17]. Therefore, reliability has to be managed by having the best steps to address, administer, and ensure reliability activities are appropriately conducted. The system of addressing reliability in the companies is core to defining and outlining every functional address to ensure continued management of reliability mentions to ensure success at all times.

Companies can use different metrics to understand the investment in reliability. Tabulating the return on investment and payback period for monies invested in reliability training, hardware, and software illuminates the influence of reliability on the organization, bringing about a better understanding of every functional entity in managing reliability in the company.

Innovations and Costs in Reliability

Different innovations can help manage the costs associated with reliability. These innovations can be provided within companies to ensure they can address reliability mentions in critical ways that assist in detailing and ensuring successful outcomes in handling the companies' everyday demands.

Blockchain

Blockchain continually works to enhance digital systems by introducing secure, tamper-proof, and transparent ways of handling transactions. A block contains transactions and cannot be altered once added, providing a safe level of functionality at all levels. Regarding reliability, blockchain acts in different ways to ensure a remarkable advancement in achieving optimal levels. The main ways blockchain enhances reliability include

- **Data Integrity:** Using blockchain leads to a high level of addressing data integrity by providing immutable data blocks that lead to high reliability.

- **Decentralization:** Blockchain distributes data across multiple nodes, ensuring optimal functionality since no single node is overloaded by administering service to the required portions.

- **Smart Contracts:** Blockchains can ensure that contracts can be coded to be self-executing, leading to reliable automation and risk reduction by identifying desired changes and addressing them.

- **Consensus Mechanisms:** Blockchain offers the chance to work with Proof of Work (PoW) and Proof of Stake (PoS), encouraging consistency by ensuring every node agrees on the blockchain's state. These measures remarkably advance to achieve the best level of blockchain functionality in all instances.

Internet of Things (IoT)

IoT are interconnected devices that exchange data and communicate. They have different applications, from household items to industrial and leisure materials, for various purposes. They help in advancing reliability in other measures. The advent of IoT leads to higher reliability through

- **Predictive Maintenance:** IoT monitors and evaluates equipment, helping to predict failures and schedule their occurrence at any given point. This proactive maintenance approach reduces downtime and makes industrial systems reliable in providing the right services that individual organizations desire.

- **Real-Time Monitoring:** IoT devices constantly collect and analyze data, helping to adjust the system beneficially. Real-time monitoring helps detect anomalies and schedule quick responses, avoiding downtime at any given point.

- **Automated Response:** IoT systems allow for the automation of responses, each scheduled to assist in shutting down equipment and ensuring that they can respond and address various advances in marking the development of critical needs to achieve appropriate outcomes in every provisional situation.

- **Scalability:** IoT devices can help organizations scale by adding devices and ensuring they can maintain performance, monitor systems, and handle increasing loads. Every step ensures that they can handle the constantly growing demand to achieve sustainable outcomes in whatever available instance.

- **Redundancy and Failover:** The development of IoT devices enables them to have a considerable level of management in terms of growing attention to work and desired practices. The devices help to ensure that the multiple sensors and tools have redundant data paths, each working to ensure consistent data provision. The model of handling these IoT devices marks the development of a redundancy that acts as a failover in the case of any operational failure on the system.

Figure 13-3. *Mechanisms in IoT that enhance reliability*

Figure 13-3 indicates various mechanisms in IoT that help advance reliability in networks and systems. Predictive maintenance, real-time monitoring, automated response, scalability, redundancy, and failover mechanisms help advance reliability as a functional aspect of networks and systems.

Summary

Reliability is an essential component in modern organizational networks and systems. Ensuring reliability starts with understanding the demand for uptime, downtime, and redundancy on the system. Investing in reliability within the company begins with costs on designs in the initial phase, which have to be conducted to create an understanding of every mention, leading to better ways to carry out activities. Direct and indirect costs associated with reliability ensure that an organization selects whatever approach is essential and suitable for achieving the right outcome in every provision. In the first instance, reliability is addressed by ensuring that the direct costs, like having backup systems with redundancy in terms of more machines and software, have to be administered to avoid indirect costs associated with loss of consumers, loss of brand reputation, and installation of systems that demand continuous address. Companies can evaluate their investment in reliability by looking at returns on investment and administering critical handling of innovations to ensure lesser expenses on reliability systems. Using blockchain and AI will increase company reliability at lower costs, implying better performance for the company. Thus, these approaches enable a better understanding of the costs associated with reliability in the company.

Bibliography

1.   K. Nagiya, A. Kumar, M. Ram, and A. Anand, "Reliability, evaluation system connected MTTF, of a computer and in star sensitivity network topology," in *The Handbook of Reliability, Maintenance, and System Safety through Mathematical Modeling*, p. 457, 2021

2.   O. Adegboye, "Reliability Culture—The Key to a Reliable and Sustainable Asset Uptime," in *2024 Annual Reliability and Maintainability Symposium (RAMS)*, 2024, pp. 1–5

3.   M. A. Mellal, S. Al-Dahidi, R. B. Patil, B. S. Kothavale, and R. S. Powar, "System reliability-redundancy optimization with high-level of subsystems," *Materials Today: Proceedings*, vol. 77, pp. 627–630, 2023

4.   A. Amin and K. M. Hasan, "A review of fault tolerant control systems: advancements and applications," *Measurement*, vol. 143, pp. 58–68, 2019

5.   M. Bennis, M. Debbah, and H. V. Poor, "Ultrareliable and low-latency wireless communication: Tail, risk, and scale," *Proceedings of the IEEE*, vol. 106, no. 10, pp. 1834–1853, 2018

6.   H. Chen et al., "Ultra-reliable low latency cellular networks: Use cases, challenges and approaches," *IEEE Communications Magazine*, vol. 56, no. 12, pp. 119–125, 2018

7.   P. Asghari, A. M. Rahmani, and H. H. S. Javadi, "Internet of Things applications: A systematic review," *Computer Networks*, vol. 148, pp. 241–261, 2019

8.   S. S. Prabhu and H. L. Shashirekha, "Effectiveness of Software Metrics on Reliability for Safety-Critical Real-Time Software," in *Congress on Intelligent Systems: Proceedings of CIS 2020, Volume 1*, Singapore: Springer, 2021, pp. 713–724

9.   H. Chen et al., "Ultra-reliable low latency cellular networks: Use cases, challenges and approaches," *IEEE Communications Magazine*, vol. 56, no. 12, pp. 119–125, 2018

10.   M. N. Alam, S. Chakrabarti, and A. Ghosh, "Networked microgrids: State-of-the-art and future perspectives," *IEEE Transactions on Industrial Informatics*, vol. 15, no. 3, pp. 1238–1250, 2018

11.   N. Adhikari, G. Ramesh, and V. Aravindarajan, "An innovation development of reliable redundancy of data backup in big data servers using RAID arrays," *ICTACT Journal on Data Science and Machine Learning*, 2022

12.   Y. Yang, H. Wang, A. Sangwongwanich, and F. Blaabjerg, "Design for reliability of power electronic systems," in *Power Electronics Handbook*, Butterworth-Heinemann, pp. 1423–1440, 2018

13.   J. Li, W. Liang, M. Huang, and X. Jia, "Reliability-aware network service provisioning in mobile edge-cloud networks," *IEEE Transactions on Parallel and Distributed Systems*, vol. 31, no. 7, pp. 1545–1558, 20

14.   K. Antosz, J. Machado, D. Mazurkiewicz, D. Antonelli, and F. Soares, "Systems Engineering: Availability and Reliability," *Applied Sciences*, vol. 12, no. 5, p. 2504, 2022

15.   H. Karimi et al., "Automated distribution networks reliability optimization in the presence of DG units considering probability customer interruption: A practical case study," *IEEE Access*, vol. 9, pp. 98490–98505, 2021

16.   V. Netes and M. Kusakina, "Reliability challenges in software-defined networking," in *Conference of Open Innovations Association, FRUCT*, no. 24, pp. 704–709, FRUCT Oy, 2019

17.   Hirsch, Y. Parag, and J. Guerrero, "Microgrids: A review of technologies, key drivers, and outstanding issues," *Renewable and Sustainable Energy Reviews*, vol. 90, pp. 402–411, 2018

CHAPTER 14

Organization Structure and Skill Set Challenges

Authors:

Sriram Panyam

Praveen Gujar

Reviewer:

Fardin Quazi

Introduction

The Imperative of Reliability: Why It's the Cornerstone of Modern Software

In the digital age, software is interwoven into the fabric of our lives. From communication and entertainment to healthcare and finance, software systems power critical operations and drive innovation. In this landscape, reliability is paramount. Unreliable software not only disrupts daily activities but can also have severe consequences, including financial losses, data breaches, safety hazards, and reputational damage.

© Saurav Bhattacharya 2024
M. Kuppam, *Enterprise Digital Reliability*, https://doi.org/10.1007/979-8-8688-1032-9_14

Modern software systems are complex, often distributed across multiple platforms and environments, with numerous dependencies and interactions. Ensuring their reliability is a challenging endeavor that requires a holistic approach, encompassing technical practices, organizational structures, and a culture of continuous improvement.

Evolution of Reliability: From Ad Hoc Practices to Strategic Initiatives

Historically, software reliability was often an afterthought, addressed through reactive measures and firefighting. Development teams focused on building new features, while operations teams struggled to keep systems running. This siloed approach led to misaligned incentives, finger-pointing, and delayed incident resolution.

The rise of agile development and DevOps methodologies marked a turning point. These movements emphasized collaboration, automation, and continuous feedback loops, laying the foundation for more proactive and systematic approaches to reliability. Organizations began to recognize that reliability was not merely a technical problem but a cultural and organizational one as well.

The Human Element: Recognizing the Role of People in Reliable Systems

While technology plays a crucial role in achieving reliability, it's essential to acknowledge the human element. Software systems are built, maintained, and operated by people. Their skills, expertise, collaboration, and decision-making processes significantly impact the overall reliability of the system.

Research has shown that factors like communication, psychological safety, and shared ownership are critical for building high-performing teams that deliver reliable software. Effective communication ensures that information flows smoothly between team members, enabling them to identify and address potential issues promptly. Psychological safety creates an environment where individuals feel comfortable raising concerns, admitting mistakes, and learning from failures. Shared ownership fosters a sense of collective responsibility, motivating team members to work together toward common goals.

In the following sections, we will delve into the historical context of reliability, explore best practices for team and organizational design, examine how these practices apply to SRE and DevOps teams, discuss adaptations for different organization sizes and domains, and outline key metrics for measuring success. We will also address the challenges organizations face in transforming their structures for reliability and conclude with a vision for the future of reliable software.

Historical Perspectives on Team Setup and Organization for Reliability and DevOps

The Siloed Past: Traditional Development vs. Operations Teams

In the early days of software development, a rigid division existed between development and operations teams. Development teams focused on writing code and building new features, while operations teams were responsible for deploying, monitoring, and maintaining systems in production. This siloed approach often led to friction, misunderstandings, and slower delivery cycles.

Development teams, driven by the pressure to release new features quickly, sometimes prioritized speed over stability. This resulted in code that was not thoroughly tested or optimized for production environments, leading to frequent outages and disruptions. Operations teams, on the other hand, were primarily concerned with maintaining system stability. They often viewed new features as potential sources of instability and resisted changes, causing delays and frustration for development teams.

The lack of communication and collaboration between these two groups hindered the ability to identify and resolve issues promptly. When problems arose in production, finger-pointing and blame games were common, further exacerbating the divide. This adversarial relationship between development and operations teams was a major obstacle to achieving reliability and agility in software delivery.

The Rise of DevOps: Bridging the Gap for Faster, More Reliable Delivery

The emergence of DevOps in the late 2000s marked a significant shift in the way software was developed and delivered. DevOps aimed to break down the silos between development and operations, fostering a culture of collaboration, shared responsibility, and continuous improvement.

The core principles of DevOps include

- **Collaboration:** Development and operations teams work together throughout the software development life cycle, from planning and design to deployment and monitoring.

- **Automation:** Repetitive tasks, such as testing, deployment, and infrastructure provisioning, are automated to reduce human error and increase efficiency.

- **Continuous Integration and Continuous Delivery (CI/CD):** Code changes are frequently integrated and tested, enabling faster and more reliable releases.

- **Monitoring and Feedback:** Systems are continuously monitored to detect and address issues proactively, and feedback loops are established to inform future improvements.

By embracing these principles, organizations were able to achieve faster delivery cycles, improved reliability, and greater customer satisfaction. The DevOps movement sparked a cultural transformation, encouraging teams to work together toward common goals and share ownership of the entire software delivery process.

Site Reliability Engineering (SRE): Google's Blueprint for High-Availability Systems

Google, with its massive scale and complex infrastructure, faced unique challenges in maintaining the reliability of its services. To address these challenges, Google developed a new discipline called Site Reliability Engineering (SRE).

SRE combines software engineering expertise with operational knowledge to create highly reliable systems. SRE teams are responsible for

- **Defining and Measuring Reliability:** Setting service-level objectives (SLOs) and tracking key metrics to ensure that systems meet or exceed reliability targets

- **Balancing Innovation and Reliability:** Establishing error budgets to allow for experimentation and new features while maintaining acceptable levels of risk

- **Automating Operations:** Developing tools and processes to automate repetitive tasks and reduce the need for manual intervention

- **Responding to Incidents:** Investigating and resolving incidents quickly and efficiently to minimize downtime

- **Building Resilient Systems:** Designing systems that can withstand failures and recover gracefully

SRE has become a widely adopted model for achieving high availability and reliability in large-scale systems. Its principles and practices have influenced the way organizations approach reliability, not only in the tech industry but also in other sectors like finance, healthcare, and government.

Organizational Models

As SRE and DevOps gained traction, organizations experimented with different ways to integrate these practices into their existing structures. Three primary organizational models emerged.

Centralized SRE Teams

In this model, SREs form a separate, specialized team responsible for the reliability of multiple services or products across the organization. This team acts as a center of excellence, providing expertise, guidance, and support to development teams. The centralized model offers several advantages:

- **Deep Expertise:** SREs can develop specialized knowledge in areas like performance optimization, incident management, and capacity planning.

- **Consistency:** Centralized teams can establish and enforce standardized practices and tools across the organization.

- **Resource Optimization:** Resources can be allocated efficiently based on the organization's overall needs.

However, this model also has some drawbacks:

- **Siloed Knowledge:** The separation between SRE and development teams can hinder knowledge sharing and collaboration.

- **Limited Context:** SREs may lack in-depth understanding of specific services or products, leading to slower response times and less effective solutions.

- **Bottlenecks:** Centralized teams can become overwhelmed with requests, slowing down development and innovation.

Embedded SREs Within Development Teams

In this model, SREs are embedded directly within development teams, working alongside software engineers and other team members. This approach fosters closer collaboration, shared ownership, and a deeper understanding of the specific service or product. The benefits of this model include

- **Faster Feedback Loops:** SREs can provide immediate feedback on reliability issues, enabling quicker resolution.

- **Contextual Expertise:** SREs develop a deep understanding of the service or product, leading to more effective solutions.

- **Increased Agility:** The integrated team can move faster and adapt more easily to changing requirements.

However, this model also presents some challenges:

- **Diluted Expertise:** Embedded SREs may have less opportunity to specialize in specific areas of reliability engineering.

- **Inconsistent Practices:** Different development teams may adopt varying practices and tools, leading to inconsistencies across the organization.

- **Resource Constraints:** Smaller teams may not have the resources to dedicate a full-time SRE to their team.

Hybrid Approaches

Many organizations have adopted hybrid approaches that combine elements of centralized and embedded models. For example, a central SRE team might provide overall guidance and support, while embedded SREs work within specific development teams to address their unique needs. This approach aims to leverage the benefits of both models while mitigating their drawbacks.

The choice of organizational model depends on various factors, including the size and complexity of the organization, the maturity of its DevOps and SRE practices, and the specific needs of its services or products. There is no one-size-fits-all solution, and organizations may need to experiment with different models to find the one that best suits their needs.

The evolution of team structures and organizational models in the realm of reliability and DevOps reflects a growing recognition of the importance of collaboration, shared responsibility, and continuous improvement. By breaking down silos, fostering communication, and empowering teams, organizations can create a culture of reliability that enables them to deliver high-quality software at speed.

General Best Practices on Team and Organization Design

Collaboration and Communication: The Lifeblood of Reliable Systems

The foundation of any successful team, especially those tasked with building and maintaining reliable software, is effective collaboration and communication. Open, transparent, and frequent communication ensures that everyone is on the same page, aware of potential issues, and aligned on solutions.

- **Regular Stand-Ups:** Daily or weekly stand-up meetings provide a forum for team members to share updates, discuss roadblocks, and coordinate efforts.

- **Shared Documentation:** Comprehensive documentation, including system architecture diagrams, runbooks, and postmortem reports, serves as a single source of truth and facilitates knowledge sharing.

- **Communication Channels:** Utilize a variety of communication channels, such as instant messaging, video conferencing, and project management tools, to cater to different communication styles and needs.

- **Blameless Postmortems:** When incidents occur, conduct blameless postmortems to analyze the root causes and identify areas for improvement, fostering a culture of learning and continuous improvement.

Shared Ownership: Fostering a Culture of Responsibility

In traditional organizations, responsibilities are often siloed, with development teams focusing on building features and operations teams handling production issues. This separation can lead to a lack of ownership and accountability, as teams may not feel responsible for the overall reliability of the system.

To overcome this, organizations should foster a culture of shared ownership, where everyone feels responsible for the success of the system. This can be achieved through

- **Cross-functional Teams:** Create teams that include members from different disciplines, such as development, operations, security, and quality assurance. This encourages collaboration and shared understanding of the entire software delivery process.

- **On-Call Rotations:** Implement on-call rotations that involve both development and operations teams. This ensures that everyone has a stake in keeping the system running smoothly and is incentivized to build reliable software.

- **Shared Metrics:** Establish shared metrics that measure the performance and reliability of the system. This aligns incentives and encourages collaboration toward common goals.

Autonomy and Empowerment: Enabling Teams to Make Decisions

Micromanagement and excessive oversight can stifle innovation and slow down decision-making. Instead, organizations should empower teams to make decisions autonomously, within a defined scope and set of guidelines. This allows teams to move faster, experiment, and learn from their mistakes.

To empower teams, organizations can

- **Define Clear Goals and Objectives:** Clearly articulate the desired outcomes and provide teams with the autonomy to determine the best way to achieve them.

- **Provide Necessary Resources:** Ensure that teams have access to the tools, training, and support they need to succeed.

- **Encourage Experimentation:** Create a safe environment where teams can experiment with new ideas and technologies without fear of reprisal.

- **Celebrate Successes and Learn from Failures:** Recognize and reward teams for their achievements, and use failures as opportunities for learning and growth.

Continuous Improvement: Learning and Adapting from Successes and Failures

In the ever-evolving world of software engineering, continuous improvement is essential for staying ahead of the curve and delivering reliable systems. Organizations should embrace a culture of learning, where teams are encouraged to experiment, gather feedback, and iterate on their processes.

To foster continuous improvement:

- **Retrospectives:** Conduct regular retrospectives to reflect on past successes and failures, identify areas for improvement, and implement changes.

- **Knowledge Sharing:** Encourage knowledge sharing through presentations, workshops, and documentation.

- **Training and Development:** Invest in training and development programs to help team members acquire new skills and stay up-to-date with the latest technologies.

- **Experimentation:** Allocate time and resources for experimentation, allowing teams to try new approaches and learn from their experiences.

Psychological Safety: Creating an Environment Where Mistakes Are Opportunities

Psychological safety is the belief that one will not be punished or humiliated for speaking up with ideas, questions, concerns, or mistakes. In a psychologically safe environment, team members feel comfortable taking risks, admitting errors, and asking for help. This is crucial for building trust, fostering collaboration, and encouraging innovation.

To create psychological safety:

- **Lead by Example:** Leaders should model vulnerability and openness by admitting their own mistakes and encouraging others to do the same.

- **Active Listening:** Listen attentively to team members' concerns and feedback, and respond with empathy and respect.

- **Constructive Feedback:** Provide feedback that is specific, actionable, and focused on improvement rather than blame.

- **Celebrate Learning:** Emphasize the importance of learning from mistakes and create a culture where failures are seen as opportunities for growth.

These best practices, when implemented effectively, can create a high-performing team that is collaborative, innovative, and focused on delivering reliable software. In the next section, we will explore how these practices can be applied specifically to SRE and DevOps teams.

Applying Best Practices to SRE and DevOps Teams

SRE Team Structures: Balancing Expertise and Integration

Given the multifaceted nature of SRE work, structuring teams effectively is crucial for success. There are a few common models:

1. **Service-Aligned Teams:** Each SRE team focuses on a specific service or product. This allows for deep expertise and context but can lead to knowledge silos if not managed carefully. It's best suited for large organizations with distinct product lines.

2. **Functional Teams:** Teams specialize in areas like performance, reliability, or tooling. This promotes expertise in specific domains but can create handoffs and coordination challenges. It's suitable when you need to tackle specific reliability bottlenecks.

3. **Mixed Teams:** A combination of service and functional alignment. Some SREs are dedicated to specific services, while others focus on cross-cutting concerns. This offers a balance but requires careful coordination to avoid duplication of effort. It's often the most flexible model for growing organizations.

Regardless of the structure, embedding SREs within product development teams, even partially, helps bridge the gap between development and operations, fostering a culture of shared responsibility.

DevOps Team Topologies: Matching Structures to Organizational Goals

DevOps team structures are diverse, mirroring the varying needs and goals of organizations. Common topologies include

1. **Fully Embedded DevOps:** DevOps engineers are fully integrated into development teams. This promotes collaboration and ownership but can dilute focus on specific DevOps tasks.

2. **Centralized DevOps Team:** A dedicated DevOps team supports multiple development teams. This allows for specialization and standardization but can create bottlenecks and slow down delivery cycles.

3. **DevOps as a Service:** DevOps teams act as internal consultants, providing expertise and tools to development teams on demand. This offers flexibility but may lack deep integration and shared ownership.

4. **SRE-Driven DevOps:** SRE teams take the lead on DevOps initiatives, focusing on reliability and automation. This ensures a strong focus on reliability but may require close collaboration with development teams to avoid friction.

The ideal DevOps team structure depends on factors like the organization's size, maturity, and culture. It's essential to align the team structure with the organization's strategic goals and ensure that it supports a culture of collaboration and continuous improvement.

Roles and Responsibilities

In the world of SRE and DevOps, roles often overlap and evolve. However, some core responsibilities can be defined:

- **SREs:** Define SLOs, monitor system health, respond to incidents, automate toil, and work with development teams to improve reliability.

- **DevOps Engineers:** Build and maintain CI/CD pipelines, automate infrastructure provisioning, and manage cloud environments.

- **Software Engineers:** Design, develop, and test software, with a focus on reliability and performance.

- **Product Managers:** Prioritize features, define requirements, and communicate with stakeholders.

- **Engineering Managers:** Lead teams, set goals, manage resources, and foster a culture of collaboration and continuous improvement.

Clear communication and collaboration between these roles are essential for ensuring that reliability is built into the software development process from the start.

Tooling and Automation: Enabling Efficiency and Reliability

Tooling and automation are the backbone of SRE and DevOps practices. They help to reduce human error, increase efficiency, and improve reliability. Key tools include

- **Monitoring and Observability Tools:** Collect metrics, logs, and traces to provide visibility into system health and performance.

- **CI/CD Platforms:** Automate the build, test, and deployment process to enable faster and more reliable releases.

- **Infrastructure as Code (IaC) Tools:** Manage infrastructure using code, allowing for versioning, reproducibility, and automation.

- **Configuration Management Tools:** Ensure consistent configuration across different environments.

- **Incident Management Tools:** Streamline incident response and communication.

Choosing the right tools and implementing effective automation can significantly improve the productivity and reliability of SRE and DevOps teams.

Adapting to Different Organization Sizes and Domains

Startups: Agility and Rapid Growth

Startups are characterized by their agility, rapid growth, and limited resources. In this environment, reliability might not be the top priority initially, as the focus is on building the product and acquiring customers. However, as the startup scales and its customer base grows, reliability becomes increasingly important to maintain customer satisfaction and prevent revenue loss.

For startups, the following approaches can be effective:

- **Prioritize Automation:** Automate as many tasks as possible, from testing and deployment to infrastructure provisioning and monitoring. This frees up valuable time and resources that can be focused on building new features and improving the product.

- **Embrace Cloud-Based Solutions:** Cloud platforms offer scalability, flexibility, and cost-effectiveness, allowing startups to focus on their core business rather than managing infrastructure.

- **Adopt a DevOps Culture:** Encourage collaboration, shared responsibility, and continuous improvement from the start. This helps to build a strong foundation for reliability as the startup grows.

- **Focus on Key Metrics:** Track key metrics like uptime, error rates, and customer satisfaction to identify and address potential reliability issues early on.

Mid-Sized Companies: Scaling Reliability Practices

As companies grow, they face the challenge of scaling their reliability practices to accommodate a larger customer base, more complex systems, and increased traffic. This requires a more structured and systematic approach to reliability.

Mid-sized companies can benefit from

- **Establishing Dedicated SRE or DevOps Teams:** As the organization grows, it becomes necessary to have dedicated teams responsible for reliability and operations. These teams can provide expertise, guidance, and support to development teams, ensuring that reliability is built into the software development life cycle.

- **Standardizing Tools and Processes:** Establish standardized tools and processes for monitoring, incident management, and capacity planning. This ensures consistency across the organization and helps to identify and address potential reliability issues proactively.

- **Investing in Training and Development:** Provide training and development opportunities for team members to enhance their skills in SRE, DevOps, and related disciplines. This helps to build a culture of continuous learning and improvement.

- **Creating a Reliability Roadmap:** Develop a roadmap for reliability initiatives, outlining priorities, milestones, and metrics for success. This ensures that reliability efforts are aligned with the organization's overall goals and objectives.

Large Enterprises: Navigating Complexity and Legacy Systems

Large enterprises often have complex, distributed systems with numerous dependencies and legacy components. This makes achieving reliability a significant challenge, requiring a comprehensive and coordinated approach.

Large enterprises should consider:

- **Adopting a Hybrid SRE Model:** A hybrid model, combining centralized SRE teams with embedded SREs within development teams, can provide the necessary expertise and support while ensuring close collaboration and contextual understanding.

- **Modernizing Legacy Systems:** Gradually modernize legacy systems to improve their reliability, scalability, and maintainability. This may involve refactoring code, adopting cloud-based solutions, and automating manual processes.

- **Implementing a Service-Oriented Architecture (SOA):** SOA allows for greater flexibility, modularity, and resilience, making it easier to isolate and address failures.

- **Establishing a Reliability Center of Excellence:** A central team can provide leadership, guidance, and support for reliability initiatives across the organization. This team can also facilitate knowledge sharing and promote best practices.

- **Building a Culture of Reliability:** Foster a culture where reliability is everyone's responsibility. This involves promoting collaboration, shared ownership, continuous improvement, and psychological safety.

Domain-Specific Considerations

Different industries have unique reliability requirements and challenges. For example:

- **Ecommerce:** Reliability is critical for maintaining customer trust and preventing revenue loss. Downtime during peak shopping periods can be catastrophic. Ecommerce companies must invest in robust infrastructure, efficient incident management processes, and proactive monitoring to ensure uninterrupted service.

- **Finance:** Financial institutions must adhere to strict regulatory requirements and ensure the security and integrity of sensitive data. Reliability is essential for maintaining customer confidence and preventing financial losses due to system failures.

- **Healthcare:** Reliability is paramount in healthcare, where software systems are used for critical tasks like patient monitoring, diagnosis, and treatment. Downtime or errors in healthcare systems can have serious consequences for patient safety and well-being.

- **Gaming:** Gaming companies rely on real-time, low-latency systems to provide an immersive and enjoyable experience for players. Downtime or performance issues can quickly lead to player frustration and churn.

To address these domain-specific challenges, organizations should

- **Understand Industry-Specific Regulations:** Be aware of and comply with any relevant industry-specific regulations and standards related to reliability and security.

- **Tailor Reliability Practices:** Adapt reliability practices to the specific needs and constraints of the industry.

- **Partner with Experts:** Collaborate with industry experts and consultants to ensure that reliability initiatives are aligned with best practices and regulatory requirements.

By understanding the unique challenges and requirements of different organization sizes and domains, businesses can tailor their reliability strategies and build systems that meet the needs of their customers and stakeholders.

In the next section, we will explore how to measure the success of these reliability initiatives.

Measuring Success: Key Metrics for Reliable Teams and Organizations

Effectively measuring the success of SRE and DevOps initiatives is crucial for demonstrating their value, identifying areas for improvement, and justifying continued investment. While the specific metrics may vary depending on the organization and its goals, some key metrics widely applicable are detailed below.

Service-Level Objectives (SLOs): Defining Acceptable Levels of Performance

SLOs are specific, measurable targets for the reliability and performance of a service. They define the acceptable level of service that customers can expect and provide a clear benchmark for measuring success. SLOs should be based on a variety of factors, including

- **Customer Expectations:** What level of reliability and performance do customers expect from the service? This can be determined through customer surveys, focus groups, and user experience research.

- **Business Goals:** How does reliability impact the organization's business goals? For example, a high availability SLO for an ecommerce website might be critical for meeting sales targets during peak shopping periods.

- **Technical Feasibility:** What level of reliability and performance is realistically achievable given the current infrastructure, technology stack, and team capabilities?

SLOs should be clearly defined and documented in a Service-Level Agreement (SLA) that is communicated to all stakeholders, including customers, development teams, and operations teams. SLAs should also specify the consequences of failing to meet SLOs, such as financial penalties or service credits for customers.

Here are some common SLO examples, along with considerations for setting them:

- **Availability:** This SLO is typically expressed as a percentage of uptime, such as "99.9%" or "three nines" of availability. When setting availability SLOs, it's important to consider the trade-off between uptime and the cost of achieving that level of reliability. For example, a service that requires very high availability (e.g., an online banking platform) may need to invest in redundant infrastructure and disaster recovery plans, which can be expensive.

- **Latency:** This SLO measures the time it takes for a request to be processed and a response to be returned. Latency SLOs are particularly important for real-time applications, such as video conferencing or online gaming. When setting latency SLOs, it's important to consider factors like network bandwidth, server response times, and geographical distribution of users.

- **Error Rate:** This SLO measures the percentage of requests that result in errors. Error rates can be caused by a variety of factors, such as software bugs, hardware failures, and network issues. When setting error rate SLOs, it's important to consider the severity of errors and the impact they have on users.

- **Throughput:** This SLO measures the number of requests that a service can handle per unit of time. Throughput SLOs are important for ensuring that a service can scale to meet demand. When setting throughput SLOs, it's important to consider factors like the capacity of the underlying infrastructure and the average processing time for requests.

Error Budgets: Balancing Innovation and Reliability

Error budgets are a powerful tool for enabling innovation while maintaining reliability. An error budget is the maximum acceptable level of unreliability for a service, expressed as a percentage of time (e.g., error budget of 1%) or a number of errors (e.g., error budget of 100 errors per day). By setting an error budget, teams can make informed decisions about how much risk they are willing to take with new features and deployments.

For example, a team responsible for a social media platform might set an error budget of 0.1% downtime per month. This means that the service can be unavailable for a maximum of 43 minutes per month. The team can then allocate this error budget across different types of incidents, such as planned maintenance downtime, unplanned outages, and errors introduced by new feature deployments.

Error budgets empower teams to experiment and innovate without compromising the overall reliability of the service. If the error budget is exceeded, the team must take corrective action, such as fixing bugs, rolling back deployments, or improving monitoring and alerting. This approach encourages a data-driven decision-making process and ensures that reliability remains a top priority throughout the software development life cycle.

Mean Time to Detection (MTTD) and Mean Time to Recovery (MTTR): Measuring Incident Response

MTTD and MTTR are key metrics for measuring the effectiveness of incident response processes. MTTD is the average time it takes to detect an incident, while MTTR is the average time it takes to recover from an incident.

By tracking these metrics, organizations can identify bottlenecks in their incident response processes and implement improvements to reduce downtime and minimize customer impact.

Customer Satisfaction: The Ultimate Indicator of Reliability

Ultimately, the success of reliability initiatives should be measured by their impact on customer satisfaction. Satisfied customers are more likely to continue using a service, recommend it to others, and provide positive feedback.

Organizations can measure customer satisfaction through surveys, feedback forms, and social media monitoring. They can also track metrics like churn rate (the percentage of customers who stop using the service) and net promoter score (NPS), which measures customer loyalty and willingness to recommend the service.

Employee Engagement and Retention: The Importance of Team Morale

The success of SRE and DevOps teams depends heavily on the engagement and morale of their members. Engaged employees are more productive, innovative, and committed to their work. High retention rates also save the organization the cost and disruption of recruiting and training new employees.

Organizations can measure employee engagement through surveys, feedback sessions, and one-on-one meetings. They can also track metrics like absenteeism, turnover rate, and employee satisfaction.

By tracking these key metrics, organizations can gain valuable insights into the effectiveness of their reliability initiatives and make data-driven decisions to improve their systems and processes.

Additional Considerations for Measuring Success

- **Business Impact Metrics:** In addition to technical metrics, it's important to track the business impact of reliability initiatives. This could include metrics like revenue, customer acquisition, and market share.

- **Leading Indicators:** Leading indicators, such as the number of incidents detected before they impact customers or the number of automated test runs, can provide early warning signs of potential reliability issues.

- **Qualitative Feedback:** Gather qualitative feedback from customers, employees, and stakeholders to gain a deeper understanding of their experiences and perceptions of reliability.

By taking a holistic approach to measuring success, organizations can ensure that their reliability initiatives are delivering value to customers, employees, and stakeholders.

Challenges in Transforming Organizational Structures for Reliability

Transforming organizational structures to prioritize and embed reliability is not without its hurdles. Many organizations face the following challenges.

Cultural Resistance: Overcoming Traditional Mindsets

One of the most significant challenges is overcoming ingrained cultural resistance to change. Traditional mindsets that prioritize individual heroics over teamwork, or that view operations as a separate concern from development, can hinder the adoption of SRE and DevOps practices.

To address this, organizations need to

- **Lead from the Top:** Leadership must champion the change and clearly communicate the benefits of a reliability-focused culture.

- **Invest in Education and Training:** Provide training programs and workshops to educate employees about SRE and DevOps principles, practices, and tools.

- **Foster a Culture of Learning:** Encourage experimentation, risk-taking, and learning from failures. Celebrate successes and recognize individuals and teams who contribute to reliability initiatives.

- **Create Incentives for Collaboration:** Align incentives with the desired outcomes, rewarding collaboration and shared responsibility.

Organizational Inertia: Dealing with Legacy Systems and Processes

Many organizations are burdened with legacy systems and processes that were not designed with reliability in mind. These systems may be difficult to monitor, automate, and scale, making it challenging to implement SRE and DevOps practices.

To overcome this, organizations can

- **Gradually Modernize Legacy Systems:** Rather than attempting a complete overhaul, start by identifying the most critical components and gradually modernize them. This could involve refactoring code, adopting cloud-based solutions, and automating manual processes.

- **Isolate Legacy Systems:** If possible, isolate legacy systems from newer, more reliable components. This can help to contain the impact of failures and reduce the risk of cascading outages.

- **Invest in Tooling and Automation:** Use tooling and automation to compensate for the limitations of legacy systems. For example, implement monitoring and alerting systems to detect issues early on and automate repetitive tasks to reduce human error.

Skills Gaps: Building Expertise in SRE and DevOps

The demand for SRE and DevOps expertise often outstrips the supply. This can make it difficult for organizations to find and retain qualified professionals.

To address this, organizations can

- **Invest in Training and Development:** Provide training and development opportunities for existing employees to upskill them in SRE and DevOps. This can include internal training programs, external courses, and certifications.

- **Hire External Talent:** Recruit experienced SREs and DevOps engineers from outside the organization. This can be expensive, but it can also accelerate the adoption of best practices and help to build a culture of reliability.

- **Partner with Consultants:** Engage with external consultants who specialize in SRE and DevOps. They can provide guidance, support, and training to help organizations implement these practices effectively.

Leadership Buy-In: Securing Support for Change

Transforming organizational structures for reliability requires strong leadership buy-in and support. Without it, initiatives may lack the necessary resources, authority, and momentum to succeed.

To secure leadership buy-in, organizations can

- **Clearly Articulate the Benefits:** Present a compelling business case for reliability, highlighting the potential cost savings, revenue growth, and customer satisfaction improvements that can be achieved through SRE and DevOps practices.

- **Demonstrate Quick Wins:** Start with small, achievable projects that can demonstrate the value of reliability initiatives early on. This can help to build momentum and secure further investment.

- **Communicate Progress and Results:** Regularly communicate the progress and results of reliability initiatives to stakeholders, highlighting the positive impact on the organization.

Measuring Progress: Demonstrating the Value of Reliability Initiatives

Measuring the progress and impact of reliability initiatives is crucial for demonstrating their value to stakeholders and justifying continued investment.

To measure progress, organizations can

- **Track Key Metrics:** Monitor key metrics like uptime, error rates, incident response times, and customer satisfaction. Use these metrics to assess the effectiveness of reliability initiatives and identify areas for improvement.

- **Conduct Regular Reviews:** Hold regular reviews with stakeholders to discuss progress, challenges, and next steps. This helps to keep everyone aligned and ensures that reliability remains a top priority.

- **Celebrate Successes:** Celebrate successes and recognize individuals and teams who contribute to reliability initiatives. This helps to build momentum and reinforce the importance of reliability.

By addressing these challenges head-on and implementing the best practices outlined in this chapter, organizations can successfully transform their structures for reliability, creating a culture of continuous improvement and delivering high-quality software that meets the needs of their customers and stakeholders.

Conclusion: Building a Future of Reliable Software

The journey toward building reliable software is ongoing, requiring constant adaptation, learning, and a commitment to excellence. As technology evolves and customer expectations rise, the definition of reliability itself will continue to shift.

The Ongoing Journey of Reliability: Continuous Learning and Improvement

Reliability is not a destination but a continuous journey. Organizations must embrace a culture of continuous learning and improvement, constantly seeking ways to enhance their systems, processes, and practices. This involves staying up-to-date with the latest technologies, investing in training and development, and fostering a culture of experimentation and innovation.

One key aspect of continuous improvement is the practice of blameless postmortems. When incidents occur, instead of assigning blame, teams should focus on understanding the root causes and identifying actions to prevent similar incidents in the future. This creates a learning environment where mistakes are viewed as opportunities for growth and improvement.

The Competitive Advantage of Reliability: Delivering Value to Customers and Stakeholders

Reliability is a key differentiator in today's competitive landscape. Customers expect seamless, uninterrupted experiences, and businesses that fail to deliver risk losing their trust and loyalty. By prioritizing reliability, organizations can

- **Enhance Customer Satisfaction:** Reliable systems meet or exceed customer expectations, leading to increased satisfaction and loyalty.

- **Reduce Costs:** Reliability initiatives can help to reduce downtime, minimize errors, and optimize resource utilization, resulting in significant cost savings.

- **Increase Revenue:** Reliable systems can drive revenue growth by enabling businesses to offer new services, expand into new markets, and attract new customers.

- **Improve Brand Reputation:** A reputation for reliability can enhance a company's brand image and attract top talent.

The Role of Leaders in Fostering a Culture of Reliability

Leaders play a crucial role in establishing and maintaining a culture of reliability. They must set the tone from the top, clearly communicating the importance of reliability and demonstrating their commitment through their actions. Leaders should also empower their teams, providing them with the autonomy, resources, and support they need to succeed.

Some specific actions leaders can take to foster a culture of reliability include

- **Establishing Clear Expectations:** Define clear expectations for reliability and communicate them throughout the organization.

- **Providing Resources and Support:** Invest in the tools, training, and infrastructure necessary to support reliability initiatives.

- **Recognizing and Rewarding Success:** Celebrate successes and recognize individuals and teams who contribute to reliability efforts.

- **Leading by Example:** Demonstrate a commitment to reliability through their own actions and decision-making.

- **Creating a Safe Environment for Learning:** Encourage experimentation, risk-taking, and learning from failures.

Emerging Trends and Technologies in Reliability Engineering

The field of reliability engineering is constantly evolving, with new trends and technologies emerging to address the growing complexity of modern software systems. Some of the most promising developments include

- **Chaos Engineering:** Chaos engineering is a disciplined approach to testing the resilience of systems by intentionally injecting failures. This helps to identify weaknesses and vulnerabilities before they cause real-world outages.

- **AIOps:** AIOps (artificial intelligence for IT operations) uses machine learning and artificial intelligence to automate and enhance IT operations tasks, such as anomaly detection, root cause analysis, and incident response.

- **Observability:** Observability is the ability to understand the internal state of a system by examining its external outputs. This is essential for detecting and diagnosing issues quickly and effectively.

- **Service Mesh:** Service mesh is a dedicated infrastructure layer that facilitates communication between services in a microservice architecture. Service mesh can provide features like load balancing, traffic management, and security, which can improve the reliability and resilience of distributed systems.

By staying abreast of these emerging trends and technologies, organizations can ensure that their reliability practices remain at the forefront of the industry and that they are well-positioned to meet the challenges of the future.

In conclusion, building reliable software is a complex and ongoing endeavor that requires a holistic approach, encompassing technical practices, organizational structures, and a culture of continuous improvement. By implementing the best practices outlined in this chapter and embracing emerging trends and technologies, organizations can create a future of reliable software that delivers value to customers, employees, and stakeholders.

PART V

Future Outlook

CHAPTER 15

Leveraging Automation and Artificial Intelligence for Enterprise Reliability

Author:

Madan Mohan Tito Ayyalasomayajula

Abstract

This chapter delves into the transformative impact of automation and artificial intelligence (AI) on enterprise reliability, specifically maintenance and asset management. Integrating innovative technology into conventional operations can bring significant benefits for organizations, including enhanced operational efficiency, reduced downtime, improved safety, and optimized resource utilization. The chapter explores various applications of automation and AI, such as predictive maintenance, condition monitoring, anomaly detection, root cause analysis, and workforce optimization. Real-world examples illustrate the advantages and challenges of adopting these innovative solutions, providing valuable insights into optimal methods and upcoming trends shaping corporate dependability.

Introduction

Background and Context

Enterprise reliability has seen substantial modifications in recent decades due to technological breakthroughs and increased worldwide market competitiveness. Traditionally focused on ensuring equipment availability and limiting unexpected downtime, today's enterprise reliability specialists confront a more complex environment shaped by digitalization, networked systems, and changing consumer expectations. This transformation needs a more complete approach to maintaining dependability, incorporating cutting-edge technology and sophisticated processes. In this setting, automation and artificial intelligence (AI) have emerged as potent technologies that provide new ways to simplify processes, manage resources, and improve overall performance. These tools assist with predictive maintenance and give insights that drive strategic decision-making, enhancing company competitiveness (Bury et al., 2014).

Automation uses control systems or computer programs to handle industrial operations without human interaction. It covers a broad spectrum of applications, from primary process controllers to advanced robots and machine learning methods. Automation improves operational efficiency by decreasing human error and ensuring process uniformity. Conversely, AI is a branch of computer science that focuses on creating intelligent computers capable of doing activities that typically need human intellect, such as reasoning, problem-solving, perception, and language comprehension. When coupled, automation and AI allow businesses to analyze massive volumes of data produced by their assets and infrastructure, detect trends, make predictions, and take remedial measures in real time (Lee, 2020). This synergy enhances reliability and allows for ongoing development and innovation in industrial operations.

The Evolution of Automation and AI in Enterprise Reliability

As businesses transition digitally, there is an increasing interest in using sophisticated technologies such as automation and artificial intelligence (AI) to improve maintenance and asset management activities. Automation has been used in industrial processes for decades, but recent advances in AI, machine learning, and edge computing have increased its capabilities and uses. These developments improve conventional automation and open new opportunities for intelligent decision-making and predictive capabilities. This section describes the growth of automation and AI in enterprise reliability, including historical milestones, present situations, and upcoming trends (Kulkarni et al., 2023). Furthermore, a discussion of how these technologies are transforming the landscape of corporate dependability, their enormous benefits, and the implications they carry for the industry as it adapts to these technological breakthroughs is presented.

Historical Perspective: Automation in enterprise reliability originated in the late 19th century with the introduction of electromechanical devices for managing steam engines and assembly lines. These early developments set the framework for increasingly complex control methods, resulting in increased industrial efficiency. Later advancements included programmable logic controllers (PLCs) in the mid-1960s and distributed control systems (DCS) in the 1980s. These technologies transformed industrial automation, allowing more accurate and flexible control over complicated operations. With the advent of computers and information systems, preventative maintenance routines became automated, allowing for proactive maintenance planning and execution. This move considerably decreased the number of unexpected equipment failures and increased asset lifespans. However, it wasn't until the turn of 2000 that AI gained momentum in enterprise reliability, first via expert systems specialized for applications. Despite their limitations, these early AI systems proved machine intelligence's promise for streamlining industrial processes. Since then, AI has evolved, moving from rule-based systems to statistical models, neural networks, and deep learning algorithms. This progress has significantly expanded the breadth and efficacy of AI applications in enterprise reliability.

Current State and Trends: Automation and AI are being integrated into enterprise reliability systems, providing benefits such as reduced downtime, increased productivity, safety, and better decision-making. These technologies allow businesses to foresee problems before they arise and react accurately and quickly. Some critical developments include using predictive maintenance based on machine learning algorithms, expanding condition monitoring systems that use sensor data analytics, and introducing smart maintenance platforms driven by edge computing and cloud services. These improvements enable real-time data processing and faster reactions to operational changes. Furthermore, the confluence of automation and AI enables advanced applications such as self-healing systems, adaptive control systems, and autonomous maintenance robots. These systems can do complicated operations with minimum human interaction, enhancing operational efficiency and dependability. Looking forward, automation and AI in corporate dependability offer even greater efficiency, more autonomy, and stronger human–machine cooperation. As these technologies advance, they are anticipated to cause substantial changes in how industries approach maintenance and asset management, resulting in a more robust and responsive operating environment (Kulkarni et al., 2023).

Predictive Maintenance with Automation and AI

Overview

Predictive maintenance offers a big step forward in industrial asset management by shifting the emphasis from reactive repair to proactive intervention, reducing downtime and increasing overall efficiency. This technique reduces the expenses associated with unexpected equipment breakdowns and increases the longevity of essential assets. Predictive maintenance systems, which combine automation and AI technology, can successfully analyze big datasets, identify aberrant behavior, and prescribe maintenance tasks before breakdowns. These technologies help companies move from planned maintenance to a more dynamic and responsive one. Understanding these principles allows companies to comprehend better the advantages of predictive maintenance and the technological improvements driving its adoption.

Predictive maintenance utilizes sensors and data-gathering systems to acquire real-time data from assets and infrastructure. These sensors include vibration monitors, temperature gauges, pressure sensors, and sound detectors, giving a complete picture of asset health. Machine learning methods, especially those based on artificial neural networks and support vector machines, are critical for evaluating this data and revealing hidden patterns and correlations (Rossini et al., 2021). These algorithms are trained on past data to recognize typical operating circumstances and detect variations that may signal a breakdown. Advanced analytical approaches, including regression analysis, time series analysis, and anomaly identification, can identify possible failure sites and propose ideal maintenance intervals. These strategies improve forecast accuracy while also providing helpful information for maintenance scheduling. Predictive maintenance has applications in various sectors, including

501

energy, transportation, manufacturing, and healthcare. Each industry
uses predictive maintenance to handle distinct operational difficulties
and increase efficiency. For example, predictive maintenance helps
monitor pipeline integrity in the oil and gas industry, lowering the chance
of leaks and spills, which may have severe environmental and financial
consequences. In aviation, it improves engine performance and lifespan,
making air travel safer and more dependable. Predictive maintenance
in manufacturing helps with continuous improvement by optimizing
production cycles and decreasing waste. It may be used in healthcare
to maintain vital medical equipment, maintain reliability, and limit
downtime that might impact patient care. The adaptability of predictive
maintenance makes it a beneficial tool in various industries, improving
reliability and performance (Rossini et al., 2021).

GE Predix Platform

GE's Predix technology perfectly illustrates how automation and AI
transform predictive maintenance at scale. This industrial Internet
platform combines several data sources to provide a comprehensive
picture of asset performance. Predix, designed to link and analyze data
from industrial assets, uses machine learning algorithms to detect aberrant
activity and forecast possible breakdowns. The platform's capacity to
handle massive volumes of data in real time allows for fast and accurate
forecasts. It also provides customized apps for specific sectors and use
cases. Predix ServiceMax, for example, helps to simplify field service
operations by delivering real-time information on technician location,
work progress, and inventory management. This helps to guarantee that
maintenance workers are appropriately deployed and that parts and
equipment are readily accessible when required. Another application,
Predix Asset Performance Management, tracks asset health and offers
maintenance schedules based on historical data and real-time sensor
inputs. This proactive strategy helps to avoid expensive, unexpected

downtime and extends asset life. GE's Predix platform exemplifies the enormous potential of merging automation and AI to create predictive solid maintenance solutions that increase corporate value and operational efficiency. The Predix platform's performance demonstrates how digital technologies have transformed industrial maintenance and asset management (www.ge.com, n.d.).

Condition Monitoring Using Automation and AI

Concepts and Challenges

Condition monitoring involves continuously evaluating the state of equipment or systems to ensure they operate reliably and promptly detect any possible problems. By adopting this proactive strategy, the occurrence of unforeseen failures is minimized, resulting in improved operating efficiency and safety. Contemporary methods for condition monitoring make use of automation and artificial intelligence (AI) technology. These technologies allow for more efficient and effective techniques by using sensor data and sophisticated analytics to assess the health of assets and provide practical insights constantly. These technologies enable real-time data monitoring and include prediction powers beyond previous approaches. However, despite its benefits, significant problems are associated with adopting condition monitoring utilizing automation and AI. Gathering and analyzing large quantities of sensor data requires a robust infrastructure and advanced data handling methods, which may be expensive and challenging to implement. Effectively analyzing intricate datasets necessitates using sophisticated analytical tools and skills, typically needing knowledge in data science and domain-specific experience. It is crucial to prioritize data security and privacy while handling sensitive information from vital industrial assets. Moreover, integrating new technologies with the current infrastructure may provide compatibility challenges, necessitating meticulous design and implementation.

Implementation Examples

The process of implementing condition monitoring via automation and AI requires numerous crucial phases, starting with the selection of suitable sensors and the establishment of communication protocols. The placement of these sensors must be carefully chosen to acquire pertinent data points and guarantee thorough monitoring. After data is gathered, it goes through preprocessing to remove errors, standardize, and combine important characteristics, ensuring the precision and uniformity of the data before analysis. Subsequently, sophisticated analytics tools, such as machine learning algorithms and signal processing procedures, extract significant insights from the data. These methodologies may discern patterns and trends that suggest problems, enabling timely action. Visualization technologies facilitate the presentation of data comprehensibly, allowing the users to make well-informed choices on maintenance tasks. These solutions often include dashboards and real-time notifications that improve situational awareness (Pimenov et al., 2022).

An exemplary instance of implementation may be seen in the power-generating industry, where condition monitoring plays a crucial role in maintaining the stability and efficiency of electrical networks. In this system, sensors are mounted on turbines, generators, and transmission lines to gather data on temperature, vibration, and other factors. This data is then evaluated in real time using machine learning algorithms to identify abnormal behavior. Implementing proactive monitoring aids in preventing outages and extends the lifespan of essential infrastructure. Early warning alerts empower operators to proactively implement remedial actions before the occurrence of faults, reducing downtime and enhancing the grid's resilience. This method improves dependability and streamlines maintenance schedules, hence decreasing operating expenses. Another instance may be found in the predictive maintenance domain in industrial settings. Manufacturers may enhance their productivity and prevent unexpected downtime by strategically placing sensors across their factory floors and using sophisticated analytics to monitor the condition of their equipment and predict potential breakdowns. Additionally, these systems can provide valuable information about operating efficiency and propose potential enhancements. Empirical evidence showcases substantial reductions in maintenance expenses and improved operational availability, underscoring the need for condition monitoring in the current competitive environment. These advantages enhance the efficiency and flexibility of the production process, enabling it to adjust to changing requirements and reduce inefficiencies quickly.

Siemens MindSphere: Siemens MindSphere is a fascinating case study demonstrating the effective incorporation of automation, artificial intelligence, and condition monitoring in industrial environments. This cloud-based Internet of Things (IoT) operating system facilitates smooth communication across goods, factories, systems, and consumers, creating opportunities for cutting-edge digital solutions. Due to its scalability and versatility, it is appropriate for a broad spectrum of applications, ranging from small-scale implementations to large industrial complexes (Petrik & Herzwurm, 2019).

MindSphere leverages data from interconnected devices and uses sophisticated analytics to provide essential insights. Condition monitoring provides real-time insight into assets' health and performance indicators, enabling prompt reaction to any concerns. Users get advantages from comprehensive diagnostic reports, proactive maintenance suggestions, and the ability to troubleshoot remotely. These characteristics aid in decreasing maintenance expenses and enhancing the dependability of assets. MindSphere also facilitates collaborative problem-solving among stakeholders, enhancing transparency and improving operational

505

efficiency. This cooperative approach guarantees that all stakeholders are well-informed and can participate in decision-making. Practically, MindSphere has produced notable achievements in several sectors. Siemens Gamesa Renewable Energy used MindSphere in the wind energy industry to improve blade inspection processes. In the past, blades needed to undergo physical examinations every six months, resulting in significant expenses and restricted coverage due to weather limitations. The fusion of artificial intelligence and automation enables ongoing surveillance and timely identification of possible problems. Automated data analysis allows for early damage identification, significantly decreasing downtime and maintenance costs. This proactive strategy has resulted in substantial cost reductions and enhanced operational effectiveness.

Similarly, in the railways field, Deutsche Bahn used MindSphere to enhance the efficiency of train repair scheduling. Engineers used data collected from sensors mounted on trains and tracks to get valuable insights about the deterioration of components, allowing them to make precise adjustments to maintenance schedules. Consequently, there were significant cost reductions and enhanced customer satisfaction due to more efficient journeys and reduced delays. MindSphere's predictive capabilities have effectively reduced interruptions and improved service quality by enabling timely repair. Siemens MindSphere represents the significant influence of automation, artificial intelligence, and condition monitoring on industrial processes. By synergistically combining these technologies, firms may achieve more efficiency, reduce downtime, and gain a long-lasting competitive edge. The platform's performance showcases digital transformation's capacity to completely overhaul asset management and operational procedures in diverse sectors (Petrik & Herzwurm, 2019).

Anomaly Detection Through Automation and AI

Anomaly identification is an essential component of data analysis that focuses on discovering atypical or aberrant observations, events, or behaviors that deviate considerably from the expected norms within a dataset. An efficient anomaly detection system enhances reliability, security, and adaptability, making it an essential tool in several sectors. Anomaly detection can help applications such as intrusion detection, problem diagnostics, fraud protection, and predictive maintenance. Rapid and precise identification of abnormalities helps avert the escalation of minor difficulties into significant ones. With the increasing volume and complexity of data generated daily, it is crucial to have accurate and efficient anomaly detection methods. Expanding intricacy necessitates inventive solutions that adjust to varied and ever-changing circumstances. As enterprises increasingly adopt digital technologies, anomaly detection becomes even more crucial in ensuring the reliability and effectiveness of systems (Krishna Parimala, 2024).

Methodologies and Algorithms

Various approaches, including statistical and machine learning frameworks, are used to ease the identification of anomalies. Statistical approaches use probability distributions to evaluate whether specific observations surpass acceptable thresholds, offering a statistical framework for detecting outliers. Mean shift clustering, ARIMA models, and statistical process control charts are often used as statistical methods for detecting anomalies. These strategies are incredibly efficient in contexts where data patterns are readily understood and reasonably consistent. Machine learning algorithms, on the other hand, acquire knowledge from past data to identify irregularities by comparing them to established patterns. This technique is very flexible and capable of managing intricate, nonlinear interactions in data. K-nearest neighbors (k-NN), local outlier factor (LOF), isolation forest, and one-class support vector machines (OC-SVM) are often used machine learning techniques for detecting anomalies. Each algorithm has unique capabilities, such as the capability to process data with many dimensions or the aptitude to operate well with tiny datasets. Thoroughly evaluating the advantages and disadvantages of each method is crucial to identifying the best option for a particular use case. In addition, combining different methodologies might sometimes provide superior outcomes by capitalizing on their complementing benefits (Wang, 2024).

Use Cases

Anomaly detection, which involves automation and AI, has many applications, including cybersecurity, healthcare, finance, and manufacturing. Anomaly detection algorithms in cybersecurity are often used to monitor network traffic and safeguard against unauthorized access and data breaches. Healthcare professionals use anomaly detection to identify rare diseases and monitor patients' vital signs for early identification, perhaps preventing fatalities via prompt treatments (Wang, 2024). Financial institutions use anomaly detection techniques to protect against fraudulent transactions and mitigate possible financial losses, bolstering economic systems' security. In industrial businesses, predictive maintenance tactics use anomaly detection to forecast equipment problems and arrange repairs before severe failures. This proactive strategy reduces the time that operations are halted and decreases the expenses associated with maintenance. Amid the continuous growth of data, there is an increasing need for accurate and adaptable anomaly detection solutions. As these methods progress, they become more available and valid for various uses, stimulating innovation in many industries.

Case Study: IBM Watson: IBM Watson is a compelling example of how anomaly detection drives the development of advanced AI applications. Watson combines natural language processing, machine learning, and cognitive computing abilities to handle large amounts of organized and unorganized data. The "anomaly findings" function detects and identifies abnormal data stream trends, patterns, or values. It also gives clear explanations, enabling users to comprehend and act based on these insights quickly. IBM Watson's anomaly detection capabilities have shown encouraging outcomes, showcasing its adaptability and efficacy in several sectors. Insurance firms have used Watson to analyze claims data and uncover suspected fraudulent activity, reducing costs and enhancing service quality. Retailers have used Watson to enhance pricing schemes by identifying inconsistencies, resulting in more competitive pricing strategies and profitability (Quiroz-Vázquez, 2023). Healthcare experts have collaborated with Watson to accurately diagnose complex medical diseases by identifying concealed abnormalities in patients' electronic health information. IBM Watson demonstrates the significant impact of automation and AI in generating groundbreaking advancements and optimizing decision-making processes across several industries by integrating anomaly detection with its vast array of AI services. This integration demonstrates the potential of AI-powered anomaly detection to revolutionize businesses by improving precision, productivity, and strategic capabilities.

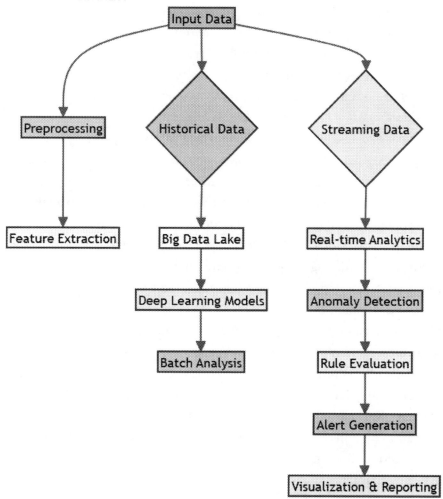

Figure 15-1. *Architecture diagram of IBM Watson's Anomaly
Detection System*

Figure 15-1 shows the major components of IBM Watson's Anomaly Detection System. This technology allows businesses to spot strange patterns and unexpected behaviors in massive datasets, offering early warnings of possible concerns before they become major problems. The input data, which includes streaming and historical data sources, is at the heart of this system. These data points are first preprocessed and then feature extracted to turn raw data into meaningful representations that can be efficiently studied. Next, the real-time analytics engine examines incoming streaming data using machine learning models designed to discover abnormalities in real time. In addition, historical data is saved in a large data lake, enabling deep learning models to perform batch analysis and pattern detection. Once abnormal occurrences are recognized, rule engines assess the severity and context of each event to determine the appropriate alert generation. Finally, visualization and reporting tools expose the data to end users, allowing them to examine and resolve the root causes of discovered abnormalities. This integrated strategy enables firms to manage their assets and systems proactively, increasing operational efficiency and effectiveness (Www.ibm.com, 2021).

Root Cause Analysis with Automation and AI

Root cause analysis (RCA) utilizes automation and AI to quickly discover and address the fundamental causes of issues, surpassing the effectiveness of conventional approaches. Automated RCA systems may expediently analyze extensive information, identify patterns, and provide practical insights using AI and machine learning. This process minimizes human mistakes and improves precision. This contemporary methodology allows for expedited incident resolution, enhanced operational efficiency, and the ability to efficiently manage intricate and extensive data environments.

Conventional Approaches

Conventional root cause analysis (RCA) procedures include a methodical
examination to ascertain the fundamental reasons for issues or
occurrences. These investigations often adhere to a systematic process
that involves collecting data, identifying the issue, formulating hypotheses,
conducting tests to verify the hypotheses, and implementing remedial
measures. Standard methodologies used for many years in many sectors
include the five whys, fishbone diagrams, and Failure Mode and Effects
Analysis (FMEA). Although traditional RCAs are successful, they may
be time-consuming, require much effort, and are susceptible to human
mistakes, especially when working with large datasets or intricate systems.
The manual approach often requires significant skill and experience, and it
might pose difficulties in maintaining uniformity and neutrality. Moreover,
conventional RCA techniques may face challenges in keeping up with
the fast data creation and intricacy of contemporary industrial settings,
hindering the timely and accurate identification of underlying causes.

Automated RCA (ARCA)

Automated RCA offers notable advantages compared to conventional
techniques using artificial intelligence (AI) and machine learning
algorithms to accelerate and enhance the RCA process. The primary
advantages of automated root cause analysis (RCA) include quicker
resolution times for incidents, less reliance on human involvement,
improved precision, and the capacity to handle bigger and more intricate
datasets. AI-driven RCA systems can efficiently analyze large volumes of
data, enabling them to promptly find correlations, trends, and causal links
that may otherwise remain unnoticed. These systems can continually
learn and enhance their diagnostic skills, improving their efficacy as time
goes on. Moreover, automated root cause analysis (RCA) can combine
data from several sources, resulting in a more extensive perspective on the

elements contributing to an issue. This comprehensive method detects
urgent factors and reveals underlying systemic problems, resulting in more
efficient long-term remedies. Furthermore, automated RCA mitigates
the potential for human mistakes and biases by reducing dependence on
manual analysis, resulting in more objective and dependable outcomes
(Soualhia & Wuhib, 2022).

Table 15-1. *Comparison of traditional root cause analysis vs.
automated RCA*

Criteria	Traditional Root Cause Analysis (TRCA)	Automated Root Cause Analysis (ARCA)
Data Collection	Manual collection of data through interviews, logs, etc.	Automatic data collection using sensors, logs, etc.
Time Consumption	Can take days to weeks to complete	Quicker identification of root cause
Accuracy	Depends on investigator skills and experience	Improved accuracy due to automated analysis
Scalability	Limited scalability for larger systems	Capable of analyzing multiple incidents simultaneously
Complexity	Suitable for simple issues	Effective for complex failure scenarios
Cost	Lower cost for small-scale incidents	Higher cost for implementation and maintenance
Continuous Improvement	Manual effort required for continuous improvement	Automatically updated with new data and algorithms
Human Error Reduction	Minimal reduction in human error	Significantly reduces human errors

(*continued*)

Table 15-1. *(continued)*

Criteria	Traditional Root Cause Analysis (TRCA)	Automated Root Cause Analysis (ARCA)
Expertise Required	Requires domain expertise	Leverages machine learning and AI algorithms
Integration with Other Tools	Limited integration options	Seamless integration with other systems and tools
Adaptability	Infrequent updates	Quickly adapt to changes in infrastructure and processes

This comparison (Table 15-1) shows the differences between traditional root cause analysis (TRCA) and automated root cause analysis (ARCA) methodologies for resolving problems in industrial systems. TRCA depends on manual data gathering via interviews, logs, and other methods, which may be time-consuming depending on the investigators' abilities and expertise. ARCA, on the other hand, collects data automatically from sensors, logs, and other sources, allowing for faster root cause identification and increased overall accuracy. Although ARCA has greater installation and maintenance costs, it dramatically decreases human error and integrates well with other technologies (Soualhia & Wuhib, 2022). Furthermore, ARCA responds swiftly to changes in infrastructure and procedures, making it an invaluable tool for dealing with complicated failure situations at a scale. However, it is essential to remember that neither technique covers every circumstance adequately and the best plan often combines both strategies to maximize their distinct advantages.

Tools and Methods: Multiple tools and methods are used to carry out automated root cause analysis (RCA). Data mining, predictive modeling, and machine learning algorithms are the fundamental components of automated RCA systems. These platforms are capable of effectively analyzing vast amounts of data. Natural language processing (NLP) and text analytics facilitate extracting semantic information from unstructured data sources such as emails, chat logs, and consumer feedback. This process yields valuable insights into possible difficulties. Graph databases and network analysis visualizations facilitate the comprehension of intricate interdependencies and interactions inside systems, enabling the identification of the underlying causes of issues. Continuous monitoring and alerting techniques enable the timely discovery of issues and rapid action, ensuring that problems are swiftly handled (Azimi & Pahl, 2020). In addition, sophisticated analytics systems can model different situations to forecast probable future failures and suggest proactive remedies. By integrating these many methodologies into a unified RCA platform, firms may address issues with unparalleled speed and accuracy, significantly improving operational efficiency.

Dynatrace ARCA: Dynatrace ARCA is a fascinating example demonstrating the efficiency of automated root cause analysis (RCA) in real-world situations. Dynatrace ARCA is a software solution mainly created for customers in the process sector. It utilizes sophisticated analytics, machine learning, and expert knowledge to provide fast and precise root cause investigation. This software uses data from many plant floor instruments and external sources for real-time anomaly detection and predictive failure analysis. The system constantly analyzes equipment and processes, identifying deviations from standard operating conditions and forecasting any breakdowns in advance. It then employs sophisticated diagnostic tools and expert databases to suggest likely reasons and suggested courses of action, offering plant managers practical insights that can be acted upon. Dynatrace ARCA enables plant managers to promptly resolve problems, reduce downtime, and enhance operational efficiency. The platform's capacity to combine and analyze data from many sources guarantees thorough and precise identification of the underlying causes, resulting in more efficient problem-solving and ongoing enhancement. Moreover, using AI and machine learning enables

Dynatrace ARCA to adjust to evolving circumstances and enhance its
diagnostic precision as time progresses. It is a beneficial instrument for
upholding elevated dependability and efficiency in industrial operations
(`Www.dynatrace.com`).

Workforce Optimization Through Automation and AI

Implementing automation and AI technologies for workforce optimization
is a revolutionary method to improve productivity, employee engagement,
and safety across many sectors. Organizations may optimize their
operations, enhance staff capabilities, and realize financial benefits using
cutting-edge technology such as robotic process automation, predictive
analytics, and virtual assistants. To fully optimize the advantages of this
comprehensive strategy, engaging in strategic planning, maintaining
effective communication, and providing continuous training are
necessary. This will help overcome skills deficiencies and privacy concerns
(Sathya et al., 2023).

Benefits and Challenges

Incorporating automation and artificial intelligence (AI) into workforce
optimization offers many advantages, including heightened efficiency,
expanded employee involvement, improved safety, and cost reduction.
Automation can assume control of monotonous and tedious jobs,
liberating people to concentrate on more valuable endeavors that need
creativity and critical thought. Enhanced employee engagement arises
from diminishing the tedium of repetitive duties and empowering
employees to make more strategic contributions toward corporate
objectives. AI-driven monitoring systems are used to enhance safety by
accurately predicting and preventing dangerous circumstances, resulting

in a decrease in workplace accidents (Jain et al., 2023). Nevertheless, the problems linked to this transformation include opposition to change, skill deficiencies, apprehensions about job displacement, and worries over privacy and security. Employees may feel anxious about emerging technology as they worry about automation replacing human employment. To address the skill shortages, it is essential to make substantial investments in training and development initiatives. These programs aim to enable the workforce to effectively collaborate with sophisticated technology. Furthermore, upholding data privacy and safeguarding sensitive information in an increasingly automated setting is essential to prevent breaches and foster trust among workers.

Strategies and Best Practices

To achieve successful workforce optimization through automation and AI, it is crucial to implement clear communication regarding technology objectives, offer training opportunities, encourage collaboration between humans and machines, establish performance metrics, and address ethical and social considerations. Efficient communication reduces anxieties and fosters a favorable sense of technological progress. Offering ongoing training and chances for skill enhancement guarantees that staff stay current and competent in the ever-changing digital environment. Facilitating cooperation between people and machines entails creating processes in which AI enhances human endeavors rather than supplanting them. Defining precise performance indicators enables firms to evaluate the influence of automation and AI on productivity and make necessary adjustments to their plans. Addressing ethical and social problems entails guaranteeing that AI upholds privacy, mitigates prejudice, and promotes equitable labor practices. A progressive implementation strategy for technology enables workers to slowly adapt to changes and acquire essential skills, facilitating easier transitions (Jain et al., 2023).

Periodic assessment and modification of policies guarantee that they align with changing business requirements and technical progress, facilitating a durable integration of AI in the workforce.

Tools and Solutions

Several kinds of tools and solutions enable workforce optimization via automation and AI. Robotic process automation (RPA) software automates monotonous operations, allowing staff to concentrate on activities that provide value, thereby improving total productivity. Intelligent agents and virtual assistants provide individualized support and direction to workers, enhancing their overall experience and productivity by effectively managing questions and regular chores. Predictive analytics and machine learning algorithms provide proactive administration of processes and resources, resulting in the reduction of bottlenecks and the minimization of waste via the prediction of demand and the optimization of resource allocation. Augmented reality (AR) and virtual reality (VR) technologies provide engaging training experiences and remote collaboration, enabling workers to acquire practical skills and interact effortlessly regardless of physical location. Furthermore, these technologies facilitate intricate repair and maintenance operations by offering immediate aid and direction (Jain et al., 2023). By incorporating these technologies into their operations, businesses may establish an adaptable, quick-to-respond, and highly productive workforce, effectively addressing contemporary difficulties.

Microsoft Azure for Manufacturing: Microsoft Azure for Manufacturing showcases the use of automation and artificial intelligence (AI) to enhance the efficiency of workforces in the manufacturing industry. This platform combines Internet of Things (IoT) devices, edge computing, and artificial intelligence (AI) services to create intelligent production environments that can adapt and react to changing circumstances in real time. Microsoft Azure for Manufacturing utilizes real-time data analysis from factory floors to allow predictive maintenance, enhance quality, and optimize energy use. This results in substantial savings in downtime and operating expenses. The platform's sophisticated analytics capabilities enable manufacturers to anticipate equipment problems in advance, guaranteeing prompt repair and reducing interruptions. In addition, it provides configurable apps designed for industrial situations, enabling firms to integrate automation smoothly and AI into their operations while meeting the distinct needs of their workforce. The technology also facilitates sophisticated quality control procedures by examining manufacturing data to detect flaws and promptly apply remedial measures. In addition, Microsoft Azure for Manufacturing improves energy efficiency by improving resource use via predictive analysis, hence supporting sustainability objectives. Microsoft Azure for production showcases the potential of using artificial intelligence and automation to revolutionize conventional production methods. This integration enhances efficiency, resilience, and sustainability while enabling the workforce to perform better (`www.microsoft.com`).

Security Considerations

Ensuring security in automation and artificial intelligence (AI) systems is crucial for protecting against cyberattacks, data breaches, and unauthorized access. Organizations should use strong security measures, such as access restriction, encryption, and constant monitoring, to effectively reduce threats. Adhering to legislation such as GDPR and HIPAA is crucial to guarantee legal compliance and safeguard people's privacy rights (Suter, 2019).

Threat Landscape

The threat landscape associated with automation and artificial intelligence (AI) systems includes risks originating from insider threats, cyberattacks, data breaches, and vulnerabilities resulting from third-party collaborations and integrations. Malicious individuals might intentionally focus on AI systems to gain illegal entry, manipulate data, cause disruptions, or steal sensitive information, substantially damaging organizations' integrity and reputation. Moreover, there is a potential danger of AI being deliberately or accidentally abused, leading to prejudices, discriminatory actions, or other undesirable consequences. This emphasizes the intricate ethical and sociological factors involved in the development and implementation of AI. With AI systems' increasing prevalence and interconnectivity, the possibility for attacks also grows, necessitating strong security measures to counter new threats successfully. In addition, the fast advancement of AI technology brings forth new security concerns, such as targeted assaults on machine learning models and the manipulation of audiovisual material via deepfakes. This calls for a constant need for alertness and creativity in defensive methods.

Mitigation Strategies

To address these threats, businesses must implement stringent security measures specifically designed for the unique attributes of AI and automation technologies. Access control rules must be carefully and systematically maintained, ensuring only authorized personnel are given permission. This is achieved using least privilege and role-based access control, which help minimize the risk of possible security breaches. Encryption and secure data transmission techniques protect sensitive data while it is being stored and sent, guaranteeing confidentiality and integrity throughout all phases of data processing. Multifactor authentication and continuous monitoring enhance the security of user accounts by promptly identifying and addressing any suspicious activity, preventing illegal access or data theft. Consistent updates and patches maintain the security and stability of the system, reducing the chances of being targeted by hackers who want to exploit system weaknesses and gain unauthorized access to sensitive data. Efforts to educate employees about AI technology aim to increase their understanding and encourage responsible use (Suter, 2019). These campaigns create a security-conscious atmosphere, enabling users to promptly identify and report security issues. Additionally, employees are expected to follow established security policies and best practices.

Compliance Regulations

Compliance rules are of utmost importance in guiding the deployment and administration of automation and AI systems. They provide a framework for assuring responsible and ethical usage while protecting people's rights and privacy. Complying with guidelines set by regulatory bodies builds confidence with stakeholders. It assures legal conformity, which helps firms prevent reputational harm, financial fines, and legal liabilities that may arise from not following the rules. For example, the General Data Security Regulation (GDPR) of the European Union enforces strict regulations on the security and privacy of data. It mandates enterprises to adopt methods like data anonymization, consent management, and breach reporting to ensure the safety of people's personal information. The Health Insurance Portability and Accountability Act (HIPAA) enforces regulations on managing protected health information in the US healthcare industry. It imposes stringent security and privacy standards on organizations that handle sensitive medical data to safeguard patient confidentiality and prevent unauthorized access or disclosure. It is crucial to stay informed about new regulations and adjust accordingly to remain compliant and avoid possible penalties. This requires continuously monitoring regulatory changes and actively consulting legal and compliance professionals to ensure alignment with evolving legal and industry norms.

Future Directions and Emerging Trends

The future of automation and AI is moving toward sophisticated analytics, machine learning, edge computing, and blockchain integration. These technologies have the potential to completely transform companies by allowing for more profound understanding, immediate decision-making, and distributed data management. Adopting these trends will improve efficiency and production and stimulate innovation and competitiveness in the changing digital environment.

Advanced Analytics

Advanced analytics will continue to influence the future of automation and artificial intelligence (AI) by extracting more profound insights from more varied and intricate datasets. Organizations may enhance their decision-making, streamline operations, and anticipate future trends using predictive modeling, prescriptive analytics, and big data analytics. Improved analytical skills also enable the development of advanced applications, such as self-driving cars, intelligent urban areas, and targeted advertising campaigns, transforming several sectors and fostering innovation. As companies gather more data from different sources, the need for sophisticated analytics tools and skills will grow, driving further research and development in this area (Baker & Ellis, 2020). In addition, integrating sophisticated data analysis with artificial intelligence systems allows for immediate decision-making, allowing firms to quickly adapt to shifting market circumstances and client needs, thus establishing a competitive advantage in dynamic contexts.

Machine Learning and Deep Learning

Machine learning (ML) and deep learning (DL) are crucial catalysts for innovation in automation and artificial intelligence (AI). ML algorithms facilitate computer learning by using data inputs and gradually enhancing performance. On the other hand, DL models imitate the structure and functionality of the human brain to identify patterns and categorize data. As these technologies develop and become more advanced, they will result in progress in fields such as voice recognition, picture classification, natural language processing, and robotics. This will eventually bring about significant changes in the healthcare, finance, transportation, and manufacturing sectors. The widespread use of machine learning (ML) and deep learning (DL) frameworks and libraries, together with the availability of extensive computer resources, democratizes the creation of artificial intelligence (AI). This enables enterprises of all scales to leverage the potential of machine intelligence to achieve desired business results. Moreover, the progress in explainable AI and reinforcement learning allows AI systems to provide clear and flexible decision-making procedures, improving confidence and usefulness in AI applications in many fields (Baker & Ellis, 2020).

Edge Computing

Edge computing is crucial in advancing automation and AI as it brings computational capabilities closer to the source of data generation, processing, and action. Edge computing improves the dependability and speed of AI systems by lowering latency and bandwidth requirements. This makes them well-suited for scenarios that need real-time decision-making and quick replies. Sectors such as agriculture, oil and gas, and logistics have the potential to gain significant advantages from the capability of edge computing to analyze data and provide insights almost instantaneously locally. Furthermore, edge computing architectures facilitate decentralized AI models and federated learning methods, enabling devices to cooperate and exchange knowledge while maintaining data confidentiality and protection. The widespread usage of edge computing devices and platforms allows for implementing artificial intelligence (AI)-powered applications in distant and resource-limited situations. This expands the scope of automation and AI to hitherto unexplored areas and scenarios (Alnemari & Bagherzadeh, 2019).

Blockchain Technology

Blockchain technology has the potential to completely transform the way automation and artificial intelligence (AI) interact with data and processes. The use of distributed ledgers, consensus processes, and cryptographic security features in blockchains may effectively prevent unauthorized alterations to data, guarantee the accuracy and reliability of data, and uphold transparency. Integrating blockchain technology with AI systems may establish safe, decentralized, and reliable networks for exchanging data. This can provide new opportunities for innovative applications in supply chain management, financial services, and identity verification. Smart contracts are pieces of self-executing code that are recorded on blockchains. They allow for the automated and transparent execution of business operations, reducing reliance on intermediaries and making transactions more efficient. In addition, blockchain-based AI markets and federated learning frameworks enable the collaborative creation of AI and the sharing of models, all while ensuring data ownership and privacy. This promotes innovation and cooperation within the AI ecosystem (Patwe, 2022). As blockchain technology advances and becomes more widely used, its incorporation into automation and AI systems will facilitate the creation of digital ecosystems that are safer, more transparent and efficient. This will enable the establishment of decentralized and autonomous organizations in the future.

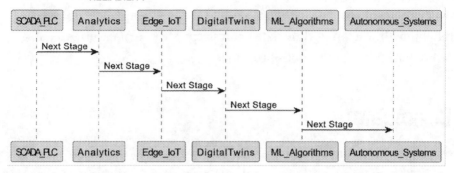

Figure 15-2. *Roadmap of emerging technologies in automation and
AI for enterprise reliability*

Figure 15-2 presents a roadmap illustrating the emerging automation
and artificial intelligence (AI) technologies that significantly enhance
enterprise reliability. Starting at the foundation, we have supervisory
control and data acquisition (SCADA) and programmable logic controllers
(PLC), which are traditionally used for monitoring and controlling
industrial processes. As we move forward, advanced analytics techniques
like predictive maintenance analytics and condition monitoring emerge,
enabling early detection of potential failures and reducing downtime. The
next stage involves the integration of edge computing and IoT sensors into
the system, providing real-time data collection and analysis closer to the
source. This leads us to digital twins, virtual replicas of physical assets,
allowing for simulating complex scenarios, optimizing performance,
and facilitating predictive maintenance. Advanced machine learning
algorithms, such as deep learning and reinforcement learning, come
into play at the subsequent level, enabling more accurate predictions,
anomaly detection, and automated decision-making. Finally, the pinnacle
of the roadmap is autonomous systems and self-healing networks, where
AI capabilities enable fully autonomous processes and self-correcting
networks, resulting in increased efficiency, improved safety, and enhanced

overall enterprise reliability. This evolution aims to enhance enterprise
reliability through real-time data analysis, predictive maintenance, and,
ultimately, fully autonomous processes.

Conclusion

Integrating automation and artificial intelligence (AI) into corporate
operations is rapidly gaining popularity, presenting a plethora of benefits
and challenges across industries. Automation and artificial intelligence
may transform conventional operations by enhancing operational
efficiency, minimizing downtime, and increasing resource utilization.
However, to ensure successful implementation and benefit realization,
it is necessary to carefully examine organizational effects such as labor
adjustments and skill upgrades. Numerous ways for using automation and
AI to increase enterprise reliability have been looked into, and potential
applications include predictive maintenance, condition monitoring,
anomaly detection, root cause analysis, and labor optimization. Real-
world examples demonstrate both the advantages and limitations of
these innovative technologies, providing valuable insights into optimal
approaches and developing trends that define the corporate dependability
environment.

Furthermore, security remains a primary issue in employing
automation and AI, and strict access restrictions, encryption measures,
regular upgrades, and intensive staff training are needed to limit risks
and ensure compliance with regulations. Strategically, businesses must
embrace data-driven decision-making, encourage agility and flexibility,
and invest in training staff capable of effectively managing and working
with AI technology. Looking forward, further research should delve deeper
into AI's ethical components, assess the long-term economic impact of
automation on employment, and analyze the relationship between AI and
sustainability. Collaboration among academia, government, and industry

will be crucial in increasing our understanding of automation and AI. At
the same time, measures to democratize AI technology and eliminate
biases in AI algorithms are vital for ensuring fair outcomes and removing
societal inequities. To summarize, the path to using automation and AI for
increased enterprise reliability is complex and dynamic, needing continual
study, adaptation, and collaboration to reach their full potential in shaping
the future of company operations.

Bibliography

1.   Alnemari, M., & Bagherzadeh, N. (2019). Efficient deep
      neural networks for edge computing. 2019 IEEE International
      Conference on Edge Computing (EDGE). doi:10.1109/
      edge.2019.00014

2.   Azimi, S., & Pahl, C. (2020). Root cause analysis and
      remediation for quality and value improvement in machine
      learning driven information models. Proceedings of the 22nd
      International Conference on Enterprise Information Systems.
      doi:10.5220/0009783106560665

3.   Baker, D., & Ellis, L. (2020). Future directions in digital
      information: Predictions, practice, participation. Chandos
      Publishing

4.   Bury, S. J., Sharda, B., Agarwal, A., & Leonard, C. (2014, April 17).
      Enabling technologies for enterprise reliability. 2014 Reliability
      and Maintainability Symposium

5.   Framework for Digital Business Transformation. (2022). The
      Human Side of Digital Business Transformation, 165–212

6.   Jain, P., Tripathi, V., Malladi, R., & Khang, A. (2023). Data-driven
     artificial intelligence (AI) models in the workforce development
     planning. Designing Workforce Management Systems for
     Industry 4.0, 159–176. doi:10.1201/9781003357070-10

7.   Krishna Parimala, V. (2024). Introductory chapter: Anomaly
     detection – Recent advances, AI and ML perspectives
     and applications. Artificial Intelligence. doi:10.5772/
     intechopen.113968

8.   Kulkarni, V., Reddy, S., Clark, T., & Proper, H. (2023). The AI-
     enabled enterprise. The Enterprise Engineering Series, 1–12.
     doi:10.1007/978-3-031-29053-4_1

9.   Lee, J. (2020). Bringing building automation systems under
     control. Web Based Enterprise Energy and Building Automation
     Systems, 91–100. doi:10.1201/9781003151234-12

10. Patwe, S. S. (2022). Blockchain-enabled decentralized
     traceability in the automotive supply chain. The Role of IoT and
     Blockchain, 177–186. doi:10.1201/9781003048367-17

11. Petrik, D., & Herzwurm, G. (2019). IIoT ecosystem development
     through boundary resources: A siemens MindSphere case study.
     Proceedings of the 2nd ACM SIGSOFT International Workshop
     on Software-Intensive Business: Start-ups, Platforms, and
     Ecosystems. doi:10.1145/3340481.3342730

12. Pimenov, D. Y., Bustillo, A., Wojciechowski, S., Sharma, V. S.,
     Gupta, M. K., & Kuntoğlu, M. (2022). Artificial intelligence
     systems for tool condition monitoring in machining: Analysis
     and critical review. Journal of Intelligent Manufacturing, 34(5),
     2079–2121. doi:10.1007/s10845-022-01923-2

13.   Quiroz-Vázquez, C. (2023, December 15). Anomaly detection in
       machine learning: Finding outliers for optimization of business
       functions. Retrieved from `https://www.ibm.com/blog/anomaly-
       detection-machine-learning/`

14.   Rossini, R., Prato, G., Conzon, D., Pastrone, C., Pereira, E.,
       Reis, J., ... Goncalves, G. (2021). AI environment for predictive
       maintenance in a manufacturing scenario. 2021 26th IEEE
       International Conference on Emerging Technologies and Factory
       Automation (ETFA). doi:10.1109/etfa45728.2021.9613359

15.   Sathya, V., Jayashree, K., & Malathi, S. (2023). Robotic
       process automation (RPA) applications and tools for the
       workforce management system. Designing Workforce
       Management Systems for Industry 4.0, 251–264.
       doi:10.1201/9781003357070-16

16.   Soualhia, M., & Wuhib, F. (2022). Automated traces-based
       anomaly detection and root cause analysis in cloud platforms.
       2022 IEEE International Conference on Cloud Engineering
       (IC2E). doi:10.1109/ic2e55432.2022.00034

17.   Suter, R. (2019). Artificial intelligence security threats. Artificial
       Intelligence and Machine Learning for Business for Non-
       Engineers, 37–43. doi:10.1201/9780367821654-4

18.   Wang, R. (2024). AI-powered predictive cybersecurity in
       identifying emerging threats through machine learning.
       2024 IEEE 3rd International Conference on Electrical
       Engineering, Big Data and Algorithms (EEBDA). doi:10.1109/
       eebda60612.2024.10485789

19.   `www.dynatrace.com`. (2024, March 18). Root-cause analysis.
       Retrieved from `https://www.dynatrace.com/monitoring/
       platform/root-cause-analysis`

20.   www.ge.com. (n.d.). What is Predix platform? Retrieved from
https://www.ge.com/digital/documentation/predix-
platforms/c_what_is_predix_platform.html

21.   www.ibm.com. (2021, February 8). IBM Watson IoT platform -
Message gateway 5.0.0. Retrieved from https://www.ibm.com/
docs/en/wip-mg/5.0.0.1?topic=overview-architecture

22.   www.microsoft.com. (n.d.). Microsoft cloud for manufacturing.
Retrieved from https://www.microsoft.com/en-us/industry/
manufacturing/microsoft-cloud-for-manufacturing

CHAPTER 16

Reliability Outlook in the Digital Age

Authors:

Fardin Quazi

Praveen Gujar

Real-Time Scenarios in Different Industries

Reliability has become the foundation across all industries with an increased focus on customer centricity. A reliable and efficient system ensures seamless operation and effectiveness in the delivery of services. It enables the system to consistently perform its intended functions without failure while maintaining high-quality standards. This makes the system secure, usable, and dependable.

With the advancement of technology, reliability has become even more important. As industries keep up the pace of innovation in design and application while leveraging the latest technology in their operations, the reliability of the systems has become even more critical. Be it finance using blockchain for leverage, healthcare turning to telemedicine, or the manufacturing sector with automation and AI, in every system, reliability is paramount. In a digital environment where industries continue to evolve and adapt to market dynamics and customer requirements, there is more

© Saurav Bhattacharya 2024
M. Kuppam, *Enterprise Digital Reliability*, https://doi.org/10.1007/979-8-8688-1032-9_16

relevance attached to reliability. It is not all about keeping the lights on but rather ensuring that services can be provided in a consistent, secure, and efficient manner regardless of outside influences.

Reliability is key to seamless operations across industries. This is the silent engine that keeps systems running and makes sure the services are delivered without any glitches. As integration and cross-collaboration among industries continue to evolve, the importance of reliability will only gain eminence in the coming years.

Reliability in Healthcare

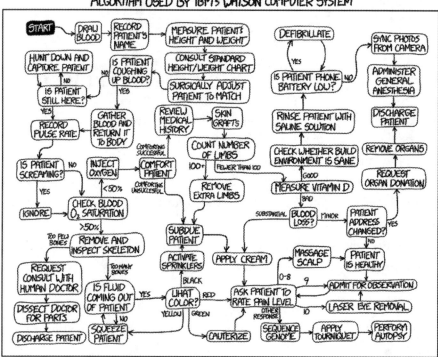

Image source: https://xkcd.com/1619/

Reliability in healthcare is of great essence, ensuring seamless operations of healthcare systems. Reliability forms the base for successful healthcare delivery, thereby assuring high-quality patient care and healthcare services. Healthcare administration involves a comprehensive range of services like patient registration, scheduling appointments, and historical medical records and bills. All these services have to work reliably to achieve effective healthcare delivery. For example, where there is an efficient system of appointment scheduling, it enhances the patient experience, while dependable Electronic Health Records management assures that there is accurate and accessible information on patients. This chapter will discuss a real-time case to emphasize the importance of reliability in healthcare. It will also show practical implications of reliability in healthcare settings that could contribute to patients' outcomes and operational efficiencies.

In healthcare, reliability is not to be considered just for the continuous performance of systems and processes. It's all about building a culture of improvement that each member of the team, from physicians and nurses to administration and IT professionals, commits to bettering the quality of care. This reinforces the belief that errors and failures are not a result of individuals' faults, but are actually learning opportunities. This encourages a nonpunitive attitude since staff members are encouraged to identify and report errors and near misses and be able to manage risks while improving patient data safety. This next case study showcases the role of reliability in a healthcare administration setup, providing insights into the practical implications.

Case Study: High-Reliability Organizing in Healthcare

In 2022, a significant study was conducted at Johns Hopkins University, a leading institution known for its groundbreaking research in healthcare [1]. The study was focused on an important and emerging concept of high-reliability organizing (HRO) in healthcare, in the face of growing demands for quality care and patient safety. High-reliability organizations (HROs) are setups operating in complex, high-risk environments while managing to maintain very high levels of safety for very long periods.

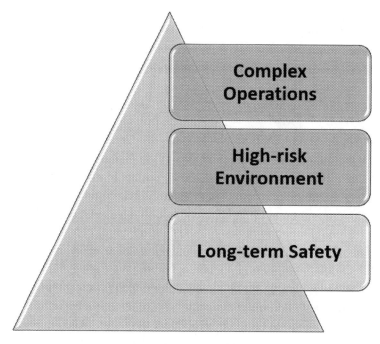

Figure 16-1. *High-reliability organizations*

The HROs are characterized by five principles as given in the diagram below. The study at Johns Hopkins University sought to explain how these principles can be applied in a healthcare setting. Its main intent was to reveal potential benefits and analyze the challenges in the implementation of a high-reliability framework among health organizations with a focus on patient safety and quality of care.

1. **Preoccupation with failure** empowers practitioners to be continuously vigilant and proactive in their risk management in healthcare settings. It enables healthcare staff to learn even from the so-called near misses and adverse events, thereby adopting a culture of continuous improvement.

2. **Reluctance to simplify interpretations** discourages generalizing the patients' symptoms or medical data collected as it may lead to incorrect diagnosis and inept treatment.

3. **Sensitivity to operations** involves awareness and alertness to frontline operations in the healthcare organization, establishing a regular interaction with frontline staff and patients while stressing the importance of timely identification, notification, and resolution of any potential issue.

4. **Commitment to resilience** ensures the development of a robust healthcare system and processes that can adapt to change and recover from disruptions. This involves implementing strong contingency plans and redundant systems to guarantee the continuity of services, even in challenging circumstances.

5. **Deference to expertise** involves cultivating and leveraging the specialized knowledge of health professionals. This principle identifies the most competent professional and expert who can take the right decision and may not be the highest in rank.

Figure 16-2. *Five pillars of HROs (source: `https://qualitysafety.`*
`bmj.com/content/qhc/31/12/845.full.pdf`)

Thus, the 2022 study published by Johns Hopkins University provided valuable insights in the implementation of the principles of HROs in healthcare. The research found an inconsistent understanding of the principles and enactment of HRO across participants. However, the suite of interventions, known as "Caring Safely," showcased the case potential for successful implementation of HRO principles even in a challenging scenario.

The study conclusively depicted that HRO principles would deliver a positive impact for health organizations. For example, it was observed that hospitals do have higher levels of staff engagement if backed by dependable leadership. This is a vital ingredient in the effective adoption of HRO principles while nurturing the values of steadfast progression and risk management foresight.

Implementation and Challenges

Despite the promise of excellence offered by the principles of HRO in healthcare, there are inherent challenges in adoption and implementation. We will discuss a few of these here:

1. **Challenges in Adopting Organization-Level Safety Culture:** The basic principles of HROs involve an organization-wide change management and collective effort toward a culture of continuous improvements and risk preparedness.

2. **Balancing Priorities:** In a health organizational transformation initiative, where patient care and data security take precedence, such as the digitalization of EHRs, the implementation of HROs takes a backseat.

3. **Development and Deployment of Process Improvement Tools and Techniques:** Large and complex healthcare organizational setup requires significant effort, resources, and time for the development and deployment of effective working process improvement tools and methods.

4. **Inconsistent Understanding and Enactment of HRO Principles:** There is variability in understanding and applying HRO principles among different healthcare professionals.

5. **Lack of Established Collaboration Across Shared Geographic Regions:** Healthcare organizations are grappling with the flow of patient loads to balance against staffing and resourcing, particularly during a public health crisis such as the COVID-19 pandemic.

Overcoming these challenges requires strong leadership, engaged staff, and a focus on continuous learning and improvement.

Outcomes and Analysis

The 2022 case study of Johns Hopkins University is a classic example of the successful implementation of HRO principles in the healthcare system. It provided conclusive evidence of attainable benefits from the implementation of HRO principles, especially the drastic improvement in patient safety and quality of care.

Over five years, the study identified and participated in a total of 3,184 process improvement projects—from enhancing patient safety protocols to boosting operational efficiency. This resulted in a huge return on investment (ROI) of US $2.8 million. This underscores how, from an economic point of view, the adoption of HRO principles led to another level of improvement in patients' results.

This case study provides powerful evidence for the potential of HRO principles to make a difference in healthcare administration. The case exemplifies that healthcare organizations might apply HRO principles effectively to derive marked enhancements not only in patient safety but also in quality of care and operational efficiency.

Reliability in healthcare administration goes beyond consistent performance to develop systems and processes that are designed to function. It's about fostering a culture of steadfast development at every level of the team, from the frontline staff to the top leadership.

As healthcare systems continue to evolve in response to technological advancements and changing patient needs, the importance of reliability will only become more pronounced. With the increasing digitization of healthcare services, ensuring the reliability of digital health solutions, such as Electronic Health Records and telemedicine platforms, has become a pressing concern.

Emerging Trends and Advancement in Reliability Engineering

The pursuit of reliability in the modern industry is the cornerstone for the successful design and operation of a system. This chapter on "Emerging Trends and Advancements in Reliability Engineering" offers a panorama of today's technologies and methodologies that are transforming every sector. Innovations such as artificial intelligence (AI), blockchain technology, and advanced data analytics are improving system dependability and system efficiency. As industries become more complex and place increased performance needs on the systems, these technologies offer engineers and administrators invaluable tools to predict, understand, and mitigate potential failures. Examples and case studies reviewed in this discussion not only serve as illustrative guides for the real-world application of these innovations but also show strategic importance in integrating advanced reliability practices for competitive advantage and operational excellence. Modern reliability engineering solutions ensure that the latest trends as discussed in this chapter are underscored and important to critical impacts forming the practice of global industry.

Figure 16-3. *Measuring the risk levels of failures and financial and regulatory impact (image source:* `Freepik.com`*)*

Reliability is a major pillar that holds the healthcare sector which improves patient care, optimization of operational efficiency, and working under very strict regulatory standards. The increased reliance of healthcare administration on modern technologies, especially in managing the performance of core functions, like patient health records or appointment scheduling, creates a persistent need for a robust, fail-proof system. This chapter explores the major trends and developments in reliability engineering in healthcare leveraging advanced technologies, which helps enhance systems in terms of security and efficiency but also in creating a user-friendly and reliable system.

Emerging Trends and Advancement in Reliability Engineering in Healthcare

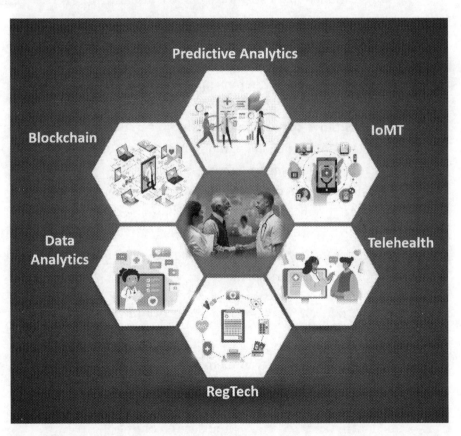

Figure 16-4. Emerging trends in reliability engineering in healthcare (image source: Freepik.com)

1. **Role of AI and Machine Learning in Healthcare Prediction**

   Artificial intelligence and machine learning tools are transforming the world of healthcare to create predictive models for enhancing patient and resource management. For example, AI algorithms can be applied to the historical database on the number of incoming patients to give more accurate predictions on admissions helping to reduce patient wait time and improve the outcome of patient care. It is also helping major hospitals to better manage bed availability and physician and administrative staff availability, which eventually translates into reduced operational overheads and efficient and streamlined processes. Predictive analytics integrated into healthcare systems would be able to predict failures or system bottlenecks and help initiate proactive assessment and mitigation strategies to maintain the reliability and performance of the system.

2. **Blockchain's Impact on Data Security and Patient Privacy**

   Blockchain technology implementation in healthcare is immensely aiding in both data security and patient privacy. Through decentralization of data in secure protocols and access controls, it helps create immutable and traceable patient records without compromising security. Blockchain technology is being increasingly used to securely exchange Electronic Health Records (EHRs) between health providers in such a way that it would allow seamless yet measured access to patient data. This technology supports authorized access, creating a transparent patient data management system which is essential for a regulated digital healthcare environment.

3. **Internet of Medical Things (IoMT) for Real-Time Monitoring**

   The Internet of Things (IoT) in healthcare, more commonly referred to as the Internet of Medical Things (IoMT), has significantly improved the reliability of health monitoring systems. IoMT devices provide real-time uninterrupted data flow useful in faster and accurate clinical decisions and resulting interventions, if needed. The use of IoMT devices in hospital networks to remotely monitor the patient's essential medical data is helping to optimize inpatient admissions. The real-time data feed to providers facilitates quick responses to medical-related abnormalities, thus leading to a drastic reduction in readmission rates and emergency interventions.

4. **Improved Operational Efficiency with Advanced Data Analytics**

   Advanced data analytics enhances operational efficiency and optimizes the reliability of healthcare systems. Health administrators rely on the vast amount of data analyzed, to find operational inefficiencies and predict future trends to create contingency and recovery plans, to manage effective resource allocations. Hospitals are using analytics to improve the scheduling of surgical procedures to reduce delays and improve the throughput of surgeries. Advanced data analytics is contributing to improvement in the reliability of healthcare systems and the overall delivery of healthcare services.

Predictive maintenance is another area in which advanced analytics helps improve health system reliability. Health facilities use such data from equipment and devices to predict a failure even before it occurs. This proactive approach has reduced downtime, extended equipment life, and ensured that critical medical devices are always available when needed. For instance, MRI machines or ventilators can be monitored in real time, with AI algorithms making predictions on potential failure events based on performance data, allowing for proper maintenance scheduling to minimize disruption to patient care.

5. **Telehealth and Remote Care Platforms**

    Digitalization of healthcare supported by state-of-the-art communication networks and connectivity has helped in global outreach and the adoption of reliable telehealth services. Physicians, especially in the rural or underserved regions are relying on telehealth consultations and follow-up care for their patients. Round-the-clock medical services and consultation through the screens of their smartphones has immensely improved patient satisfaction and reduced the number of appointment no-shows, thus making the remote care solutions more reliable and effective.

6. **Compliance and Regulatory Technology (RegTech)**

   RegTech supports healthcare organizations in ensuring compliance with changing regulations efficiently and effectively. RegTech in healthcare offers automated compliance systems that help healthcare providers adhere to changing regulations without manual oversight, which eventually helps reduce human effort and improve system reliability. Another usage of RegTech is automated Medicare reporting to significantly reduce the long laborious hours spent in the manual reporting process.

7. **Advanced Cybersecurity Measures for Healthcare Systems**

   As healthcare systems become increasingly digitized, robust cybersecurity measures are paramount. Advanced cybersecurity is essential for maintaining the reliability and integrity of healthcare systems, safeguarding sensitive patient data, and ensuring seamless healthcare services. Key advancements in healthcare cybersecurity include

   1. **AI-Powered Threat Detection:** Machine learning algorithms that can identify and respond to new and evolving cyber threats in real time

2. **Zero Trust Architecture:** A security model that mandates strict identity verification for every person and device attempting to access network resources, regardless of their location

3. **Secure Cloud Solutions:** Utilizing advanced encryption and access controls for cloud-based healthcare data and applications

4. **Endpoint Security:** Protecting all devices connected to the healthcare network, including IoMT devices, from potential security breaches

5. **Continuous Security Training:** Regular cybersecurity awareness training for healthcare staff to mitigate human factor risks

As cyber threats continue to evolve, investing in advanced cybersecurity measures is crucial for maintaining the reliability and trustworthiness of healthcare systems.

Emerging trends and advancements in reliability engineering are transforming the healthcare landscape. Technologies like AI, machine learning, blockchain, IoMT, edge computing, and enhanced cybersecurity are boosting the reliability of healthcare systems while improving patient outcomes and operational efficiency. As healthcare continues to evolve, integrating these innovative solutions will be essential for creating more resilient, efficient, and patient-centered systems. The future of healthcare reliability hinges on successfully adopting and integrating these technologies, coupled with strong security measures and adherence to evolving regulations.

Generative AI and LLMs Reshape Reliability's Future

Big data surges, the Internet of Things connects, and artificial intelligence evolves at breakneck speed—reliability engineering stands at a crossroads. This discipline, the guardian of system dependability, faces a revolution fueled by Generative AI and Large Language Models (LLMs). These technologies, creators of content, analysts of vast information landscapes, and predictors of intricate events, are not mere tools; they are catalysts for transformation.

As big data continues its relentless growth, the Internet of Things connects more devices than ever before, and artificial intelligence advances at an unprecedented pace, reliability engineering finds itself at a pivotal moment. The field, which has long served as the custodian of system dependability, is on the brink of a transformation driven by the rise of generative AI and Large Language Models (LLMs). These cutting-edge technologies, capable of generating content, analyzing vast information landscapes, and predicting complex events, are not just tools—they are the engines of change.

Reliability engineers, once reliant on models and intuition, now embrace a data-driven future. AI's power unlocks insights, predicts breakdowns, and optimizes upkeep, changing how we ensure resilience and longevity in complex systems. This chapter delves into this paradigm shift, where AI enhances existing practices and forges new frontiers in reliability.

In this new era, reliability engineers who once depended solely on traditional models and intuition are now stepping into a data-driven future. The immense power of AI is revolutionizing the field by unlocking deep insights, predicting potential failures, and optimizing maintenance strategies, fundamentally altering the way we achieve resilience and longevity in complex systems. This chapter explores this paradigm shift in detail, highlighting how AI not only enhances existing reliability practices but also creates entirely new frontiers for the discipline.

The Data-Driven Dawn of Reliability

In the realm of asset maintenance and optimization, traditional practices have historically relied on well-established models, statistical analysis, and the invaluable expertise of seasoned engineers. These approaches, while undeniably valuable, often encountered limitations due to the scarcity of available data and the intricate, often elusive patterns hidden within complex systems. As a result, decision-making was frequently based on incomplete information and educated guesses, which could lead to suboptimal outcomes. These challenges were further compounded by the need for constant human intervention and the reliance on historical data, which sometimes failed to capture the dynamic nature of modern systems. Consequently, while traditional methods provided a solid foundation, they often fell short of delivering the level of precision and foresight necessary to optimize asset performance fully.

The advent of the Internet of Things (IoT) revolutionized this landscape by ushering in an era of unprecedented data abundance. Sensors embedded within critical assets began generating a deluge of information, providing real-time insights into their health, performance, and operational conditions. This newfound wealth of data held the promise of unlocking a deeper understanding of asset behavior, enabling more accurate predictions of failures, and facilitating proactive maintenance strategies. However, the sheer volume and complexity of this data presented a new challenge: how to effectively harness its potential to drive meaningful improvements. The need for advanced tools and methodologies to process, analyze, and act on this data became apparent, as traditional methods were quickly overwhelmed by the scale and speed of information generated.

Generative AI and Large Language Models (LLMs) have emerged as the key to unlocking the value hidden within this sea of sensor data. With their exceptional ability to process and analyze vast amounts of information, these AI-powered models have rapidly surpassed traditional methods in terms of both speed and accuracy. By leveraging their data-crunching prowess, engineers can now detect subtle anomalies, predict failures with remarkable precision, and generate comprehensive reports that provide a holistic view of asset health. This transformative capability has fundamentally shifted the paradigm of asset maintenance from a reactive approach to a proactive one. No longer do engineers have to wait for signals of failure; they can now anticipate issues and take preventive action well before any actual damage occurs, leading to a significant reduction in unexpected downtime and maintenance costs.

The implications of this shift are profound. Instead of waiting for failures to occur and then scrambling to address them, engineers can now identify potential problems before they escalate, minimizing downtime, reducing costs, and enhancing overall system reliability. Furthermore, the ability to predict failures with high accuracy allows for more efficient resource allocation, ensuring that maintenance activities are prioritized based on actual needs rather than on arbitrary schedules. In essence, generative AI and LLMs are empowering engineers to make data-driven decisions that optimize asset performance, maximize uptime, and extend the lifespan of critical equipment. This proactive approach not only enhances the operational efficiency of assets but also contributes to the sustainability of operations by extending the lifecycle of machinery and reducing waste.

The integration of generative AI and LLMs into asset management represents a paradigm shift that promises to reshape the future of industries that rely on complex systems and infrastructure. By harnessing the power of data and AI, organizations can move beyond the limitations of traditional approaches and embrace a new era of proactive, predictive, and data-driven asset maintenance. This transformative capability has the potential to not only improve operational efficiency and reduce costs but also enhance safety, increase productivity, and ultimately drive greater value across the entire enterprise. As the technology continues to evolve and mature, the possibilities for innovation and optimization are virtually limitless. The adoption of these advanced tools signals a move toward a future where asset management is not just about maintaining the status quo but about driving continuous improvement and innovation, paving the way for smarter, more resilient industrial operations.

Anomaly's Whisper, Maintenance's Foresight

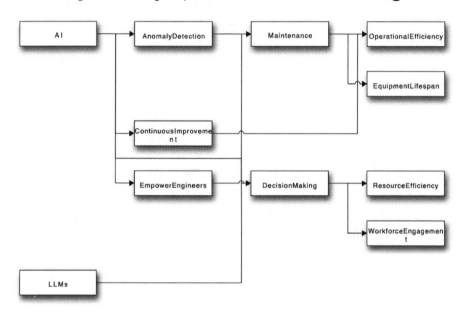

Artificial intelligence (AI) has bestowed upon the realm of reliability engineering a truly transformative gift: the ability to detect anomalies and predict maintenance needs with unparalleled precision. This evolution is not merely a technological upgrade but a revolutionary shift that reshapes the very foundations of how we approach asset management and operational efficiency. At the heart of this capability lies the marriage of real-time sensor data with sophisticated AI models, a union that enables a level of insight previously thought impossible. Sensors embedded within critical assets continuously monitor a wide array of parameters, from temperature and vibration to pressure and acoustic emissions.

These advanced sensors act as the nervous system of machinery, providing a constant stream of data that reflects the real-time health of each component. The AI models, trained on vast amounts of historical data, possess the remarkable ability to discern even the most subtle deviations from normal operating conditions. These models don't just detect obvious faults; they have the capacity to identify minute irregularities that might indicate the earliest stages of wear and tear or potential failure. These deviations, often invisible to the human eye and undetectable by traditional methods, are the telltale signs of impending failure, the whispers of a machine crying out for attention. By identifying these anomalies in their nascent stages, engineers can intervene proactively, addressing minor issues before they escalate into catastrophic breakdowns. This proactive approach transforms maintenance from a reactive, often crisis-driven activity into a strategic process that enhances reliability and efficiency. Not only does this avert costly downtime and repairs, but it also maximizes the uptime and lifespan of valuable equipment, ensuring that assets are utilized to their full potential. In essence, AI-powered anomaly detection and predictive maintenance represent a triumph of foresight over hindsight, where data-driven insights translate into tangible improvements in both operational efficiency and the bottom line. This technological advancement is not just about preventing failures; it's about optimizing every aspect of asset performance to create a more resilient and productive operational environment.

Large Language Models (LLMs), renowned for their linguistic prowess, have also found a valuable application in the realm of asset maintenance. Their ability to analyze vast amounts of textual data, including maintenance logs, inspection reports, and even informal discussions among technicians, adds a unique dimension to the understanding of asset health. LLMs bring a new level of sophistication to data analysis by transforming unstructured text into actionable insights. By weaving together insights gleaned from both human observations and sensor data, LLMs can uncover hidden patterns and correlations that might otherwise

551

go unnoticed. This holistic view of asset performance, where qualitative data from human expertise is integrated with quantitative sensor data, enables engineers to develop maintenance strategies that are tailored to the specific needs of each individual piece of equipment. Maintenance intervals can be optimized, ensuring that interventions occur at the most opportune times, while unnecessary downtime is minimized. The ability to customize maintenance schedules based on real-time data and historical trends represents a significant leap forward in asset management. This personalized approach to maintenance not only maximizes the efficiency of resources but also extends the operational life of assets, contributing to a more sustainable and cost-effective approach to asset management. Furthermore, this approach ensures that maintenance resources are allocated where they are most needed, reducing waste and enhancing overall operational efficiency.

Moreover, the integration of AI into reliability engineering fosters a culture of continuous improvement. As AI models learn from each anomaly detected and each maintenance intervention, they become increasingly adept at predicting future events and identifying areas where further optimization is possible. This continuous learning process is akin to a feedback loop, where every piece of data enhances the model's accuracy and predictive capabilities. This iterative process of learning and refinement leads to a virtuous cycle of improved asset performance, reduced downtime, and enhanced operational efficiency. The insights generated by AI not only empower engineers to make more informed decisions but also inspire innovation in maintenance practices, ultimately driving greater value across the entire enterprise. As AI tools evolve, they can suggest new methods and strategies that may not have been previously considered, pushing the boundaries of what's possible in maintenance and reliability engineering. This culture of continuous improvement becomes embedded within the organization, leading to a more agile and responsive operational environment.

The impact of AI on reliability engineering extends beyond the technical realm, influencing organizational culture and workforce dynamics. By automating routine tasks, AI frees up engineers to focus on more strategic activities, such as root cause analysis, process optimization, and the development of new maintenance protocols. This shift in focus allows organizations to leverage the full potential of their engineering talent, fostering a more engaged and empowered workforce. Engineers can now spend more time on creative problem-solving and innovation, driving the company forward rather than merely maintaining the status quo. Furthermore, the data-driven insights generated by AI can inform broader decision-making processes, such as asset investment strategies, risk mitigation planning, and resource allocation. AI becomes a strategic tool, not just for maintenance but for overall business strategy, providing a competitive edge in asset management.

In conclusion, the integration of AI, particularly anomaly detection and predictive maintenance powered by real-time sensor data and LLMs, has ushered in a new era of reliability engineering. By harnessing the power of data and AI, organizations can move beyond the limitations of traditional approaches and embrace a more proactive, predictive, and data-driven approach to asset management. This paradigm shift not only improves operational efficiency, reduces costs, and extends asset lifespan but also fosters a culture of continuous improvement, empowers engineers, and drives greater value across the entire enterprise. The ripple effects of AI integration are felt across all levels of the organization, from the shop floor to the boardroom. As AI technology continues to evolve and mature, the possibilities for innovation and optimization in the realm of reliability engineering are boundless. Future advancements will likely include even more sophisticated models, deeper integrations with IoT, and expanded use cases that will further transform the field, solidifying AI's role as a cornerstone of modern reliability engineering.

Failure's Anatomy, AI-Augmented

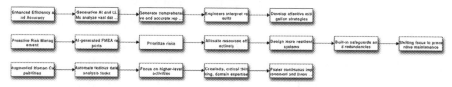

Traditional FMEA: A Laborious but Essential Endeavor

Failure Mode and Effects Analysis (FMEA) has long stood as a cornerstone in engineering, meticulously identifying and addressing potential failure points within intricate systems. Traditionally, this was a human-intensive process, where seasoned engineers would carefully sift through mountains of data, spanning historical failure records, maintenance logs, and intricate design specifications. Though comprehensive, this manual approach was inherently time-consuming and prone to the limitations of human cognitive abilities. The sheer volume of data often overwhelmed human analysis, potentially leading to oversight of critical insights and the unavoidable introduction of biases that accompany subjective interpretations.

The AI-Powered Transformation of FMEA

The advent of generative AI and Large Language Models (LLMs) has ushered in a revolutionary era for FMEA. These advanced technologies have injected a potent cocktail of speed and precision into the analytical process. LLMs, equipped with the capability to process and comprehend vast amounts of textual information, can rapidly delve into extensive datasets encompassing everything from past operational failures to nuanced design documents. By harnessing sophisticated algorithms, these models can uncover subtle patterns and correlations that might remain obscured in a traditional, human-led analysis. This not only drastically accelerates the FMEA process but also enhances the accuracy and breadth of the insights generated, offering engineers a significantly clearer and more detailed understanding of potential failure modes and their underlying causes.

Proactive Engineering: Building Robustness from the Ground Up

Armed with AI-generated FMEA reports, engineers possess a powerful tool for precision risk prioritization and strategic resource allocation. The rich insights gleaned from these AI systems illuminate the root causes of potential failures, their likelihood, and their projected impact on system integrity and safety. This deep level of understanding empowers engineers to take a proactive stance, designing systems that are inherently more robust from the early stages of development. By strategically incorporating safeguards and redundancies, organizations can shift from a reactive, damage-control approach to one that is preventative and proactive, ultimately resulting in systems that boast enhanced reliability, increased safety, and superior performance.

The Augmented Engineer: AI As a Catalyst for Innovation

The incorporation of generative AI and LLMs in FMEA transcends mere operational streamlining; it redefines the role of human engineers. Far from diminishing human expertise, AI tools amplify it, liberating engineers from the time-consuming and repetitive tasks of data analysis. This reallocation of human resources empowers engineers to dedicate themselves to higher-level activities that demand creativity, critical thinking, and strategic decision-making. By delegating the labor-intensive data processing to AI, engineers can engage more deeply with the intellectual core of their work, exploring innovative design solutions and tackling complex problems with greater efficiency and focus.

FMEA in the AI Age: Pioneering a Future of Engineering Excellence

The evolution of FMEA from a manual, expert-driven process to an AI-augmented, data-driven one marks a pivotal advancement in the engineering field. The fusion of generative AI and LLMs into this critical practice not only streamlines and refines failure analysis but also cultivates a culture of perpetual innovation and improvement. As these technologies continue to mature, we can anticipate FMEA becoming even more sophisticated, enabling engineers to confront increasingly complex systems and challenges with unprecedented confidence and capability. The future of FMEA, powered by AI, holds the promise of redefining the boundaries of engineering achievement, paving the way for a world filled with safer, more reliable, and highly efficient systems across industries.

Words into Wisdom: NLP Decodes Root Causes

Failures, even the most catastrophic ones, rarely occur in a vacuum. They leave behind a trail of linguistic breadcrumbs scattered throughout the intricate tapestry of communication and documentation that permeates complex systems and organizations. Maintenance logs, incident reports, emails, even seemingly casual exchanges—all these disparate sources of unstructured textual data hold the potential to reveal the underlying causes of failures, whether they stem from technical malfunctions, human error, or systemic organizational deficiencies.

Traditionally, unraveling the root cause of a failure has been a Herculean task. It involved countless hours of manual analysis, with human experts meticulously sifting through mountains of documents, seeking patterns, and connecting dots. This approach was not only time-consuming and laborious but also susceptible to human limitations. Subtle nuances in language, implicit connections across disparate sources, and unconscious biases could easily skew the analysis, leading to incomplete or inaccurate conclusions.

The advent of natural language processing (NLP) and Large Language Models (LLMs) has ushered in a new era of root cause analysis. These AI-powered models are capable of processing and comprehending vast amounts of unstructured textual data with remarkable speed and accuracy. They delve into the linguistic intricacies, identify recurring themes, and discern patterns that might elude even the most seasoned human analysts. LLMs can map the complex web of interactions and dependencies within a system, revealing hidden connections and tracing the origins of a failure back to its root cause.

The benefits of NLP-powered root cause analysis extend far beyond mere speed and efficiency. It's not just about identifying the immediate trigger of a failure; it's about gaining a deep and comprehensive understanding of the underlying factors that contributed to it. By analyzing the language used to describe failures, LLMs can shed light on systemic issues, recurring patterns of error, communication breakdowns, or even ingrained organizational practices that might be increasing the risk of future failures.

Armed with this deeper level of insight, organizations can take decisive corrective action. They can address the immediate problem, implement preventive measures to avoid similar failures in the future, and embark on a journey of continuous improvement. By proactively identifying and addressing systemic issues, organizations can create more resilient systems, foster a culture of safety and accountability, and enhance their overall performance.

The applications of NLP-powered root cause analysis are as diverse as the systems and organizations it can be applied to. In engineering, it can help identify design flaws or manufacturing defects before they result in catastrophic failures. In healthcare, it can analyze patient records and identify systemic issues that contribute to adverse events. In finance, it can detect early warning signs of fraud or market instability. From transportation to telecommunications, from energy to environmental protection, NLP is transforming the way we learn from failures and shaping a more resilient, reliable, and safe future.

In essence, NLP-powered root cause analysis empowers us to turn failures into opportunities for growth and innovation. It's not just about fixing what's broken; it's about understanding why it broke in the first place. By harnessing the power of language and AI, we can transform data into knowledge, knowledge into action, and action into continuous improvement. In this way, NLP is more than just a tool for analysis; it's a catalyst for progress, driving us toward a future where failures become stepping stones to greater success.

Real-World Echoes: AI in Action

The impact of artificial intelligence (AI) is no longer confined to the realm of theoretical speculation or futuristic visions; it's actively reshaping industries and revolutionizing the way we work and live. This transformation is unfolding across a multitude of sectors, where AI-powered solutions are driving unprecedented levels of efficiency, safety, and cost savings. The convergence of AI with traditional industrial practices is ushering in a new era of intelligent automation, predictive maintenance, and data-driven decision-making.

On the modern factory floor, the marriage of robotics and AI is giving rise to a new generation of smart machines. An assembly line robot, equipped with an array of sensors that continuously monitor its performance, generates a constant stream of data. This data is then fed into a Large Language Model (LLM), a sophisticated AI system trained on vast amounts of operational data. The LLM, with its ability to discern patterns and anomalies within the data, acts as a vigilant sentinel. It detects a subtle vibration—a barely perceptible tremor that might escape human notice—but to the LLM, it's an early warning sign of a worn bearing. This seemingly minor issue, if left unaddressed, could lead to a catastrophic failure, resulting in costly downtime, production delays, and potential safety hazards. Armed with this AI-driven insight, maintenance teams can intervene proactively, replacing the faulty component before it causes disruption. In this way, AI transforms from a passive observer to an active participant, ensuring the smooth and uninterrupted operation of complex manufacturing processes.

In the realm of renewable energy, AI is proving to be an invaluable asset in optimizing the performance and reliability of wind farms. Each turbine, a towering testament to human ingenuity, is equipped with a network of sensors that capture a wealth of data on wind speed, temperature, vibration, and other critical parameters. This data deluge is then channeled into a generative AI model, an AI system capable of not just analyzing data but also generating new insights. The model meticulously sifts through the information, searching for patterns and anomalies. It might, for instance, detect a recurring irregularity in the data, revealing a subtle design flaw in the blade control system. Such a flaw, if left unaddressed, could lead to premature wear and tear, reduced energy production, and costly repairs. By identifying and rectifying this issue early on, engineers can enhance the performance, reliability, and longevity of the wind farm. The financial and environmental benefits of such AI-driven insights are substantial, contributing to a more sustainable and efficient energy future.

The aerospace industry, where safety is of paramount importance, is also undergoing a profound transformation thanks to AI. Natural language processing (NLP) systems, capable of understanding and analyzing human language, are now being deployed to analyze vast quantities of unstructured data, such as maintenance logs, pilot reports, and incident records. These systems, trained on a corpus of aviation-related text, can identify subtle patterns, correlations, and linguistic cues that might escape human attention. In one remarkable case, an NLP system flagged a recurring complaint from pilots about a particular component, prompting engineers to investigate further. The investigation revealed a latent defect in the component that, if left unaddressed, could have led to a catastrophic in-flight failure. By uncovering this hidden risk, AI played a crucial role in averting disaster and ensuring the safety of countless passengers.

These real-world examples underscore the tangible and transformative impact that AI is having on industrial reliability, efficiency, and safety. By detecting anomalies, predicting failures, and revealing root causes, AI is empowering engineers and operators to make data-driven decisions, optimize performance, and prevent costly disruptions. It's important to note that AI is not replacing human expertise; rather, it's augmenting it, providing valuable insights and tools that enable humans to make better, faster, and more informed decisions. The collaboration between humans and AI is proving to be a powerful combination, unlocking new levels of efficiency, productivity, and safety across a wide range of industries. As AI continues to evolve and mature, its impact on industrial operations is only set to grow, paving the way for a future where machines and humans work seamlessly together to achieve unprecedented levels of performance and reliability.

A Glimpse into Reliability's AI-Powered Future

The integration of generative AI and LLMs with the torrent of data flowing from the Internet of Things is already transforming the landscape of asset management. The ability to analyze, predict, and optimize performance in real time has led to unprecedented levels of efficiency and cost savings. But this is just the first step in a journey with limitless horizons. As artificial intelligence continues to mature and expand its capabilities, we stand on the cusp of a new era in asset management, one where the possibilities seem boundless.

The Emergence of Digital Twins: A Virtual Playground for Innovation
One of the most promising developments on this horizon is the rise of digital twins—virtual replicas that faithfully mirror their physical counterparts in intricate detail. These digital doppelgangers allow engineers to venture into a risk-free environment where they can experiment, test, and optimize without the fear of real-world consequences. Within this virtual realm, stress tests can be conducted, potential failures simulated, and designs iterated upon to achieve peak performance and resilience. By harnessing the power of digital twins, engineers gain an invaluable understanding of asset behavior under a wide array of conditions, enabling them to proactively identify vulnerabilities and fine-tune designs for optimal outcomes.

AI As a Trusted Advisor: Enhancing Decision-Making

The role of AI in asset management extends far beyond mere analysis. AI-powered decision support systems are emerging as trusted advisors, capable of sifting through mountains of data, extracting meaningful insights, and offering recommendations. These systems can identify potential risks, evaluate the effectiveness of various strategies, and provide valuable guidance to human decision-makers. The symbiotic relationship between human judgment and machine intelligence promises to enhance decision-making accuracy, speed, and overall efficacy.

Generative Design: Unleashing Creative Potential

Beyond analysis and prediction, AI is now venturing into the realm of creation. Generative design algorithms are capable of conceiving entirely new components, systems, and processes that are optimized for performance, reliability, and resilience. These AI-generated designs often push the boundaries of what is possible, leading to innovative solutions that transcend the limitations of human imagination. By tapping into the creative potential of AI, engineers can unlock new levels of efficiency, performance, and sustainability, forging a path toward a future where assets are not just managed but optimized to their fullest potential.

Data-Driven Proaction: From Failures to Learning Opportunities

The future of asset management is one of data-driven proaction. Failures, once dreaded events, will be transformed into valuable learning opportunities. AI-powered systems will continuously monitor and analyze asset performance, detecting anomalies, predicting potential failures, and enabling proactive interventions before problems escalate. The lessons gleaned from these near-misses and actual failures will feed into a virtuous cycle of continuous improvement, driving the iterative refinement of systems, processes, and practices.

The Synergy of Human Ingenuity and Artificial Intelligence

As AI continues to evolve, it will become an indispensable ally to human ingenuity. The future of asset management is not about replacing humans with machines; it's about empowering engineers and operators with the tools and insights they need to make informed decisions, design resilient systems, and proactively manage assets to achieve optimal outcomes. The synergy between human expertise and artificial intelligence will pave the way for a future where failures are minimized, efficiency is maximized, and innovation flourishes.

A Balanced Path: Challenges and Ethics

Here's the expanded content, doubling the original paragraphs while preserving the core messages and adding some additional insights.

The boundless potential of artificial intelligence (AI) to revolutionize industries and enrich our lives is undeniable. From healthcare to finance, transportation to manufacturing, AI's transformative power is poised to optimize processes, streamline decision-making, and unlock new frontiers of innovation. However, as with any technological leap, the path to realizing AI's full potential is paved with challenges. These challenges, while significant, are not insurmountable, and addressing them thoughtfully will be key to ensuring AI's responsible and beneficial integration into society.

Data, often hailed as the lifeblood of AI, fuels its learning and decision-making processes. The quality, quantity, and integrity of this data are paramount. Incomplete, inaccurate, or biased data can cripple even the most sophisticated AI models, leading to flawed outputs and potentially harmful consequences. Furthermore, the insatiable appetite of modern AI systems for vast troves of data raises concerns about privacy and security. As we entrust AI with increasingly sensitive personal and proprietary information, the risks associated with data breaches and misuse become ever more pressing. Safeguarding data and ensuring its ethical use will be crucial to building public trust in AI technologies.

As AI models grow in complexity, their inner workings often become shrouded in opacity, raising concerns about explainability and trust. The "black box" nature of many AI algorithms makes it difficult to understand how they arrive at their conclusions. This lack of transparency can be particularly problematic in high-stakes domains like healthcare or finance, where understanding the rationale behind AI-driven decisions is critical for accountability and informed decision-making. Building trust in AI systems necessitates the development of techniques that render their decision-making processes more interpretable and comprehensible to both experts and the general public.

Bias, an insidious and pervasive issue in society, can also infiltrate AI models, perpetuating and even amplifying existing prejudices and discrimination. If the data used to train AI algorithms is biased, the resulting models will inherit and propagate those biases, leading to discriminatory outcomes and reinforcing societal inequalities. Addressing this challenge requires a multipronged approach. Ensuring diversity and inclusivity in data collection and model development is crucial. Additionally, ongoing monitoring and evaluation are necessary to detect and mitigate biases as they arise, promoting fairness and equity in AI systems.

The prospect of AI automating tasks and displacing jobs is a legitimate concern that must be addressed with foresight and empathy. While AI has the potential to boost productivity and efficiency, it also raises questions about the future of work and the need to equip the workforce with the skills necessary to thrive in an AI-driven world. Striking a balance between automation and human employment requires careful consideration of ethical, social, and economic factors. The goal should be to leverage AI to complement and augment human capabilities, creating a future where humans and machines work together synergistically, each contributing their unique strengths.

In navigating the complexities of AI integration, ethical considerations must remain at the forefront. It is imperative to ensure that AI systems are designed and deployed in ways that align with human values, respect privacy and autonomy, and promote fairness and equity. Developing robust ethical frameworks and governance mechanisms will be vital to

563

ensure that AI serves humanity's best interests and does not exacerbate existing societal problems. The delicate balance between harnessing AI's transformative power and mitigating its risks demands ongoing dialogue, collaboration, and a steadfast commitment to ethical principles that prioritize human well-being and societal benefit.

Conclusion

Generative AI and Large Language Models (LLMs) have transcended their roles as mere tools; they are emerging as the architects of a new paradigm in reliability engineering. These advanced technologies are empowering engineers and organizations to move beyond reactive maintenance and embrace a proactive, predictive approach to asset management. By harnessing the vast amounts of data generated by IoT sensors and other sources, generative AI and LLMs can identify subtle patterns and anomalies that often go unnoticed by human experts. This enables them to predict potential failures with remarkable accuracy, allowing for timely interventions that prevent costly downtime and catastrophic events.

Moreover, these AI-powered tools are not just reactive; they are also proactive. They can design systems with resilience in mind, anticipating potential vulnerabilities and suggesting design modifications that mitigate risks before they materialize. This ability to create inherently reliable systems is a game-changer, particularly in industries where safety and continuous operation are paramount. While challenges undoubtedly remain, such as ensuring data quality, addressing ethical concerns, and navigating the complexities of AI integration, the potential rewards are too significant to ignore. Safer systems, more efficient operations, and a more sustainable future are all within reach.

To fully realize these benefits, we must embrace this data-driven dawn, harnessing the power of AI while respecting and valuing human ingenuity. AI is not a replacement for human expertise but rather a powerful complement. By combining the analytical capabilities of AI with the creativity and problem-solving skills of human engineers, we can forge a new era of collaboration that leads to truly resilient and reliable systems. Together, we can weave reliability into the very fabric of our technological tapestry, creating a world where failures are not just minimized but anticipated and prevented. This is not just a vision of the future; it is a reality that is unfolding before our eyes, and it is up to us to seize this opportunity and shape it for the betterment of society.

Blockchain Principles: Immutability and Consensus

Across various industry sectors, the prominence of blockchain technology in enabling reliability is pivotal to creating a secure and scalable system, leveraging the principles of immutability and consensus. Immutability ensures that once the data is added to the blockchain, it is protected from being tampered with or vulnerable to fraud. This is an essential feature in areas, but not limited to healthcare, legal, or finance, where stability and accuracy of the stored and transactional data records are essential for operational integrity.

On the other hand, consensus requires the agreement of data validity among all parties within the network before updating new entries to the blockchain. This decentralized verification process avoids the possibility of the data being manipulated by any single person, hence enhancing the security and transparency of the system.

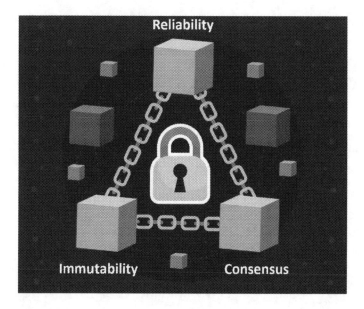

Figure 16-5. *The strong dependency of Reliability with Immutability and Consensus*

Blockchain Concept of Immutability and Consensus

Together, both the immutability and consensus approach to data management not only ensures a secure operational setup but also assists in making it transparent and democratic for data validation and modification. These characteristics of a blockchain have a direct influence on the overall reliability of a system. Blockchain guarantees a trusted environment because it is reasonably necessary to ensure that the entered data has not been fiddled with and that it has been verified by all stakeholders before any change. A system equipped with blockchains enables an organization to operate at very high levels of confidence and assurance. This ensures that the organizations operate efficiently and reliably, in an optimized workflow, with minimal or zero risks of data leakage or corruption.

Technical Overview of Blockchain

Before delving into the impact of blockchain on healthcare reliability, it is crucial to gain a comprehensive understanding of the technical underpinnings that contribute to the security and reliability of this transformative technology.

Key Components of Blockchain

1. **Blocks**: These are units of information, each containing a set of transactions and a unique identifier known as a hash.

2. **Chaining**: Each block includes the hash of the previous block, forming an unalterable chain of blocks.

3. **Distributed Ledger**: This decentralized database is shared across a network of computers, called nodes.

4. **Consensus Mechanisms**: These protocols ensure all nodes agree on the blockchain's state. Common methods include

   - Proof of Work (PoW)

   - Proof of Stake (PoS)

   - Delegated Proof of Stake (DPoS)

5. **Cryptography**: This secures transactions using public and private key pairs.

6. **Smart Contracts**: These are self-executing contracts with terms directly written into code.

How Blockchain Achieves Immutability

1. **Cryptographic Hashing**: Each block contains a unique hash based on its contents and the previous block's hash.

2. **Timestamp**: Each block is timestamped, establishing a chronological order that can't be altered.

3. **Consensus**: New blocks are added only after network verification, making it extremely difficult to modify existing blocks.

Grasping these technical elements is vital for understanding how blockchain ensures data integrity and security in healthcare systems.

Impact of Blockchain Principles on Healthcare Reliability

Blockchains have emerged as the central transformational agents within the domain of reliability engineering. Healthcare systems grapple with some critical challenges related to data integrity and security, wherein blockchain addresses these concerns. Blockchain ensures that patient records are immutable and fully traceable, thereby maintaining high data security and integrity standards. This key aspect of blockchain enormously fortifies the system's defense against fraud and unauthorized access to sensitive health information, creating an insurmountable barrier to potential security breaches.

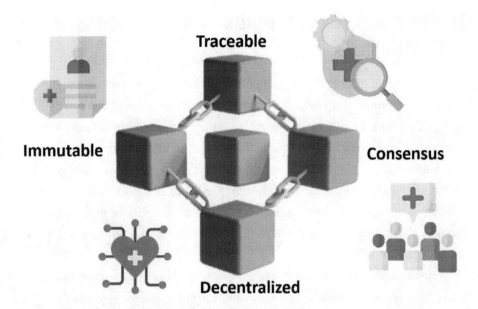

Figure 16-6. *Tenets of healthcare reliability*

Key Attributes of Blockchain in a Healthcare Ecosystem

One of blockchain's most valuable features is its capability to provide
permanent and verifiable records for every transaction. Every transaction
in the patient health record is diligently logged and tracked, enabling a
highly auditable system. This not only secures the data from tampering but
also ushers in a new level of transparency in healthcare processes. This will
create a reliable environment for all stakeholders involved in delivering
secure and efficient patient care, confidently relying on the authenticity
and correctness of records.

The Principle of Immutability and Consensus in Healthcare

The principle of immutability and consensus is critical in accentuating the transformational impact of blockchain on healthcare reliability. Immutability guarantees that the data recorded in the blockchain, remains the same—unchanged, thus offering essential protection to the integrity of the medical records. This is of utmost importance in healthcare where the accuracy and authenticity of patient health information directly affects clinical decisions and the outcome of treatment. It helps keep medical records from tampering and fraud, ensuring historical health data remains secure and reliable.

The consensus mechanism balances immutability by necessitating every change or addition to the blockchain to be approved by multiple verified parties before the contents are accepted as valid. The concerted verification process is done by various healthcare stakeholders, including hospitals, insurance companies, and specialist practitioners, each having an equal stake in keeping the data precise and secured. The blockchain system minimizes insecurity in the data through consensus and, therefore, offers excellent protection against unauthorized alteration of information.

In healthcare applications, certain consensus mechanisms are favored for their efficiency and security. Proof of Authority (PoA) is commonly used in private healthcare blockchains because it offers faster transaction times without the energy-intensive computations required by Proof of Work. In PoA, consensus is achieved through a set of approved validators, typically reputable healthcare institutions or regulatory bodies. This setup ensures that only trusted entities can validate transactions, adding an extra layer of security and reliability to sensitive healthcare data.

These core principles of blockchain enhance transparency in addition to strengthening the security and integrity of healthcare data. All transactions on the blockchain are visible to parties involved in verification, ensuring that the data management system is transparent and accountable. This transparency is paramount to the trust of the users and stakeholders, thereby creating a system of reliable information to be utilized in the decision-making process. The combined attributes of immutability and consensus in blockchain technology create a robust framework for data management that can improve the reliability and effectiveness of health service delivery. This would enable a more coordinated and secure healthcare environment where data integrity and security are the attributes.

Another vital application of blockchain in healthcare is the management of Electronic Health Records (EHRs). The decentralized setup in blockchain ensures that no single party has control over the complete dataset, thereby minimizing the risk of centralized data breaches. This further enhances the portability of patient data across different healthcare providers without compromising integrity or security to the same. Sharing patient data and EHRs is crucial for the effective delivery of healthcare, especially when there are several providers involved or requires coordinated care management. However, the application of blockchain in healthcare concerns more than just data security and the protection of data integrity; rather, it aids in facilitating better cohesion through secure and seamless sharing of data. This kind of interconnection could lead to improved health outcomes for the patient, as healthcare providers receive timely information on accurate and comprehensive patient health information (PHI). The improved data flow would eventually contribute to fewer diagnostic errors and improved treatment effectiveness through a complete provision of the patient's medical history.

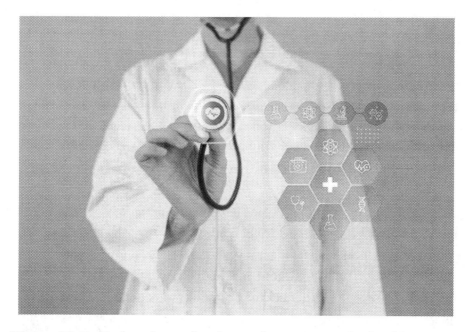

Figure 16-7. *Embracing technology advancements for healthcare (image source: Freepik.com)*

Embracing blockchain technology has drastically reduced the administrative burden on healthcare providers. A simplified, structured, and well-defined process to access and update patient records enables healthcare professionals to spend more time on care management activities and less time jumping through bureaucratic hoops. The standardization of the blockchain protocol for the use and sharing of data could bring a new era of innovation in health technologies, through higher operational efficiencies and improved quality of healthcare service delivery. The reliable and secure framework provided by the blockchain is enabling the integration and deployment of state-of-the-art digital health services, from telemedicine platforms to AI-driven diagnostic tools that can revolutionize healthcare and make services more accessible and efficient

The integration of blockchain technology and healthcare is a massive step forward for reliability engineering. It offers an opportunity for a more secure, transparent, and efficient approach to managing PHI and EHR, hence improving the overall delivery of healthcare services. The adoption of blockchain not only safeguards information but also propels the healthcare industry toward a future where data-driven decisions and interconnected services become the norm, significantly improving both patient outcomes and system resilience. This technological shift not only meets the current demands of healthcare administration but also sets a new standard for the future of healthcare operations and management.

While the potential of blockchain in healthcare is vast, widespread adoption faces several challenges. These include regulatory hurdles, the need for significant infrastructure investments, and concerns about scalability and energy consumption. However, as the technology advances and these issues are addressed, the future of blockchain in healthcare appears bright. Ongoing research in quantum-resistant cryptography and more efficient consensus mechanisms promises even more secure and scalable blockchain solutions. As healthcare continues to digitize, prioritizing data integrity and interoperability, blockchain's principles of immutability and consensus will become increasingly vital for ensuring the reliability and efficiency of healthcare systems worldwide.

Bibliography

1.   AI in Healthcare Study – A 2023 Definitive Healthcare special report; AI-in-healthcare-study-2023.pdf (definitivehc.com)

2.   Predictive analytics in healthcare; What is predictive analytics in healthcare? | Definitive Healthcare (definitivehc.com)

3.   What is Internet of Medical Things (IoMT): Explained in Detail, March 18, 2024, By Team EMB; What is Internet of Medical Things (IoMT): Explained in Detail (emb.global)

4.   Using Data Analytics to Predict Outcomes in Healthcare, By Lesley Clack, ScD, CPH, June 20, 2024; Using Data Analytics to Predict Outcomes in Healthcare (ahima.org)

5.   Yousef, E. A., Sutcliffe, K. M., McDonald, K. M., & Newman-Toker, D. E. (2022). Crossing Academic Boundaries for Diagnostic Safety: 10 Complex Challenges and Potential Solutions From Clinical Perspectives and High-Reliability Organizing Principles. Human Factors, 64(1), 6–20. https://doi.org/10.1177/0018720821996187

6.   High reliability organizing in healthcare: still a long way left to go; Crossref DOI link: https://doi.org/10.1136/bmjqs-2021-014141

7.   Can High-Reliability Organization Principles Help Transform Healthcare Delivery in the U.S.? HIMSS Roundtable Insights Review; https://www.himss.org/sites/hde/files/media/file/2023/03/23/teletracking_wp_himss-roundtable-insights-review-5.pdf

8.   Evidence Brief: Implementation of High Reliability Organization Principles; https://www.hsrd.research.va.gov/publications/esp/high-reliability-org.pdf

9.   Burns, B., & Beda, J. (2021). Kubernetes: Up & Running. O'Reilly Media

10. Turnbull, J. (2017). The Art of Monitoring. Turnbull Press

11. Red Hat. (2020). Understanding Observability. Red Hat

12. Sigelman, B. (2019). Distributed Tracing in Practice: Instrumenting, Analyzing, and Debugging Microservices. O'Reilly Media

13.   Rajan, S. (2020). AI and Machine Learning for Monitoring and Observability. Packt Publishing

14.   Williams, T. (2018). Modern Observability with Prometheus and Grafana. Packt Publishing

15.   Smith, J. (2021). Implementing Observability: Strategies for Building Reliable Systems. Addison-Wesley Professional

16.   Blockchain Revolutionizing Healthcare Industry: A Systematic Review of Blockchain Technology Benefits and Threats, By Fatma M. AbdelSalam, MBA, FISQua, CSM, CSSBB; Blockchain Revolutionizing Healthcare Industry: A Systematic Review of Blockchain Technology Benefits and Threats (ahima.org)

17.   Digital Transformation of Healthcare Administration: How can Blockchain augment Process Automation, The video was recorded during the "Global Virtual Blockchain in Healthcare Symposium" on 29th February -2024 https://www.youtube.com/watch?v=6WRCBCHtib0

Glossary

Digital Twin: A virtual replica of a physical asset, system, or process used to simulate and analyze real-world performance and conditions

ELK Stack: A suite of tools including Elasticsearch, Logstash, and Kibana, used for aggregating, analyzing, and visualizing log data

Fluentd: An open source data collector that unifies data collection and consumption for better use and understanding of data

FMEA (Failure Mode and Effects Analysis): A systematic process for identifying and addressing potential failures in a system, product, or process

Generative AI: A type of artificial intelligence that can generate new content, such as text, images, or music, based on the data it has been trained on

Grafana: A multiplatform open source analytics and interactive visualization web application that provides charts, graphs, and alerts for supported data sources

IoT (Internet of Things): A network of physical objects embedded with sensors, software, and other technologies to connect and exchange data with other devices and systems over the Internet

Jaeger: An open source end-to-end distributed tracing tool that monitors and troubleshoots transactions in complex distributed systems

LLMs (Large Language Models): Advanced AI models, such as GPT-3, capable of understanding and generating humanlike text based on vast amounts of data

© Saurav Bhattacharya 2024
M. Kuppam, *Enterprise Digital Reliability*, https://doi.org/10.1007/979-8-8688-1032-9

Logging: The process of recording discrete events that occur within a system, providing detailed information about specific actions, errors, or state changes

Metrics: Quantitative data points that reflect the performance and health of a system over time, such as CPU usage, memory consumption, and request rates

NLP (Natural Language Processing): A branch of artificial intelligence focused on enabling computers to understand, interpret, and generate human language

Observability: The ability to infer the internal state of a system from its external outputs

Predictive Maintenance: Techniques that use data analysis and AI to predict when maintenance should be performed to prevent unexpected equipment failures

Proactive Maintenance: Maintenance activities performed before a failure occurs, based on predictive insights and data analysis, to prevent potential issues

Prometheus: An open source monitoring and alerting toolkit designed for reliability and scalability, used for collecting and querying metrics

Tracing: A method for tracking the flow of requests through a system, providing a high-level view of how different services and components interact

Zipkin: A distributed tracing system that helps gather timing data needed to troubleshoot latency problems in microservice architectures

Index

A

Actionable alerts, 366, 386

Active listening, 473

Adopting a testing mindset
    benefits, 150
    confidence in releases, 152
    cost savings, 152
    cultural and organizational
        barriers, 195, 196
    customer satisfaction, 151
    empowerment of teams,
        152, 153
    faster time to market, 151
    principles of effective
        testing, 153–156
    promotion of continuous
        improvement, 152
    resistance to change
        communicate the
            benefits, 194
        leading by example, 194
        provide training and
            support, 194
        team member's concerns
            and objections, 194
    resource constraints, 194, 195
    risk of defects and errors, 151
    software quality, 151

Advanced data analytics, 236, 336,
    539–540, 544

Annual failure rate (AFR)
    levels, 120

AI, *see* Artificial intelligence (AI)

AI-generated FMEA, 555

AI/ML in software testing
    automated smart test case
        generation, 203
    benefits
        cost reduction, 204
        enhanced efficiency, 203
        improved accuracy, 204
    challenges
        model drift, 204
        training data quality, 204
        unforeseen test cases, 204
    practices
        be patient, 205
        learn prompt
            engineering, 205
        tester's capabilities, 205
        understand AI/ML
            systems, 204
    test accuracy and efficiency, 202

Printed in the United States
by Baker & Taylor Publisher Services